VLSI High-Speed I/O Circuits

VLSI High-Speed I/O Circuits

Theoretical Basis, Architecture, Modeling and Circuit Implementation

---

Hongjiang Song

To order additional copies of this book, contact:
Xlibris Corporation
1-888-795-4274
www.Xlibris.com
Orders@Xlibris.com
62861

CONTENTS

PREFACE

My early exposure to the so-called I/O circuits can probably be dated back to many years ago when I was working on a circuit design project of the computer interface between a PC8088 personal computer and the scanning electron microscope (SEM) at Yunnan University. Some of the I/O circuit terms, such as the data bus, the strobe, the handshaking, and the synchronization, came to me the first time. The 8-bit 200kb/s data acquisition rate was considered to be high enough at that time that the DMA (direct memory access) technique was needed. The entire circuit was implemented with very little computer-aided (CAD) tool support. My later touch on the I/O circuits was during a VHDL course project at Arizona State University at around 1992. A PSK MODEM was designed and simulated based on the VHDL language. In the past decade, I have been participating in various VLSI high-speed I/O circuit developments in the industry, such as the original 480Mbps USB2.0, 1.6Gbps SATA I/O interfaces, the SRAM interface, and the first (2.5Gbps) and second (5Gbps) generation PCI Express interface. In these cases, leading edge VLSI process technologies with sophisticated CAD tools are intensively used.

My above story might be used to precisely describe some of the important trends of VLSI high-speed I/O circuit developments. These trends can be characterized by the increasing data rates as driven by the demands for higher I/O bandwidth and the advancements of VLSI process technologies, by the parallel to serial I/O transition as driven by the pin count reduction and better signal integrity, by the wide adoption of sophisticated CAD tools and innovated circuit techniques as driven by the design productivity and by the increasing design challenges that are caused by the increasing data rates under the legacy I/O channel requirement for low cost.

This text was modified from the class notes of a graduate level VLSI circuit design course (EEE598) I offered at the Electrical Engineering Department at Arizona State University. The main goals of this text are to introduce to the students and beginners the theoretical framework, the VLSI circuit design techniques, and the most recent developments of VLSI high-speed I/O circuits based on the widely available public domain VLSI high-speed I/O research results contributed from both industrial and academic communities. The book is organized in twenty-one special topics, covering various aspects of VLSI high-speed I/O circuits, including the theoretical basis, architectures, modeling, and VLSI circuit implementations.

I would like to thank many of my colleagues and friends, especially Professor David R. Allee (Arizona State University) and Yan Song (Intel Corporation), for their encouragements, informative discussions, and various help during this text development. I would like to thank my lovely daughter Lucy Song for helping me design the book cover. I also would like to thank my students in EEE598 class for their valuable feedback and on whom much of the material was tested.

Hongjiang Song, Ph.D.

Mesa, Arizona, December 2008

CHAPTER 1
INTRODUCTION TO VLSI HIGH-SPEED I/O CIRCUITS

The increasing demands for more bandwidth at low cost to support the inter-chip communications as a result of continuous advancing of VLSI on-chip signal processing powers and the VLSI process technologies have been among the main driving forces for the most recent developments of VLSI high-speed I/O circuits. A typical computer platform now employs multiple multi-Gbps data rate serial I/O circuits, such as the first- and second-generation PCI Express and SATA links. Such high-speed I/O circuits are becoming standard features on the computer systems. However, this does not mean that the design challenges for faster inter-chip communication have been solved. On the contrary, it actually indicates that even higher inter-chip communication data rates have yet to come, and the design challenges continue. Consequently, deep understandings of the theoretical basis, the circuit operations, the circuit architectures, the modeling methodologies, and the circuit implementation techniques of VLSI high-speed I/O circuit are becoming increasingly crucial for the successful developments of VLSI electronic products.

1.1 OVERVIEW OF HIGH-SPEED I/O CIRCUIT TRENDS

The traditional I/O circuits that are typically characterized by wide (32-, 64-, or even 128-bit) parallel buses and the multipoint interfaces have been well established in past years employing many innovated techniques for increasing the data throughput using higher I/O interface frequency, wider I/O buses, the pipelining operation, the splitting transaction operation, and the out-of-order operation. However, the I/O circuit techniques suffer from increasingly severe design constraints under the trend of the increasing data rate throughput:

- The increasingly severe delay skews uncertainties among signals in the same bus, and the decreasing available timing budget at higher data rate results in severe limitation in the maximum achievable frequency for I/O with wide bus width.
- The high device pin count, long PCB traces, and larger connectors associated with wide bus width results in the increasing product cost and power dissipation and the limitation in the number of I/O interfaces a design can provide.
- The attachment of multiple devices to the same bus to support multipoint interface become very difficult design options at high data frequency and wide bus width.

As a result, there has been a trend in the most recent VLSI high-speed I/O circuit to focus on maximizing the overall throughput with reduced pin count to reduce the circuit complexity, design cost, and power dissipation. However, to obtain the higher data throughput with reduced pin count or data bus width, the clock frequency must be further increased. Increasing the frequency makes the channel more sensitive to the parasitic capacitance and inductance loading effects, such as bandwidth limitation, reflection, ringing, and cross talk effects. The moving to a lower pin count with higher I/O circuit frequency can be achieved but at the penalty of losing the multipoint interface ability. Typical VLSI high-speed low-pin count I/O circuits usually incorporate point-to-point arrangements in response to such issues.

As a result, modern VLSI high-speed I/O circuits provide solution to the inter-chip connection problem with several common techniques that emphasize efficiency, reliability, robustness, and cost:

- Packet switching that replaces the shared bus with switch fabric to control the packet data flow between devices
- Point-to-point unidirectional connections to release the skew and parasitic effects and allow higher operation speed
- Reduced pin count for each I/O interface standard to support lower power dissipation and better thermal characteristics

- Built-in scalabilities of I/O circuits for both frequency and data width to provide the I/O operation speed and data width flexibility that can be programmed based on application conditions.

1.2 OVERVIEW OF HIGH-SPEED I/O CIRCUIT TECHNIQUES

The cost-effective solutions for higher inter-chip data rates based on existing low-cost I/O circuit infrastructures, such as FR4 PCB, multimode fiber, and legacy connectors, in the market places result in the high-speed I/O circuit performance increasingly limited by the transmission channel nonidealities, such as the ISI, reflection, ringing, and cross talk effects. These I/O channel effects can significantly degrade the transmitted signal and impact the high-speed I/O circuit performances.

The continuous device performance improvements and reduction in on-chip device cost under VLSI technology scaling motivate VLSI I/O circuit designers to overcome these high-speed I/O transmission channel nonidealities employing innovations of on-chip circuit techniques for even increasing I/O data rates:

- Non-return-to-zero (NRZ) signaling schemes that offer the power-efficient signaling solutions are usually adopted in VLSI high-speed I/O circuits. Other signaling schemes with better spectral efficiency, such as PAM4 and duobinary signaling, are also investigated for even higher data rate beyond 10Gbps.
- Data scrambling and channel coding techniques such as the 8B/10B coding are usually employed in the VLSI high-speed I/O circuits for improving the data transmission reliability.
- CMOS low-voltage differential signaling (LVDS) circuit techniques are widely employed in multi-Gb/s VLSI high-speed I/O circuits that offer lower power and high performance solutions to high data rate operations.
- VLSI high-speed I/O transceiver circuits are usually implemented and associated with adaptively controlled transceiver impedances and signal slew rate to avoid the reflection, ringing, and cross talk effects in high-speed data transmission.
- Channel equalization circuit techniques are widely used to minimize the intersymbol interference (ISI) effects typically resulting from the transmission channel bandwidth limitation effects. Equalizations are performed in either the VLSI digital or the analog circuit modes, within the transmitter or receiver circuits, with feed-forward or feedback topologies, and in the linear or nonlinear forms. The combination of several equalization techniques are usually employed for even higher performance at very high data rates. For unknown or time-varying channel, adaptive equalization techniques are required to ensure high-speed I/O performances.
- Precise timing controlled clocking circuit techniques, such as the source synchronous clocking and the clock data recovery scheme, are commonly used for the bit

level timing recovery of VLSI high-speed I/O circuits. These circuit techniques, employing the VLSI phase-locked loop (PLL), delay-locked loop (DLL), clock recovery circuits (CRC), and data recovery circuits (DRC), provide significantly improved timing accuracy, compared with the traditional common clocking scheme based on low-speed I/O circuits, to ensure the targeted bit-error rate (BER) at high I/O data rate.

- Various VLSI circuit nonideality compensation circuit techniques, such as the signal dc offset compensation in receiver analog front-end (AFE) circuit, clock duty-cycle error, and I/Q phase error compensation in clock generation and distribution circuits, are commonly used in VLSI high-speed I/O to improve the signal and clock quality at high data rate.

- Modern VLSI high-speed I/O circuit systems also include various power-saving features, such as the power states, power and clock gating, input signal power level sensing, to ensure high power efficiency operation of the data transmission. Such features are becoming increasingly important with the trend of SOC integration that integrates multiple I/O (different types and large number of ports) with the cores (MCU or DSP) circuits.

1.3 THE ORGANIZATION OF BOOK CHAPTERS

The goals of the book are to provide the theoretical basis, the circuit architectures, the effective modeling methods, and the circuit implementation of modern VLSI high-speed I/O circuits. These related topics are covered in twenty-one chapters with the contents arranged as follows:

The introduction to VLSI high-speed I/O circuit technology is covered in chapter 1. The trends of modern VLSI high-speed I/O circuit technologies and the fundamental design challenges are discussed.

The delay timing-based I/O circuit prototype and time-domain SFG models are introduced in chapter 2. Such models provide effective methods for analyzing various VLSI high-speed I/O circuits. The basic timing constraint equation based on the I/O prototype circuit model provides the theoretical basis for the design implementations of various VLSI high-speed I/O circuits. The basic delay transformation and timing borrow techniques provide useful methods in the design and design analysis of various high-speed I/O circuits.

In chapter 3, various basic VLSI circuit delay sources are studied. Major sources of VLSI circuit delay times are penalties for charging the parasitic capacitors and inductors in the VLSI circuits. Delay models of VLSI passive, active, and I/O circuit channels are discussed.

In chapter 4, various basic VLSI circuit delay uncertainty sources are discussed. Basic VLSI circuit delay time noise sources, such as the power/ground noises, substrate coupling noise, cross talk, reflection/ringing effects, the ISI effects, and the thermal noise, are analyzed. The fundamental mechanisms of voltage noise to delay variation conversion effects of VLSI circuit components are discussed.

In chapter 5, the statistical analysis techniques of VLSI circuit delay variations are provided. The high-speed I/O link delay timing budgeting and the delay variations quantification methods are discussed. The VLSI circuit jitter signatures, such as the Gaussian-based random jitter (RJ) and the deterministic jitter (DJ) consisting of the periodic jitter (PJ), the duty-cycle distortion (DCD) jitter, and the data-dependent jitter (DDJ), are analyzed. The jitter combination and decomposition techniques and their applications are studied.

In chapter 6, three basic VLSI delay control circuit elements, including the voltage-controlled delay line (VCDL) circuits, the voltage-controlled oscillator (VCO) circuits, and the phase interpolator (PI) circuits are studied. Time-domain modeling and VLSI circuit implementations of these elements are discussed.

In chapter 7, basic VLSI synchronization circuit elements are introduced. The performance limitations of VLSI synchronization circuits are analyzed. Various VLSI synchronization circuit implementations are provided.

In chapter 8, basic VLSI delay time and phase computational circuit elements in VLSI time-domain signal processing circuits, including the delay time addition, delay time scaling, delay time integration, and frequency divider circuits, are studied. Time-domain models and VLSI circuit implementations of these circuits are discussed.

In chapter 9, basic VLSI phase detector (PD) circuits are studied. The VLSI circuit operations and implementations of two phase detector families, including the clock phase detectors (CPDs) and the data phase detectors (DPDs), are discussed.

In chapter 10 and 11, the VLSI phase-locked loop (PLL) and delay-locked loop (DLL) circuits are introduced. The SFG-based modeling and analysis methods of VLSI PLL and DLL circuits are introduced. VLSI circuit implementations of PLL and DLL circuits are discussed.

In chapter 12, VLSI on-chip high-speed I/O impedance termination circuit techniques are studied. The VLSI circuit implementation techniques of adaptive termination of high-speed I/O circuits are discussed.

In chapter 13, VLSI high-speed I/O circuit channel equalization circuit techniques are studied. The theoretical basis of VLSI equalization circuits for the NRZ data stream is provided. The circuit implementations of VLSI high-speed I/O equalizer circuits are discussed.

In chapter 14, basic VLSI I/O circuit architectures, including the common clock I/O circuits, the source synchronous clocking I/O circuits, and the clock and data recovery I/O circuits, are discussed. The timing basis and the practical strengths and limitations of these I/O circuit architectures are analyzed.

In chapter 15 and 16, the VLSI high-speed I/O transceiver (transmitter and receiver) circuit design techniques are discussed. The design trade-offs of VLSI high-speed I/O transmitters and receiver circuits are studied.

In chapter 17, the theoretical basis of VLSI high-speed I/O data recover circuits (DRC) is provided for. The loop transfer characteristics of the first- and second-order DRC structures are analyzed. The circuit implementations of the VLSI high-speed I/O DRC are discussed.

In chapter 18, the impacts of VLSI circuits to the signal delay time variation are analyzed. The jitter transfer functions of two VLSI circuit families, including the phase-domain signal circuits and filterlike circuits, are studied. The theoretical basis of jitter cancellation techniques for high performance high-speed I/O circuits is discussed.

In chapter 19, the signal integrity (SI) and power delivery (PD) design issues in VLSI high-speed I/O circuits are analyzed. The sources and solutions to the typical SI and PD issues in VLSI high-speed I/O circuits are discussed.

In chapter 20, the channel characterization techniques based on TDR and VNA techniques are discussed. The theoretical basis and reflection signature of various I/O link components are studied.

In chapter 21, the basic VLSI reference circuit implementations for high-speed I/O circuits are discussed. The current biasing circuit structures, including the Vt-based, the V_{BE}-based, and the constant Gm current bias and the voltage reference circuit structures, such as the bandgap reference and temperature sensor circuits, are studied.

References:

[1]   F. Zarkeshvari, et al, "High-Speed Serial I/O Trends, Standards, and Techniques," 2004 1st International Conference on Electrical and Electronics Engineering.

[2]   M. J. E. Lee, et al, "Low-Power Area-Efficient High-Speed I/O Circuit Techniques," IEEE JSSC, Vol. 35, No. 11, Nov. 2000.

[3]   M. J. E. Lee, et al, "CMOS High-Speed I/Os, Present and Future," ICCD2003.

[4]   J. F. Bulzacchelli, et al, "A 10-Gb/s 5-Tap DFE/4-Tap FFE Transceiver in 90nm CMOS Technology," IEEE JSSC Vol. 41, No. 12, December 2006.

[5]   H. Hatamkhani, et al, "Power-Center Design of High-Speed I/Os," DAC 2006.

[6]   J. R. Broomall, et al, "Extending the Useful Range of Copper Interconnects for High Data Rate Signal Transmission," 1997 Electronic Components and Technology Conference.

[7]   C. K. K. Yang, "Design of High-speed Serial Links in CMOS," Technical Report No. CSL-TR-98-775, December 1998.

[8]   E. Haq, et al, "JAZIO Signal-Switching Technology—A Low-Cost Digital I/O for High-Speed Applications," IEEE 2001.

[9]   B. Casper, et al, "An 8-Gb/s Simultaneous Bidirectional Link with On-Die Waveform Capture," IEEE JSSC, Vol. 38, No. 12, Nov. 2003.

[10] J. F. Bulzaccheli, et al, "A 10-Gb/s 5-Tap DFE/4-Tap FFE Transceiver in 90-nm CMOS Technology," IEEE JSSC, Vol. 41, No. 12, Dec. 2006.

[11] H. Tamura, "CMOS High-Speed I/O—Background, Circuits, and Future Trends," IEEE, Nov. 26th, 2004.

References

CHAPTER 2
I/O CIRCUIT PROTOTYPE MODELS AND TIME-DOMAIN SIGNAL PROCESSING

VLSI high-speed I/O circuits are a family of VLSI time-domain signal processing circuits that are mainly focused on information transmission. For the typical VLSI high-speed I/O circuit configurations, the information sender and receiver circuits are physically located in different chips within the electronic systems connected through electrical channels, typically consisting of the chip packages, binding wires, connectors, and the PCB traces. VLSI high-speed I/O circuits usually suffer from severe signal delay time and voltage uncertainty effects compared with the typical on-chip VLSI circuits. Consequently, VLSI high-speed I/O circuits usually adopt sophisticated circuit techniques and implementations to support reliable performance of data transmission at high data rates. Among all circuit operations, bit-level signal recovery operations are the most fundamental and most important in VLSI high-speed I/O circuits. Bit-level signal recovery operation requires that the transmitted data signals in I/O circuits maintain high-enough signal-to-noise ratio (SNR) during the data transmission, and that the data streams are sampled at the optimal time in the receivers. VLSI high-speed I/O circuits can be modeled based on the time-domain signal processing (TDSP) system approaches. Under the time-domain prototype I/O circuit model, only the delay time aspects of the VLSI high-speed I/O circuits are included in the I/O circuit performance analysis. The voltage variation effects are indirectly modeled based on the voltage to delay time variation conversion mechanisms. This method provides a highly effective way for the analysis and implementations of VLSI high-speed I/O circuits. The basic I/O delay time constraint equations and the signal flow graph (SFG) model developed based on the prototype I/O model provide the basis for the performance modeling and implementation of VLSI high-speed I/O circuits.

2.1. THE PROTOTYPE I/O CIRCUIT MODEL

A VLSI high-speed I/O physical layer (PHY) circuit example is shown in figure 2.1. Such a circuit typically consists of significant number of circuit blocks to support the targeted I/O circuit operations.

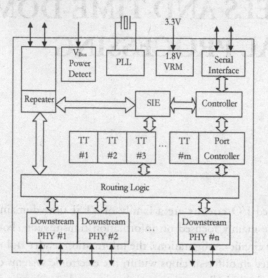

Fig. 2.1. High-speed I/O circuit example

At behavioral level, it has been observed that only a limited number of fundamental functional circuit blocks and operations play the critical roles in the high-speed I/O circuit performances. VLSI high-speed I/O circuits can therefore be significantly simplified for functional analysis, performance modeling, and circuit implementation. An I/O circuit prototype model, employing the most fundamental I/O delay time elements, is introduced in this text as effective method for the analysis and design of various VLSI high-speed I/O circuits.

For a specific VLSI high-speed I/O circuit structure, an I/O circuit prototype model can be constructed based on the following time-domain signal processing circuit concepts:

- Only time-domain parameters of the I/O circuit are used in the I/O circuit prototype model.

- Voltage variations indirectly impact the I/O circuit performance, and that can be modeled using the voltage to delay time variation transfer functions.

- High-speed I/O channel can be modeled as an equivalent delay time element (t_D).

- Data transmitter is modeled as a synchronization element, and it is specified by a clock-to-output delay time (T_{CO}) parameter.

- Data receiver is modeled as a sampler (synchronization) element, and it is specified by its input setup time (T_{Setup}) and hold time (T_{Hold}) parameters.

- Transmitter reference clock is modeled using the transmitter clock delay time (or phase) (δ_{TX}) with respect to a global time reference.
- Receiver reference clock is modeled using the receiver clock delay time (or phase) (δ_{RX}) with respect to a global time reference.
- The delay time parameters of the high-speed I/O circuits are discrete-time functions that in general should be modeled using z-domain functions. However, these functions can be simplified using s-domain functions based on the s-to-z transformation under the oversampling operation conditions.

Shown in figure 2.2 is the generic time-domain high-speed I/O circuit prototype model based on the time-domain signal processing system approach.

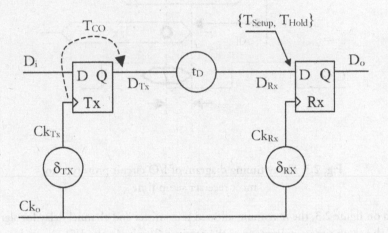

Fig. 2.2. I/O prototype circuit model

2.2. I/O CIRCUIT DELAY CONSTRAINT EQUATION

The time-domain I/O circuit prototype provides an effective way to model the fundamental data transmission operation of VLSI high-speed I/O circuits at the microscopic (or bit) level:

- The digital data bit D_i is transmitted from the transmitter synchronization unit represented by D_{TX}, synchronized by the transmitter clock and with a transmitter clock to output delay time (T_{CO}).
- The data bit propagates through the channel with an equivalent delay time (t_D).
- The data bit D_{RX} is captured by the receiver sampler, specified by the receiver setup time (T_{setup}) and hold time (T_{hold}), controlled by the receiver reference clock.
- The transmitter and receiver reference clocks are specified by their delay times (δ_{Tx} and δ_{Rx}) respectively, with respect to a given global time reference.

The above data transmission operation for meeting the setup and hold times of the receiver can be graphically described using the delay timing diagrams shown in figure 2.3 and figure 2.4.

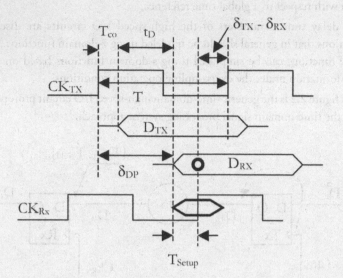

Fig. 2.3. Delay timing diagram of I/O circuit prototype to meet receiver setup time

Based on figure 2.3, the maximum allowed transmitter and channel delay for signal to meet the setup time requirement of the receiver sampler in the I/O circuit prototype can be expressed as

$$\delta_{DP} < \delta_{RX} - \delta_{TX} + T - T_{Setup}$$

$$(2.1)$$

In above equation, δ_{DP} is the equivalent datapath delay of the high-speed I/O circuit, which is the sum of the T_{CO} of the transmitter synchronization element and the effective propagation delay (t_D) of the I/O channel.

On the other hand, based on in figure 2.4, the minimum allowed transmitter and channel delay for signal to meet the hold time of the receiver sampler in I/O circuit prototype can be expressed as

$$\delta_{DP} > \delta_{RX} - \delta_{TX} + T_{Hold}$$

$$(2.2)$$

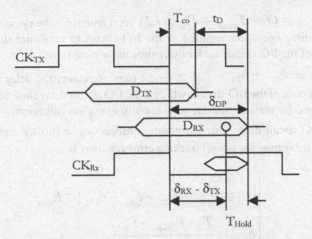

Fig. 2.4. Delay timing diagram of I/O circuit prototype to meet receiver hold time

Equations 2.1 and 2.2 can be combined into a so-called basic I/O delay time constraint equation for the valid bit-level data transmission as

$$T_{Hold} < \delta_{DP} + \delta_{TX} - \delta_{RX} < T - T_{Setup}$$

$$(2.3)$$

2.3 I/O TRACKING PHASE ERROR EQUATION AND I/O SFG MODEL

The basic I/O delay time constraint equation provides the fundamental delay time criteria for valid data transmission at the bit level. This equation can also be rewritten into the symmetric form as

$$-\frac{T - T_{Setup} - T_{Hold}}{2} < \delta_{DP} + \delta_{TX} - \delta_{RX} - \frac{T - T_{Setup} + T_{Hold}}{2} < \frac{T - T_{Setup} - T_{Hold}}{2}$$

$$(2.4)$$

These terms in the symmetric delay time constraint equation have their very specific meaning in the practical VLSI high-speed I/O circuit operations:

- The ($\delta_{DP} + \delta_{TX} - \delta_{RX}$) term represents the effective datapath delay time of the I/O, which determines the performance of the I/O circuit.

- The $(E_{max} \equiv (T - T_{Setup} - T_{Hold})/2)$ term represents the maximum available (single-side) delay time budget for an I/O circuit at targeted clock frequency under the constraint of the receiver setup and hold time penalty.

- The $(\delta_{T/2} \equiv (T - T_{Setup} + T_{Hold})/2)$ term represents the eye-centering delay time shifting operation required in the I/O circuit to maximize the delay time margin of the I/O circuit under delay time noise conditions.

- The ($E \equiv \delta_{DP} + \delta_{TX} - \delta_{RX} - \delta_{T/2}$) term represents the delay time (phase) tracking error of the I/O circuit, which is the I/O circuit delay time deviation from its optimal (or maximum delay time margin) operation condition.

The basic I/O circuit delay time constraint equation can be further expressed in the form of I/O delay time (or phase) tracking error equation as

$$
\begin{cases}
|E| \equiv |\,\delta_{DP} + \delta_{TX} - \delta_{RX} - \delta_{T/2}\,| < E_{max} \\
\delta_{T/2} \equiv \dfrac{T - T_{Setup} + T_{Hold}}{2} \\
E_{max} \equiv \dfrac{T - T_{Setup} - T_{Hold}}{2}
\end{cases}
$$

(2.5)

The above equation can be used to describe some important insights of the VLSI high-speed I/O circuits:

- The negative contribution of the receiver clock delay time element (δ_{RX}) in the effective datapath delay time implies the possibility of delay compensation in I/O circuit using properly selected receiver clock phase. Such a concept serves as the mathematical basis for various high-speed I/O timing compensation techniques that are based on correlated delay generation of receiver reference clock for high-speed I/O performance improvement.

- For any positive maximum available delay time budget term of the I/O circuit (i.e., $E_{max} \equiv (T - T_{Setup} - T_{Hold})/2 > 0$), there is a range of delay time parameter ($\delta_{DP} + \delta_{TX} - \delta_{RX} - \delta_{T/2}$) that can meet the basic I/O circuit delay time equation.

- The uncertainties in delay time elements make it impossible in practical I/O circuit to be designed with perfectly zero delay time tracking error. Instead, only the mean value of the phase tracking error can be minimized (as shown in figure 2.5) to provide a so-called maximized time margin (MxTM) operation condition, and such a delay time arrangement serves as the fundamental delay time design criteria for all types of high-speed I/O circuits:

$$
\overline{E} = \overline{\delta_{DP} + \delta_{TX} - \delta_{RX} - \delta_{T/2}} = 0
$$

(2.6)

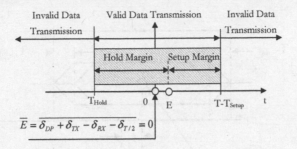

Fig. 2.5. The MxTM I/O circuit operation condition

The MxTM equation given in equation 2.6 can also be expressed in the desired receiver reference clock phase form as

$$(\delta_{RX})_{MXTM} = \delta_{DP} + \delta_{TX} - \delta_{T/2} \tag{2.7}$$

This receiver MxTM sampling condition allows the receiver sampling clock edge to be centered at the incoming data eye pattern as shown in figure 2.6.

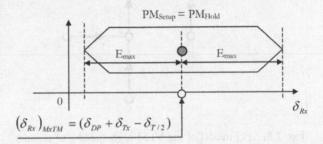

Fig. 2.6. The MxTM sampling condition

For practical high-speed I/O circuit under noise conditions, actual receiver sampling point is usually deviated from the MxTM sampling condition that can be expressed by the delay time tracking error E in terms of receiver reference clock as

$$E \equiv \delta_{RX} - (\delta_{RX})_{MxTM} = \delta_{RX} - (\delta_{DP} + \delta_{TX} - \delta_{T/2}) \tag{2.8}$$

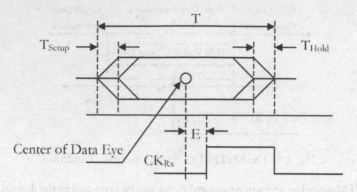

Fig. 2.7. The delay time tracking error of I/O circuits

Based on the equation 2.5, the prototype I/O circuit can also be modeled mathematically using a time-domain I/O delay time tracking error signal flow graph (SFG) as

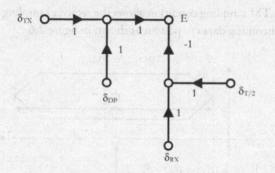

Fig. 2.8. SFG model of the VLSI high-speed I/O circuits

Under this SFG model, the performances of an I/O circuit for the bit-level data transmission can be modeled using the tracking phase error of the circuit. The basic delay parameters of the I/O circuits are specified by the equivalent datapath delay, the receiver reference clock phase, and the eye-centering delay shift. Such a model, incorporated with the statistical properties of these I/O circuit delay time parameters, provides a macroscopic performance modeling method to the high-speed I/O circuits.

The SFG model also provides the VLSI time-domain signal processing (TDSP) system partitioning of the VLSI high-speed I/O design practices. The design efforts of VLSI high-speed I/O circuit can be described as the selection of a set of timing parameters based on certain VLSI time-domain signal processing circuit techniques such that the constraint given in the SFG can be met during the data transmission operation. Such TDSP system design approach to VLSI high-speed I/O circuit design is shown in figure 2.9.

Under this TDSP system model, the key tasks for the high-speed I/O circuit design can be summarized based on the design solutions to the four prototype I/O delay parameters as follows:

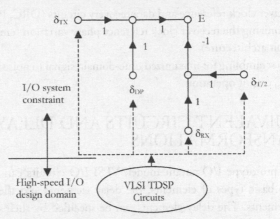

Fig. 2.9. TDSP system approach of high-speed I/O circuit design

- Design solution of δ_{DP} related to the following VLSI high-speed I/O circuits:
  - o   TCO and TCO variations minimization in transmitter driver and pre-drivers;
  - o t_D and its variation minimization with transmitter equalization and termination circuits;
  - o t_D and its variation minimization in PCB signal trace circuits;
  - o t_D and its variation minimization with receiver equalization and termination circuits

- Design solution of $\delta_{T/2}$ related to the following VLSI high-speed I/O circuits:
  - o   Receiver preamplifier and sampler design for minimizing the receiver synchronization circuit setup and hold times and variation.
  - o   High-speed I/O circuit signaling plan for maximum the available data transmission timing window and cycle-to-cycle clock period variation
  - o   I/Q clock phase error minimization

- Design solution of δ_{TX} related to the following VLSI high-speed I/O circuits:
  - o   Transmitter clock reference generation circuits (PLL, DLL, etc.) for minimizing the transmitter clock reference phase variation
  - o   Transmitter clock tree and clock buffer design for minimizing the transmitter clock reference phase variation

- Design solution of δ_{RX} related to the following VLSI high-speed I/O circuits:
  - o   Receiver clock reference generation circuits (PLL, DLL, PI etc.) for optimized clock skew and the minimized receiver clock reference phase variation (common clock I/O circuit architecture)
  - o   Receiver clock reference generation circuits (PLL, DLL, PI etc.) for minimizing the receiver clock reference phase variation (forward clock I/O circuit architecture)

o   Receiver clock reference and data recovery circuits (DRC, PLL, DLL, PI.) for minimizing the receiver clock reference phase variation (embedded clock I/O circuit architecture)

o   Data scrambling for maximized time-domain signal to noise ratio for improved DRC circuit operation

2.4 EQUIVALENT CIRCUITS AND DELAY TIME TRANSFORMATIONS

Based on the prototype I/O circuit model, VLSI I/O circuits can be fully modeled based on two basic types of elements, the delay elements and the synchronization (sampling) elements. The delay elements can be specified by their delay parameters. The synchronization elements, on the other hand, are defined by their setup and hold times (T_{Setup} and T_{Hold}) and clock-to-out delay (T_{CO}) parameters.

Under the I/O prototype circuit model, two circuits are said to be equivalent if, and only if, both have the same delay time parameters. It can be seen that the four circuits shown in figure 2.10 are equivalent since all of them have the same total delays.

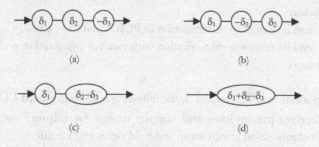

Fig. 2.10. Equivalent VLSI delay time circuits structures

The equivalent prototype circuit concept can be used to derive prototype circuit transformations that can be used to simplify the analysis of the I/O circuits.

Shown in figure 2.11 is the clock path to datapath delay transformation based on the equivalent prototype circuit concept for the prototype sampling circuit element. For the first circuit shown in figure 2.11a, the effective TCO of the circuit equals the sum of the TCO of the original synchronization circuit element and the added delay time δ in the clock path of the circuit. The equivalent setup time and hold time window of circuit equals to the setup time and hold time window of the original synchronization circuit element with delay shifted by δ in the time-domain. Similarly, the delay parameters of the second circuit shown in figure 2.11b can also be analyzed. The effective TCO of the circuit equals to the TCO of the original synchronization circuit element with the added delay time of δ, which is the same as the effective TCO of the circuit in figure 2.11a. On the other hand, the setup time and hold time window of the second circuit equals to the original synchronization circuit element with a delay shift by δ, which is also the same as the first circuit. Therefore by definition, the two circuit structures are equivalent to each other.

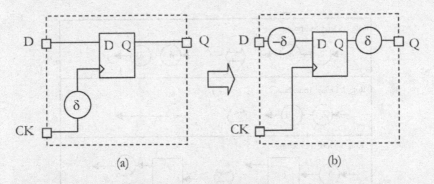

(a)   (b)

Fig. 2.11. Equivalent VLSI synchronization circuit structures

The above equivalent circuit structures can also be derived using the equivalent delay transformation shown in figure 2.12 based on the following equivalent delay transformation rules:

- The difference of the input datapath and the reference clock path should be unchanged.
- The sum of the reference clock path delay, the TCO of sequential element, and the delay time of the output datapath of circuit block should be unchanged.

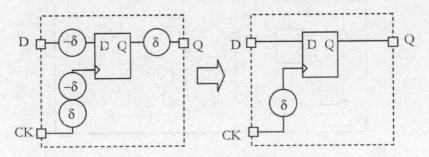

Fig. 2.12. Equivalent delay transformation of synchronization circuit elements

Shown in table 2.1 are some equivalent delay transformations derived based on the equivalent circuit structures under the prototype modeling methods.

Table 2.1. Basic equivalent VLSI circuit delay time transformations

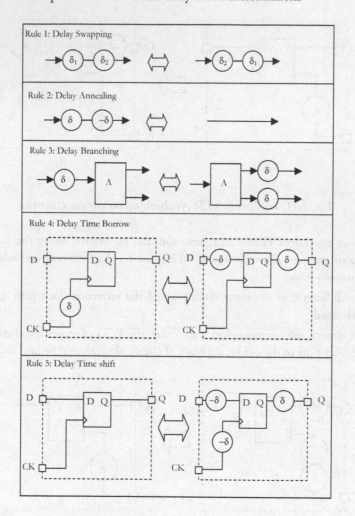

2.5 THE TIME-BORROW TECHNIQUE

Time-borrow technique based on the equivalent delay transformation can be used to maximize the delay time performance of a given high-speed I/O circuit. Such a technique allows borrowing time from the neighboring stage by purposely introduced clock path delays in either the transmitter or the receiver clock references. By applying the delay transformation rule, it can be seen that the effect of the clock path delays can be used to adjust the datapath delay of the I/O circuits as shown in figure 2.13.

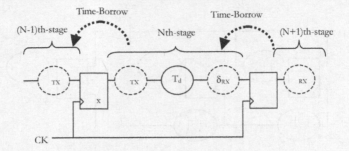

Fig. 2.13. The time-borrowing technique

An application of the time-borrow techniques is the VLSI-matched delay line sampler circuit shown in figure 2.14, where very fine sampling resolution can be achieved with coarse delay elements based on delay matching. In the matched delay line circuit structure, two sets of matched delay elements are placed inside each bit slice of the sampler circuit in the datapath and clock path respectively.

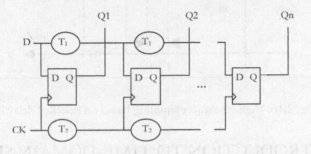

Fig. 2.14. The VLSI-matched delay line sampler circuit structure

In this circuit, delay time T_1 and T_2 are significantly higher than the targeted sampling delay time resolution, representing the delay of the matched delay elements in the data and clock path respectively. By applying the equivalent delay transformation, we have the equivalent circuit of the sampler as shown in figure 2.15.

The attractive feature of the matched delay line sampler circuit is that very fine resolution sampling can be achieved using well-matched coarse delay elements. For the circuit shown in figure 2.15, very high sampling resolution of $(T_2-T_1) \ll T_2$ or T_1 can be achieved, at some penalty of increased clock to output delay.

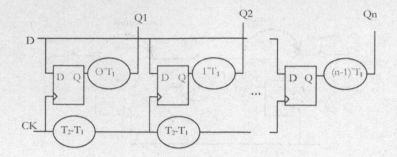

Fig. 2.15. The equivalent circuit for matched delay
line sampler circuit

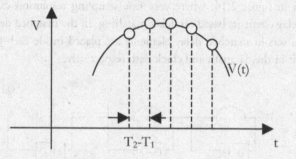

Fig. 2.16. High-resolution sampling based on matched delay line

2.6 INTRODUCTION TO TIME-DOMAIN SIGNAL PROCESSING

VLSI high-speed I/O circuits operation can be effectively modeled based on the delay (or phase) aspects of the signals. Such model suggests a special VLSI signal processing family that is fully based on the timing aspect of the signals. We can define the time-domain (TD) signal using the delay (phase) of the events (e.g., rise or fall edges) of the digital signal with respect to a time reference. Such time-domain signals and systems can be formally defined by the following statements:

- VLSI time-domain signals are the delay (phase) of the events (such as rise and fall edges of a NRZ signal) related to a given time reference.

- VLSI time-domain signals are discrete-time signals that are only defined at the discrete-time points and should be treated in z-domain. However, under the oversampling conditions, such signals can also be approximated as continuous time signals and represented in s-domain using the s-to-z transformation.

- The defined events may not occur at every defined discrete-time points. In such cases, interpolation techniques are usually used to interpolate the time-domain signal at these time points.

Shown in figure 2.17 is the time-domain representation of a NRZ signal where an event missing occurs at t = 3T and the interpolation method using previous delay value is used.

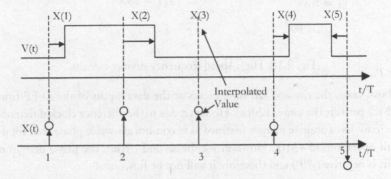

Fig. 2.17. Time-domain representation of NRZ signal

The time-domain signal can be expressed in two ways: Either using the delay time of the event with respect to the given clock reference or using the phase of the event with respect to the given clock reference. The relationship between the delay time and phase representations is given as:

$$\phi = 2\pi (t / T)$$

(2.9)

Where t is the delay time of the event and the ϕ is the phase of the event. T is the reference clock period.

Note that in applications with single reference clock period T, the delay time and the phase representations of the time-domain signals are equivalent to each other. However, for circuits with multiple reference clock periods, the delay time and phase representations may have different meanings. Such difference can be very important for the time-domain signal processing system operations.

The difference between the delay time and the delay phase can be explained using the time-domain signal processing (frequency divider) circuits shown in figure 2.18, where both circuits are identical and only have different control clock periods.

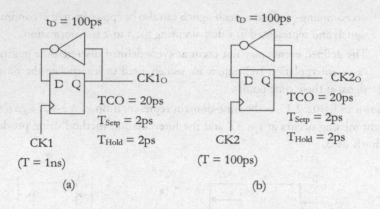

Fig. 2.18. High-speed frequency divider circuits

In above cases, the rise and fall time events at the data inputs of the D-FF from the feedback path are the same (120ps). However, due to the reference clock difference, the first circuit has a positive margin (defined as maximum allowable phase shift for proper circuit operation) of +316°. However for the second circuit, the phase margin of the circuit is negative (-79°) and therefore it will not be functional!

It can be seen that VLSI high-speed I/O circuits can be modeled based on the time-domain signal process system. Such a system can be constructed using a set of basic time-domain signal processing elements:

- TD signal generation and addition: injection of delay to the digital signal, addition of multiple delay element to a digital signal
- TD signal scaling
- TD signal integration
- Delay/phase detection
- Synchronization
- Voltage to time (V/T) and time to voltage (T/V) conversions.

APPENDIX 2

A2.1 THE TRANSMITTER CLOCK TO OUTPUT DELAY

High-speed I/O transmitter clock to output delay (T_{CO}) is defined as the delay time from the effective sampling edge (such as the rise edge) of the transmitter clock CK_{TX} to the sampled data appear at the output of the transmitter under the proper loading condition of the channel.

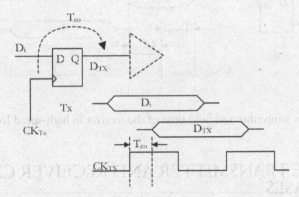

Fig. 2.19. The transmitter clock to output delay

A 2.2 THE PROPAGATION DELAY OF CHANNEL

Under the time domain circuit model, the high-speed I/O channel can be modeled using the channel propagation delay t_D as the delay of the entire channel with proper driving (signal slew rate and signal swing) from the transmitter and proper loading from the receiver circuit as shown in figure 2.20. Note that channel here includes all circuit components between the transmitter and the receiver.

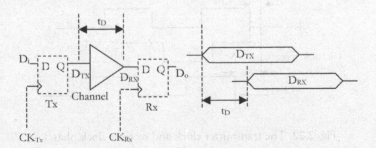

Fig. 2.20. The channel propagation delay

A2.3 THE RECEIVER SETUP TIME AND HOLD TIME

Setup time (T_{Setup}) and hold time (T_{Hold}) are essential delay time parameters for synchronization circuit elements. For a valid data transfer, it is required that a valid (proper slew rate and swing) input signal be stable for a time period defined by the setup time (T_{Setup}) before the effective sampling clock edge, and it must remain stable for a time period defined by the hold time (T_{Hold}) after the effective sampling edge of the receiver clock.

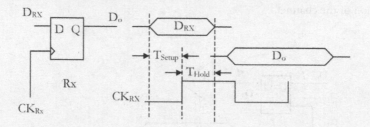

Fig. 2.21. The setup time and hold time of the receiver in high-speed I/O circuits

A2.4 THE TRANSMITTER AND RECEIVER CLOCK PHASES

In a typical high-speed I/O circuit, the transmitter and the receiver usually belong to different clock domains. The difference between the transmitter and the receiver clock phases is modeled in the basic high-speed I/O delay time model using the transmitter clock phase δ_{TX} and the receiver clock δ_{RX} as shown in figure 2.22.

The transmitter clock phase (δ_{TX}) is defined as the relative delay of the transmitter clock with respect to a global (virtual or real) clock reference.

Similarly, the receiver clock phase (δ_{RX}) is defined as the relative delay of the transmitter clock with respect to a global (virtual or real) clock reference.

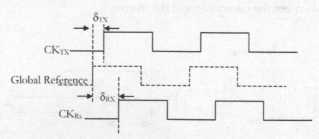

Fig. 2.22. The transmitter clock and receiver clock phases

A2.5 DELAY TIME EQUATION BASED ON UNIFIED CLOCKS

For the high-speed I/O circuit shown in figure 2.23, if both the transmitter and the receiver are clocked by the same control clock, the delay time constraint for the valid data transfer can be analyzed.

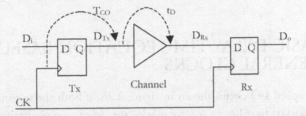

Fig. 2.23. Basic high-speed I/O circuit with unified clock reference

From the setup time requirement of the I/O circuit and considering the unified reference clock constraint as shown in figure 2.24, we have the maximum datapath delay condition as

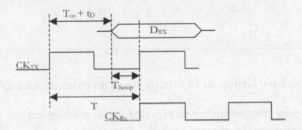

Fig. 2.24. The setup time requirement of the I/O circuit with unified clock reference

$$T_{CO} + t_D < T - T_{setup}$$

(2.10)

Based on the same way for the hold time requirement as shown in figure 2.25, we have the minimum datapath delay condition as

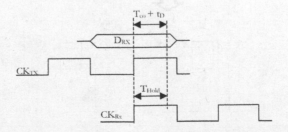

Fig. 2.25. The hold time requirement of the I/O circuit with unified clock reference

$$T_{CO} + t_D > T_{Hold} \tag{2.11}$$

By combining equations 2.10 and 2.11, we have the basic delay time equation of the high-speed I/O circuit with unified clock reference as

$$T - T_{Setup} > T_{CO} + t_D > T_{Hold} \tag{2.12}$$

A2.6 BASIC DELAY TIME EQUATION BASED ON GENERAL CLOCKS

For the high-speed I/O circuit shown in figure 2.26, if both the transmitter and the receiver are clocked by different control clocks, the delay time constraint for the valid data transfer can be analyzed.

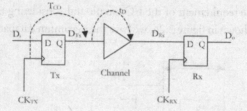

Fig. 2.26. Basic high-speed I/O circuit based on different clock references

Based on the delay transformation rule provided in the previous section, the above I/O structure can be converted to the unified clock reference scheme as shown in figure 2.27.

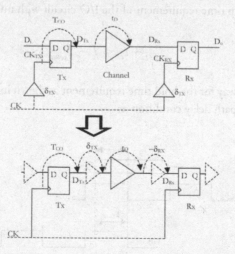

Fig. 2.27. Conversion of I/O circuit to unified clock reference

By applying equation 2.12, we have the basic delay time constraint equation for general high-speed I/O circuits as

$$T - T_{Setup} > T_{CO} + t_D + \delta_{TX} - \delta_{RX} > T_{Hold} \tag{2.13}$$

Note that the above basic I/O circuit delay time equation provides the basis of all types of I/O systems. It also leads to the delay time borrow and the jitter compensation design techniques, the two important design techniques in high-speed I/O design.

A2.7 SIGNAL FLOW GRAPH (SFG)

Signal flow graph (SFG) is a type of block diagram that is used to graphically represent the relations among the variables of a set of linear algebraic relations. With the SFG, the nodes represent variables, and are joined by branches that have assigned directions (indicated by arrows) and gains. A signal can transmit only in the direction of the arrow. When two or more branches are jointed, an addition operation is realized. SFG representation provides a simple yet powerful way for representing the signal processing systems.

Shown in figure 2.28 are the basic SFG representations of the amplification (or scaling), addition, and the integration operations.

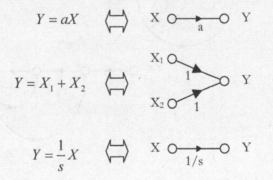

Fig. 2.28. SFG representations of basic signal processing elements

SFG representation can also be used to construct complicated linear signal processing systems. Shown in figure 2.29 are some examples of feedback system SFG representations.

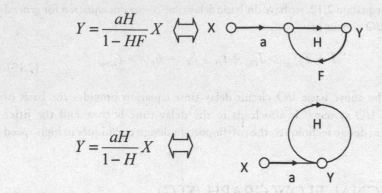

$$Y = \frac{aH}{1-HF}X$$

$$Y = \frac{aH}{1-H}X$$

Fig. 2.29. SFG representations of basic signal processing elements.

SFG representation can also be used for either s-domain or z-domain signal processing systems. Shown in figure 2.30 are two linear second-order sigma-delta modulator SFG models. One of the sigma-delta is implemented in the s-domain and the second one is in the z-domain.

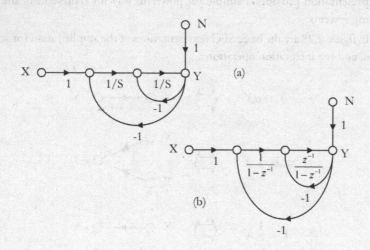

Fig. 2.30. SFG representations of sigma-delta modulators

References:

[1]   S. H. Unger, et al, "Clocking Schemes for High-Speed Digital Systems," IEEE Trans. Comput., vol. C-35, no. 10, Oct. 1986.

[2]   E. G. Friedman, "Clock Distribution Networks in VLSI Circuits and Systems," IEEE Press.

[3]   K. D. Wagner, "Clock System Design," IEEE Design Test Comput., Oct. 1988.

[4]   K. A. Sakallah, et al, "Synchronization of Pipelines," IEEE Trans. Computer-Aided Design, vol.12, no.8, Aug. 1993.

[5]   C. Thomas, et al, "Delay Time Constraints for Wave-Pipelined Systems," IEEE Trans. Computer-Aided Design, vol. 13, no.8, Aug. 1994.

[6]   A. F. Champernowne, et al, "Latch-to-Latch Timing Rules," IEEE Trans. Computers, vol.39, no.6, June 1990.

[7]   E. G. Friedman, "Clock Distribution Networks in VLSI Circuits and Systems, IEEE Press, pp.1-31.

CHAPTER 3
VLSI CIRCUIT DELAY EFFECTS

Electrical signals propagation through VLSI circuits will experience fundamental time delay effects that are contributed from the intentional or parasitic resistive, capacitive, and inductive circuit component effects. Both the on-chip and the off-chip resistor, capacitor, and inductor circuit components result in the signal delay of VLSI high-speed I/O circuits.

Signal delay effects for on-chip VLSI circuits are usually contributed from two circuit families, including the active devices, such as logic gates and buffer circuits, and the passive circuit elements, such as the VLSI interconnects. Although inductive effect begins to shown up in some global interconnects, delay times of on-chip VLSI circuit elements can typically be modeled using their effective RC time constants.

The off-chip circuit elements in VLSI high-speed I/O circuits, such as the channels that are used for inter-chip data communications, are among the dominant delay effect contributors. High-speed I/O circuit channels are typically modeled using the lossless or lossy transmission line circuits combined with lumped RLC components that are typically contributed from package and trace discontinuities. Transmission line effects, such as the reflection, ringing, cross talk, ISI, and attenuation, are among key factors that impact the performances of the VLSI high-speed I/O circuits.

3.1 BASIC VLSI CIRCUIT DELAY EFFECTS

The signal delay effects of on-chip VLSI circuits are usually contributed from the effective RC time constant of the circuit elements. Delay times of the VLSI circuits strongly depend on various circuit parameters, such as the power supply voltage, threshold value, signal swing, and loading.

3.1.1 RC MODEL OF VLSI INVERTER CIRCUIT

The delay time of the basic VLSI inverter buffer circuit can be behaviorally modeled using an effective RC circuit as shown in figure 3.1.

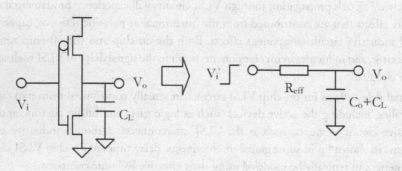

Fig. 3.1. RC behavioral model of VLSI inverter circuit

In the above model, C_L represents the external load capacitance. C_o represents effective output parasitic capacitance of the inverter. R_{eff} represents the effective output impedance of the inverter.

The effective driver impedance based on the linear MOS device model can be approximately expressed as

$$R_{eff} \approx \begin{cases} \dfrac{1}{\beta_n(V_{CC} - V_{Tn})} & \textit{fall} \\[2ex] \dfrac{1}{\beta_p(V_{CC} - |V_{Tp}|)} & \textit{rise} \end{cases}$$

(3.1)

Under this model, the delay time of the output signal to reach 50% of the maximum swing can be approximately expressed as

$$t_d \approx \begin{cases} 0.7 \cdot \dfrac{C_L + C_o}{\beta_n(V_{CC} - V_{Tn})} & fall \\[4mm] 0.7 \cdot \dfrac{C_L + C_o}{\beta_p(V_{CC} - |V_{Tp}|)} & rise \end{cases}$$

$$(3.2)$$

For a specific length of the MOS devices (usually minimal device length for digital device), the output parasitic capacitance of inverter is approximately proportional to the parasitic input capacitance of the inverter, which, in turn, is proportional to the total width of the devices:

$$C_o \approx \gamma C_g = \gamma \frac{\varepsilon_r \varepsilon_o}{t_{ox}} \sum LW = \gamma \frac{\varepsilon_r \varepsilon_o}{t_{ox}} L \sum W$$

$$(3.3)$$

where L and W are the length and width of the MOS devices. For a given CMOS process technology, power supply voltage, and P/N device ratio, it is interesting to see that the product of the effective resistance and the total parasitic capacitance of the inverter is approximately a constant, independent of the device width:

$$0.7 R_{eff} \cdot (C_g + C_o) \equiv 0.7 R_{eff} C_g (1 + \gamma) = t_o \approx const.$$

$$(3.4)$$

Such delay parameter t_o can therefore be used as a fundamental process technology parameter that represents the intrinsic delay of the inverter with a load of the same type of inverter buffer (i.e., FAN-OUT = 1).

3.1.2 VLSI ON-CHIP INTERCONNECT RC MODELS

VLSI on-chip interconnects can usually be modeled using distributed RC network. Two commonly used interconnect parasitic capacitance models, including the "Sandwich" and the "Pizza" models, are shown in figure 3.2, where the sandwich model is used for lower-level interconnect layers and the pizza model for the top-level VLSI interconnect layer.

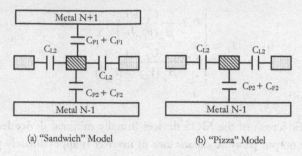

(a) "Sandwich" Model   (b) "Pizza" Model

Fig. 3.2. VLSI interconnect capacitance models

In the above model, C_{P1} and C_{P2} are the plate capacitances of specific interconnect layer that are coupled to the lower- and the upper-layer (for sandwich model) metals respectively. C_{L1} and C_{L2} are the lateral coupling capacitances to the left and right sides of interconnects respectively. C_{F1} and C_{F2} are the fringing field associated capacitances to the upper and lower plates respectively. The total parasitic capacitance of interconnect based on the above model can be expressed as

$$C = C_P + C_F + MCF \cdot C_L$$

(3.5)

where MCF is the Miller capacitor coefficient of interconnect that is used to model Miller effect resulted from the switching effects of the lateral parasitic capacitance. C_P and C_L are the total plate and lateral capacitances.

The resistance of interconnect can be modeled using the width W, length L, and the sheet resistance of interconnect as

$$R = R_s \frac{L}{W}$$

(3.6)

Under the first-order approximation, the unity length capacitance of interconnect can be expressed as function of the width W, the space S and pitch P of the routing interconnect as

$$\frac{C}{L} \approx \alpha_P \cdot W + \alpha_F + \frac{\alpha_L}{S} = \alpha_P \cdot W + \alpha_F + \frac{\alpha_L}{P - W}$$

(3.7)

The delay effect of the distributed RC interconnect can be modeled employing the piecewise RC circuit model by dividing interconnect into multiple equal length pieces with each piece represented by a lumped RC model. Shown in figure 3.3 are simulated 50% swing signals delay and 10%-90% rise times versus the number of lumped RC segment for the step signal input.

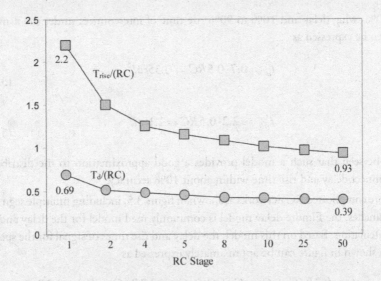

Fig. 3.3. Simulated interconnect delay and rise time

Based on the above results, the delay time and rise time of the distributed RC-based interconnect have small differences for more than four segments. The 50% delay time and the 10-90% rise time of the distributed RC-based interconnect can then be approximately expressed as

$$t_d \approx 0.39 RC = 0.39 rcL^2 \tag{3.8}$$

$$t_{rise} \approx 0.93 RC = 0.93 rcL^2 \tag{3.9}$$

In most practical application, a π lumped interconnect model as shown in figure 3.4 is widely used as a highly simplified yet fairly accurate model for the VLSI RC interconnect. Under the π model, a VLSI distributed interconnect circuit can be represented by a lumped resistor with the resistance equal to the total interconnect parasitic resistance and two identical lumped capacitors with each of capacitance equal to half of the total parasitic interconnect capacitance.

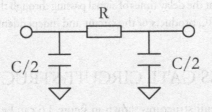

Fig. 3.4. The π lumped VLSI interconnect model

The 50% swing delay and 10% to 90% rise time of interconnect under the π model can then be expressed as

$$t_d \approx 0.7 \cdot 0.5RC = 0.35rcL^2$$

(3.10)

$$t_{rise} \approx 2.2 \cdot 0.5RC = 1.1rcL^2$$

(3.11)

It can be seen that such a model provides a good approximation to the distributed interconnect delay and rise time within about 10% accuracy.

For more complicated RC networks as shown in figure 3.5, including multiple segments and branches, the Elmore delay model is commonly used model for the delay and rise time calculation. Based on this model, the delay and rise times of signal for the specific branch shown in figure can be approximately expressed as

$$t_d \approx 0.7R_1(C_1 + C_2 + C_3 + C_4 + C_5) + 0.7R_3(C_3 + C_4) + 0.7R_4C_4$$

(3.12)

$$t_{rise} \approx 2.2R_1(C_1 + C_2 + C_3 + C_4 + C_5) + 2.2R_3(C_3 + C_4) + 2.2R_4C_4$$

(3.13)

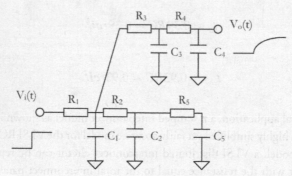

Fig. 3.5. Delay time of lumped RC circuits based on Elmore model

It is important to note that the delay times of signal passing through the passive RC network are proportional to the RC products of the circuit and independent of supply voltage.

3.1.3 VLSI PASS GATE CIRCUIT STRUCTURES

The VLSI pass gate circuit structures shown in figure 3.6 can be modeled using their effective linear resistance. The delay time of the electrical signal passing through such circuits can then be approximately expressed as

$$t_D \propto \frac{C_L}{\beta(V_{CC} - V_T)} \qquad (3.14)$$

$$t_D \propto \frac{C_L n^2}{\beta(V_{CC} - V_T)} \qquad (3.15)$$

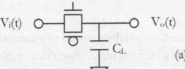

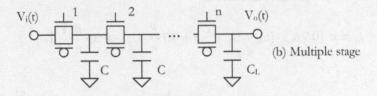

Fig. 3.6. VLSI pass gate circuit structures

Note that the delay time of a pass gate circuit increases with the square of n.

3.2 BUFFERED INTERCONNECT DELAY MODELS

For buffered interconnect shown in figure 3.7, the delay and rise time of the signal can be derived based on the equivalent driver RC model, the π interconnect model, and the Elmore delay model as

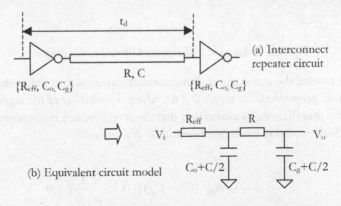

Fig. 3.7. Delay model of buffered interconnect circuit

53

$$t_d \approx 0.7 R_{eff} \cdot (C_o + C_g + C) + 0.7 R \cdot C_g + 0.35 R \cdot C$$

(3.16)

$$t_{rise} \approx 2.2 R_{eff} \cdot (C_o + C_g + C) + 2.2 R \cdot C_g + 1.1 R \cdot C$$

(3.17)

3.2.1 MINIMAL DELAY INTERCONNECT BUFFER CIRCUITS

For very long interconnect trace, repeaters are usually inserted to minimize the overall delay time of the interconnect circuits. For a buffered interconnect that employs repeaters of identical driving strength as shown in figure 3.8, signal delay time can be derived based on the equations 3.16 and 3.17 as

$$t_d \approx n \cdot [0.7 R_{eff} \cdot (C_o + C_g + \frac{C}{n}) + 0.7 \frac{R}{n} \cdot C_g + 0.35 \frac{R}{n} \cdot \frac{C}{n}]$$

(3.18)

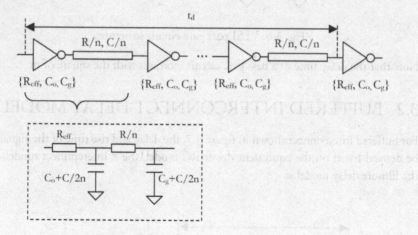

Fig. 3.8. Delay model of repeated interconnect circuit

The delay time of the entire repeated interconnect circuit path contains three major terms that are proportional to n^1, n^0, and n^{-1}, where n is number of RC segmentation. This implies that there exists a condition that the total repeated interconnect delay is minimal. Such condition can be mathematically derived as

$$\frac{\partial t_d}{\partial n} \approx [0.7 R_{eff} \cdot (C_o + C_g) - 0.35 \frac{R}{n} \cdot \frac{C}{n}] = 0$$

(3.19)

$$\Rightarrow 0.7 R_{eff} \cdot (C_o + C_g) = 0.35 \frac{R}{n} \cdot \frac{C}{n}$$

$$(3.20)$$

that is,

$$n = \sqrt{\frac{0.35 R \cdot C}{0.7 R_{eff} \cdot (C_o + C_g)}}$$

$$(3.21)$$

or

$$L_{min} = \sqrt{\frac{t_o}{0.4 r \cdot c}}$$

$$(3.22)$$

It is important to note that

- the minimal delay buffering condition is achieved when the interconnect delay time of the each segment is equal to the buffer intrinsic delay;

- equation 3.21 provides the criteria for minimal delay buffering stage number n for a given length interconnect under the targeted process technology; and

- based on equation 3.22, the minimal delay buffering interconnect length can be calculated.

On the other hand, the buffer size for minimal delay can also be derived based on equation 3.18 as

$$\frac{\partial t_d}{\partial C_g} \approx -\frac{0.7 R_{eff} \cdot C}{C_g} + 0.7 R = 0$$

$$(3.23)$$

We have that

$$\Rightarrow \frac{C_g}{R_{eff}} = \frac{C}{R} = \frac{c}{r}$$

$$(3.24)$$

or

$$\Rightarrow C_g = \sqrt{\frac{1}{0.7(1 + \gamma)} \cdot \frac{c}{r} \cdot t_o}$$

$$(3.25)$$

where t_o is the inverter intrinsic delay introduced in previous section. It can be seen that for the same VLSI process technology, the optimal buffer size for minimal delay is proportional to the square root of the interconnect capacitance to the resistance ratio. This means that high-level interconnect metal layers usually require large repeaters for signal repeating.

Under the minimal delay buffering condition, the total delay time of the buffered interconnect path can be derived as

$$t_d \approx 2 \cdot 0.7 \sqrt{R_{eff} C_g \cdot CR} = 2 \sqrt{\frac{0.7}{0.4}} \cdot \sqrt{\frac{t_o}{1+\gamma}} \cdot t_{flight} \approx 2 \sqrt{t_o \cdot t_{flight}} \propto L$$

(3.26)

As can be seen, the buffered interconnect delay is linearly proportional to the interconnect length. Such a result is therefore significantly different from the unbuffered interconnect delay, which is proportional to the square of the interconnect length.

3.2.2 LOW POWER INTERCONNECT BUFFER CIRCUITS

Since a significant portion of VLSI chip power dissipations are usually contributed from interconnect buffering circuits, it is highly desired that the power dissipation of interconnect buffering circuit can be optimized for low power dissipation. Low power interconnect repeating circuit can be implemented based on the minimization of the power delay product of the repeated interconnect circuits.

For a power supply voltage of V_{CC}, frequency f, and activity factor α, the total active power dissipation of buffered interconnect shown in figure 3.8 can be expressed as

$$P \approx n \cdot \alpha \cdot f \cdot V_{CC}^2 \cdot (C_o + C_g + \frac{C}{n})$$

(3.27)

The power-delay product of the buffered interconnect can then be expressed as

$$P \cdot t_d \propto n^2 \cdot (C_o + C_g + \frac{C}{n})[0.7 R_{eff} \cdot (C_o + C_g + \frac{C}{n}) + 0.7 \frac{RC_g}{n} + 0.35 \frac{RC}{n^2}]$$

(3.28)

The closed form minimal power-delay product condition can be solved through the equations as

$$\begin{cases} \dfrac{\partial(P \cdot t_d)}{\partial n} = 0 \\[2mm] \dfrac{\partial(P \cdot t_d)}{\partial C_g} = 0 \end{cases}$$

$$(3.29)$$

We may also use the graphical method as shown in figure 3.9 to solve for the minimal delay power product condition.

It can be seen from the figures that the optimization point for minimal power delay product curves is slightly lower than the minimal delay only curves, implying slightly longer interconnect buffering length and smaller buffer size compared with the minimal delay design can usually be selected for low power interconnect buffering:

$$\begin{cases} n^{PD} < n^{D} \\[2mm] C_g^{PD} < C_g^{D} \end{cases}$$

$$(3.30)$$

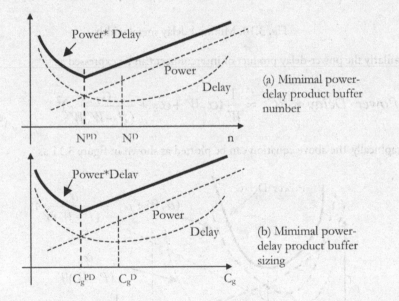

(a) Mimimal power-delay product buffer number

(b) Mimimal power-delay product buffer sizing

Fig. 3.9. Typical power-delay product versus buffer number and buffer size

3.2.3 INTERCONNECT LAYOUT PITCH TRADE-OFF

Based on the interconnect resistance and capacitance versus width relations, the delay time of the interconnect can be expressed as

$$t_d \propto RC \propto \alpha_P + \frac{\alpha_F}{W} + \frac{\alpha_L}{(P-W)W} \tag{3.31}$$

As graphically shown in figure 3.10, there is an optimal metal width for each metal pitch such that the delay time is minimized.

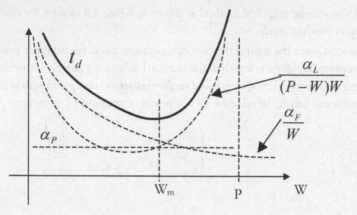

Fig. 3.10. Minimal delay metal width

Similarly the power-delay product of interconnect can be expressed as

$$Power \cdot Delay \propto RC^2 \propto \frac{1}{W}\left(\alpha_P W + \alpha_F + \frac{\alpha_L}{(P-W)W}\right)^2 \tag{3.32}$$

Graphically, the above equation can be plotted as shown in figure 3.11 as

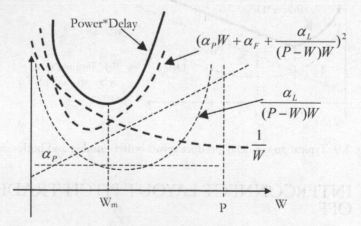

Fig. 3.11. Minimal power-delay product metal width

3.3 VLSI DELAY BUFFER CIRCUIT MODELS

For the VLSI buffered circuit shown in figure 3.12, the delay and rise time of the signal can be derived based on the Elmore delay model and the effective buffer model as

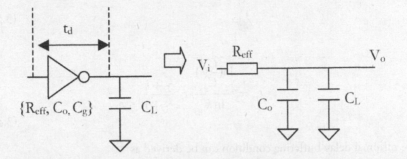

Fig.3.12. Delay model of VLSI buffer circuit

3.3.1 MINIMAL DELAY BUFFER CIRCUITS

For high fan-out (FO = C_L/C_g >>1) buffering applications, the increasingly sized (with ratio k) buffer stages as shown in figure 3.13 are commonly used to minimize the delay time of the buffer circuit.

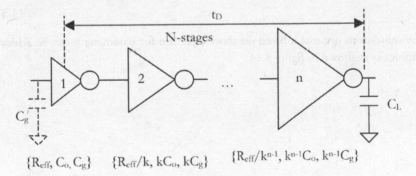

Fig. 3.13. Delay time of VLSI buffer chain

Under such circuit arrangement, each stage based on the RC delay model provides an equal delay time as

$$t_D = 0.7 R_{eff} \cdot (C_o + kC_g) = t_o \frac{\gamma + k}{\gamma + 1}$$

(3.33)

The delay time of the entire n-stage buffer chain can be expressed as

$$\begin{cases} t_D = n \cdot t_o \dfrac{\gamma + k}{\gamma + 1} \\ \dfrac{C_L}{C_g} \equiv k^n \end{cases}$$

(3.34)

or

$$t_D = \frac{\ln(\dfrac{C_L}{C_g})}{\ln k} \cdot t_o \frac{\gamma + k}{\gamma + 1}$$

(3.35)

The minimal delay buffering condition can be derived as

$$\frac{\partial t_D}{\partial k} = \frac{\partial}{\partial k} \left(\frac{\ln(\dfrac{C_L}{C_g})}{\ln k} \cdot t_o \frac{\gamma + k}{\gamma + 1} \right) = 0$$

(3.36)

or

$$\ln k = 1 + \frac{\gamma}{k}$$

(3.37)

The solution to optimal k based on above equation for parameter γ can be achieved graphically as shown in figure 3.14.

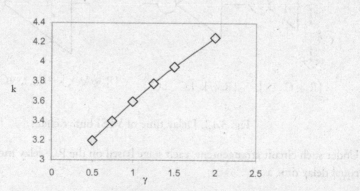

Fig. 3.14. Minimal delay fan-out versus γ for buffer chain

It can be seen that a fan-out of 3-4 is typically preferred to achieve the minimal delay buffering.

3.3.2 LOW POWER BUFFER CIRCUITS

The active power dissipation of the buffering can be approximately expressed based on the buffer parameter and buffering condition as

$$P = afV_{CC}^2[C_g + (C_o + kC_g)\sum_{i=0}^{n-1}k^i]$$

$$= afV_{CC}^2[C_g + (C_o + kC_g)\frac{\frac{C_L}{C_g}-1}{k-1}]$$

$$(3.38)$$

For n-stage buffering with equally assigned fan-out set by the input and output capacitance, the total delay can be expressed as

$$t_D = \frac{\ln(\frac{C_L}{C_g})}{\ln k}\cdot t_o\frac{\gamma+k}{\gamma+1}$$

$$(3.39)$$

The power-delay product is given as

$$P\cdot t_D \propto [1+(\gamma+k)\frac{\frac{C_L}{C_g}-1}{k-1}]\cdot\frac{\gamma+k}{\ln k}$$

$$(3.40)$$

The minimal delay power product condition for various k and C_L/C_g can be graphically solved as figure 3.15:

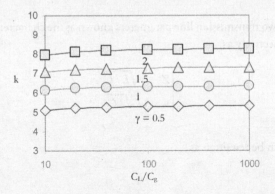

Fig. 3.15. Minimal power delay fan-out versus γ and CL/Cg for buffer chain

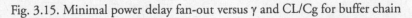

It can be seen that a slightly higher fan-out is generally desired for low power repeater circuit design.

3.4   HIGH-SPEED I/O CHANNEL DELAY EFFECTS

To the first-order, a VLSI high-speed I/O channel can be approximately modeled using a lossless transmission line circuit as shown in figure 3.16.

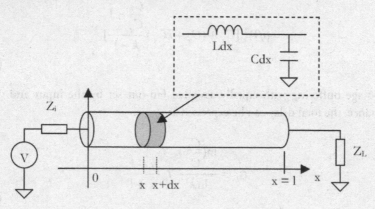

Fig. 3.16. Lossless transmission line circuit model

The voltage and current inside such transmission line circuit can be expressed using the differential voltage-current equations as

$$\begin{cases} \dfrac{\partial V}{\partial x} = -L\dfrac{\partial I}{\partial t} \\[2mm] \dfrac{\partial I}{\partial x} = -C\dfrac{\partial V}{\partial t} \end{cases}$$

$$(3.41)$$

By introducing two transmission line parameters known as the characteristic impedance and velocity respectively as

$$\begin{cases} Z_o = \sqrt{L/C} \\[2mm] v = 1/\sqrt{LC} \end{cases}$$

$$(3.42)$$

equation 3.41 can be rewritten as

$$\begin{cases} \dfrac{\partial}{\partial(x/v)}(\dfrac{V}{\sqrt{Z_o}}) = -\dfrac{\partial}{\partial t}(\sqrt{Z_o}I) \\[4mm] \dfrac{\partial}{\partial(x/v)}(\sqrt{Z_o}I) = -\dfrac{\partial}{\partial t}(\dfrac{V}{\sqrt{Z_o}}) \end{cases}$$

(3.43)

It is convenient to introduce two wave signals as

$$\begin{cases} a^+(x,t) \equiv \dfrac{1}{2}(\dfrac{V}{\sqrt{Z_o}} + \sqrt{Z_o}I) \\[4mm] a^-(x,t) \equiv \dfrac{1}{2}(\dfrac{V}{\sqrt{Z_o}} - \sqrt{Z_o}I) \end{cases}$$

(3.44)

or

$$\begin{cases} V(x,t) \equiv \sqrt{Z_o}\,(a^+(x,t) + a^-(x,t)) \\[4mm] I(x,t) \equiv \dfrac{1}{\sqrt{Z_o}}(a^+(x,t) - a^-(x,t)) \end{cases}$$

(3.45)

Equation 3.41 can be further simplified using the wave signals as

$$\begin{cases} \dfrac{\partial}{\partial\left(\dfrac{x}{v}\right)}a^+(x,t) = -\dfrac{\partial}{\partial t}a^+(x,t) \\[6mm] \dfrac{\partial}{\partial\left(\dfrac{x}{v}\right)}a^-(x,t) = +\dfrac{\partial}{\partial t}a^-(x,t) \end{cases}$$

(3.46)

It is easy to find that the general solution to equation 3.46 are two traveling waves as

$$\begin{cases} a^+(x,t) = a^+(\dfrac{x}{v} - t) \\[4mm] a^-(x,t) = a^-(\dfrac{x}{v} + t) \end{cases}$$

(3.47)

We may now get some insights into the general electrical property of the lossless transmission lines as the following:

- The two general solutions a+(x,t) and a-(x,t) represent a left and right traveling waves in the transmission line as shown in figure 3.17. The two waves are independent of each other in the transmission line. In general case, the solution to signal in the transmission line is the linear combination of the two waves.

- The velocities of the waves are the same and are given in equation as $v = 1/\sqrt{LC}$.

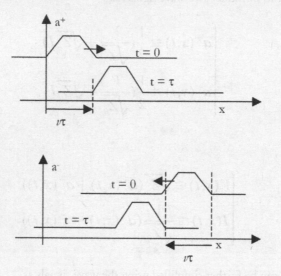

Fig. 3.17. The traveling wave solutions to the lossless transmission line

- At x = 0, we have that

$$V_s(t) = V(0,t) + Z_i I(0,t)$$

$$= \sqrt{Z_o}(a^+(0,t) + a^-(0,t)) + \frac{Z_i}{\sqrt{Z_o}}(a^+(0,t) - a^-(0,t))$$

$$= \sqrt{Z_o}\{a^+(0,t)\cdot[1 + \frac{Z_i}{Z_o}] + a^-(0,t)\cdot[1 - \frac{Z_i}{Z_o}]\}$$

$$(3.48)$$

For infinite or load matched transmission line, there is no left traveling wave. We have that

$$\begin{cases} a^-(0,t) = 0 \\ V_s(t) = \sqrt{Z_o}\{a^+(0,t)\cdot[1 + \frac{Z_i}{Z_o}] \end{cases}$$

$$(3.49)$$

The right traveling voltage and current are given as

$$\begin{cases} V(0,t) \equiv \sqrt{Z_o}\,(a^+(x,t)+a^-(x,t)) = V_s(t)\dfrac{Z_o}{Z_i+Z_o} \\[3mm] I(0,t) \equiv \dfrac{1}{\sqrt{Z_o}}(a^+(x,t)-a^-(x,t)) = V_s(t)\dfrac{1}{Z_i+Z_o} \end{cases}$$

(3.50)

It can be seen that the right traveling wave is a voltage-divided version of the source signal with division factor determined by the source impedance and the characteristic impedance of the transmission line.

• At x = l, we have that

$$V(l,t) = Z_L I(l,t)$$

$$\Rightarrow \sqrt{Z_o}\,(a^+(l,t)+a^-(l,t)) = \frac{Z_L}{\sqrt{Z_o}}(a^+(l,t)-a^-(l,t))$$

(3.51)

The reflection coefficient is given as

$$\Rightarrow \rho \equiv \frac{a^-(l,t)}{a^+(l,t)} = \frac{Z_L-Z_o}{Z_L+Z_o}$$

(3.52)

Note that the reflection will occur if there is an impedance discontinuity.

3.4.1 DELAY TIME OF LOSSLESS TRANSMISSION LINE

The time-domain response of the lossless transmission line can be analyzed based on the injection, propagation, and reflections diagram through the superposition method as shown in figure 3.18 and figure 3.19.

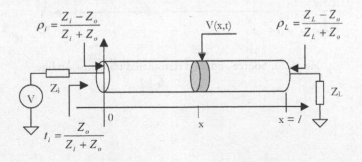

Fig. 3.18. General impedance transmission line termination configuration

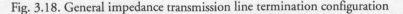

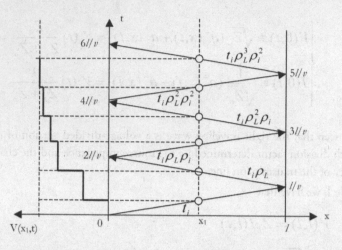

Fig. 3.19. Reflection diagram of lossless transmission line

It can be seen that the following transmission line termination schemes offer the best situation of the signal transmission in the transmission line that only depends on the electrical and physical dimension of the transmission line and not on other electrical parameters.

3.4.2 SOURCE AND LOAD MATCHED CHANNEL

- For lossless transmission line with both source and load well matched ($Z_i = Z_L = Z_O$), there are no reflection on both ends of the line ($\rho_i=0$, $\rho_L=0$). The transmission coefficient at the source based on voltage division is given as $t_i = 0.5$ as shown in figure 3.20. The reflection diagram is shown in figure 3.21.

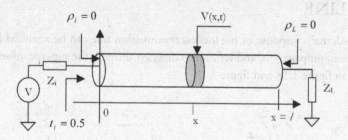

Fig. 3.20. Source and load terminated transmission line

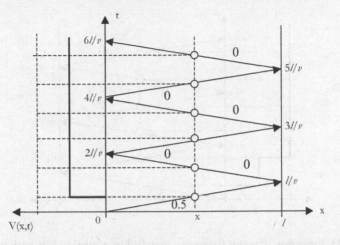

Fig. 3.21. Reflection diagram for source and load matched transmission line

The delay time of the signal traveling through the channel is given by the transmission line length divided by the velocity of the signal in the line as

$$t_D = l\sqrt{LC}$$

(3.53)

Note that the signal at the load can only reach half of the swing of the source.

3.4.3 SOURCE MATCHED/LOAD OPEN CHANNEL

- For lossless transmission line with source well matched and load open ($Z_i = Z_O \ll Z_L$), there is no reflection on source end of the line ($\rho_i = 0$) and the signal at the load will be totally reflected ($\rho_L = 1$). The transmission coefficient at the source based on voltage division is given as $t_i = 0.5$ as shown in figure 3.22. The reflection diagram is shown in figure 3.23.

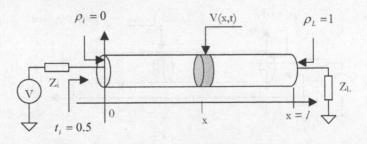

Fig. 3.22. Source matched transmission line with open load

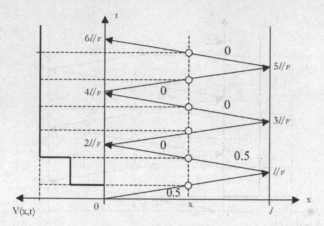

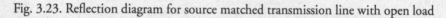

Fig. 3.23. Reflection diagram for source matched transmission line with open load

The delay time of the signal traveling through the channel is given by the transmission line length divided by the velocity of the signal in the line as

$$t_D = l\sqrt{LC}$$

(3.54)

Note that the signal at the load can reach full voltage swing of the source.

3.4.4 SHORT SOURCE/MATCHED LOAD CHANNEL

- For lossless transmission line with short source and matched load ($Z_i \ll Z_L = Z_o$), there are no reflections on the load ends of the line ($\rho_L = 1$). The transmission coefficient at the source based on voltage division is given as $t_i = 1$ as shown in figure 3.24. The reflection diagram is shown in figure 3.25.

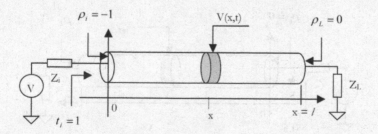

Fig. 3.24. Source short transmission line with matched load

The delay time of the signal traveling through the channel is given by the transmission line length divided by the velocity of the signal in the line as

$$t_D = l\sqrt{LC}$$ (3.55)

Note that the signal at the load can also reach full swing of the source.

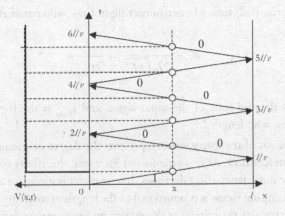

Fig. 3.25. Reflection diagram for short source and matched load lossless transmission line

3.5 VLSI ON-CHIP LRC DELAY EFFECTS

In early-generation integrated circuit (e.g., LSI circuits), the parasitic resistance and capacitance of the MOS devices in integrated circuits dominate the circuit delay performance. On-chip interconnect parasitic in these integrated circuit was usually ignored and interconnect was model as a short. Situations are changed in the VLSI circuits; interconnect capacitances and resistances have become comparable to the active devices even for long on-chip signal routes. In these cases, interconnects are required to be modeled as the lumped capacitances or even as the distributed RC networks. For global interconnects, such as the clock routes, high metal layers that have thick and wide metals are usually used. These interconnects are low resistance lines that can exhibit significant inductive effects under the fast signal rise/fall time. On the other hand, higher performance requirements are pushing the introduction of new materials such as copper interconnect for low resistance interconnect and new dielectrics to reduce interconnect parasitic capacitance. In these cases, interconnect may need to be modeled using LRC circuit networks or transmission line components.

In general, there are two circuit parameters, named the damping factor ζ and the signal rise/fall time to interconnect flight time ratio κ, that determine the significance of the inductance effects in interconnects. The two interconnect parameters are defined as follows:

- The damping factor ζ of an interconnect is defined as

$$\varsigma \equiv \frac{Rl}{2}\sqrt{C/L} \equiv \frac{1}{2Q}$$

$$(3.56)$$

where R, L, and C are the resistance, inductance, and capacitance per unit length of interconnect, respectively, and l is the length of interconnect. Q is the quality factor of the (unloaded) interconnect.

- The signal rise (fall) time to interconnect flight time ratio parameter is defined as

$$\kappa \equiv \frac{t_r}{2l\sqrt{LC}} = \frac{t_r}{2t_{flight}}$$

$$(3.57)$$

where t_r is the rise time of the input signal and t_{flight} is the flight time of the interconnect with length l.

The damping factor of the interconnect represents the degree of attenuation the wave travels down interconnect. As this attenuation increases, the effects of the reflections decrease and the inductance effect decreases. Interconnect is more like an RC network. Note that the damping factor is proportional to the length of interconnect. Therefore, the damping factor sets the criteria of the maximum interconnect that can be treated as an RLC network (or transmission line). An alternative damping factor expression is given as

$$\varsigma = \frac{(Rl)(Cl)}{2l\sqrt{LC}} = \frac{\tau_{RC}}{2\tau_{LC}}$$

$$(3.58)$$

where τ_{RC} and τ_{LC} are the RC and LC time constants of interconnect. Such equation represents the competition between the RC and LC time constant effects of interconnect.

The general criteria for using LRC network (or transmission line) to model interconnects are given as

$$\begin{cases} \varsigma \equiv \dfrac{Rl}{2}\sqrt{C/L} < 1 \\[2mm] \kappa \equiv \dfrac{t_r}{2l\sqrt{LC}} < 1 \end{cases}$$

$$(3.59)$$

Such criteria can also be expressed in terms of the interconnect length as

$$\frac{t_r}{2\sqrt{LC}} < l < \frac{2}{R}\sqrt{L/C}$$

$$(3.60)$$

Since many technology advancements have been targeted at reducing the RC time constant, such as using the copper interconnect and low K dielectric, such trend leads to increased inductance effects for the on-chip interconnects. On the other hand, the advancements of the circuit operation speed implies the continuous decrease of the signal rise time, which also leads to the increased inductance effects in interconnect circuits.

For VLSI tree interconnect structures, equivalent figures of merit can be developed to characterize the significance of the inductance effects. The criteria for using LRC interconnect model can be described using the equivalent damping factor and raise to flight time ratio parameters at node i of an interconnect tree as

$$
\begin{cases}
\varsigma_i \equiv \dfrac{1}{2} \dfrac{\sum\limits_k C_k R_{ik}}{\sqrt{\sum\limits_k C_k L_{ik}}} < 1 \\[3ex]
\kappa \equiv \dfrac{t_r}{2\sqrt{\sum\limits_k C_k L_{ik}}} < 1
\end{cases}
\tag{3.61}
$$

The rise time of signals traveling across the LRC interconnect improves as the inductance effects increase. The attenuation constant becomes less frequency dependent as inductance effects increase (lower damping factor). As a result, interconnect with lower damping factor suffer less rise time degradation effect.

Inductance effects in interconnect also impact the delay time of interconnect. A general expression for propagation delay from input to output of an LRC interconnect of length l with ideal power supply and open circuit load is given as

$$
t_d = l\sqrt{LC}[e^{-2.9\varsigma^{1.35}} + 0.74\varsigma]
\tag{3.62}
$$

In the two extreme cases, where the interconnect is either RC or LC dominated respectively, the delay time of the interconnect can be modeled as a distributed RC network or as a lossless transmission line respectively as

$$
t_d =
\begin{cases}
0.37RCl^2 \propto l^2 & L \to 0 \\[2ex]
\sqrt{LC}l \propto l & R \to 0
\end{cases}
\tag{3.63}
$$

For the LRC tree circuit structures, the 50% delay time of the signal at node i in the tree based on the equivalent Elmore delay is given as

$$t_d = 1.047 \cdot \sqrt{\sum_k C_k L_{ik}} \cdot e^{-\varsigma_i/0.85} + 0.695 \cdot \sum_k C_k R_{ik}$$

$$(3.64)$$

where ζ_i is the equivalent damping factor given in equation 3.61.

APPENDIX 3

A3.1 GENERAL DELAY MODEL OF VLSI DIGITAL CIRCUITS

For more complicated VLSI logic circuits, such as NAND, NOR gates as shown in figure 3.26, to provide the same driving capability as an inverter circuit, the device sizes must be properly upsized. Assuming all devices in the logic gates have minimal device length, the device driving strength is approximately proportional to the width of the device (which is labeled using a dimensionless ratio number next to the device in the following figures).

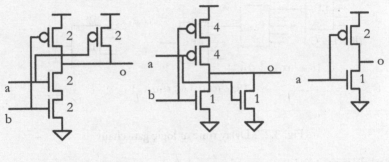

(a) 2-Input NAND   (b) 2-Input NOR   (c) Inverter

Fig. 3.26. VLSI logic gate circuit that have similar driving capability

Since the input (gate) capacitance of the device is also approximately proportional to the width of the specific device. The upsizing of the MOS device for matching the inverter gate driving strength results in the penalty of increased input gate capacitance of the gate that results in increased delay and power penalties in the VLSI circuit implementations.

Logic effort (LE) can usually be used to specify the input capacitance penalty for using complex gates, which can be defined as the ratio of input capacitance (or total width of a given input terminal) of a given complex gate to the input capacitance of the inverter with similar driving capability. Based on such definition and the circuit structures shown in figure 3.26, the LE for the 2-input NAND and the 2-input NOR gates are 4/3 and 5/3 respectively. An example of LE of some commonly used VLSI logic gates is shown in table 3.1.

Table 3.1. Logic efforts of some commonly used VLSI logic gates

Gate	Inverter	NAND2	NOR2	NAND3	NOR3	NAND4	NOR4
LE	1	4/3	5/3	5/3	7/3	6/3	9/3

It can be seen that with increased input terminals, the LE is increased, implying increased capacitive load penalty for using large input pin count devices.

The minimal delay condition for the logic gate chain with specified input gate capacitance and the final load capacitance of the chain similar to inverter chain in figure 3.13 is shown in figure 3.27.

For a given fan-out (FO = C_L/C_g) of a signal path, assume the input gate capacitance is increased at a fan-out of $\{\alpha_i\}$ for the ith stage.

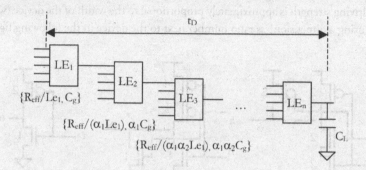

Fig. 3.27. Delay time of logic gate chain

The total delay time and the constraint are given as

$$\begin{cases} t_D = 0.7 R_{eff} C_g \sum_{i=1}^{n} Le_i \alpha_i \\ \dfrac{C_L}{C_g} \prod_{i=1}^{n} Le_i = \prod_{i=1}^{n} Le_i \alpha_i = const. \end{cases}$$

(3.65)

The minimal delay gate sizing condition can be found mathematically based on the given constraint as

$$\frac{\partial}{(Le_i \alpha_i)} [0.7 R_{eff} C_g \sum_{i=1}^{n} (Le_i \alpha_i) + \lambda(\frac{C_L}{C_g} \prod_{i=1}^{n} Le_i - \prod Le_i \alpha_i)] = 0$$

$$i = 1, 0, ..., n$$

(3.66)

$$\Rightarrow \begin{cases} (t_D)_{\min} = n 0.7 R_{eff} C_g (Le \cdot \alpha) \\ le_i \alpha_i = (\dfrac{C_L}{C_g} \prod_{i=1}^{n} Le_i)^{1/n} \equiv (le \cdot \alpha) \quad i = 1, 2, ..., n \end{cases}$$

(3.67)

A3.2 GRAPHIC DESIGN SOLUTIONS TO VLSI BUFFER CIRCUITS

Considering the linear delay versus fan-out (FO) relationship of the VLSI buffer circuits (either single-stage buffer, including the inverter or the complex buffer circuit, including multiple circuit stages) as shown in figure 3.28, design optimizations can be realized graphically based on the following technique.

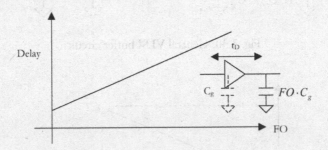

Fig. 3.28. Delay time versus fan-out of general VLSI buffer circuit

The new total delay of the above circuit with one buffer insertion under the constant fan-out constraint can be solved graphically as shown in figure 3.29.

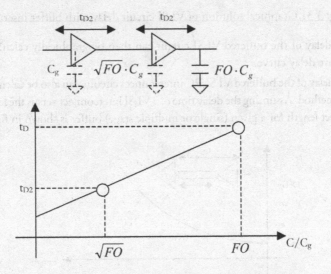

Fig. 3.29. Graphical solution of VLSI circuit delay with buffer insertion

If the new delay time of each buffered stage in the new circuit is less than half of the original circuit, the insertion of the buffer will yield shorter total circuit delay.

Similarly, the above graphical technique can also be extended into n buffer stage insertion case (n>1).

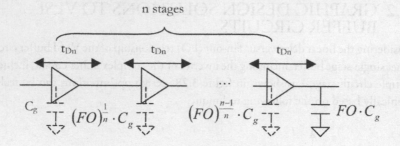

Fig. 3.30. General VLSI buffer circuit

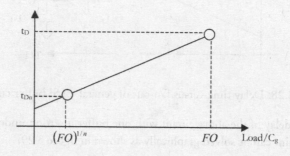

Fig. 3.31 Graphical solution of VLSI circuit delay with buffer insertion

The total delay of the buffered VLSI circuit can then be graphically calculated based on the above delay curve.

Similarly, delay of the buffered VLSI RC interconnect circuits can also be calculated using graphical method. Assuming the delay time of a VLSI interconnect versus the length of the interconnect length for a given (single or multiple stage) buffer is shown in figure 3.32.

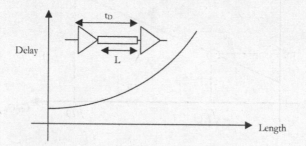

Fig. 3.32. Delay time versus interconnect length of VLSI interconnect circuit

The new total delay of the above circuit with one buffer insertion under the identical buffer constraint can be solved graphically as shown in figure 3.33.

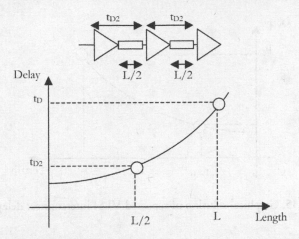

Fig. 3.33. Delay time of interconnect circuit with repeater insertion

If the delay time of each buffered interconnect segment in the new circuit is less than half of the original circuit, the insertion of the repeater will yield shorter total circuit delay.

Similarly, the above graphical technique can also be extended into multiple repeater insertion case (n>1).

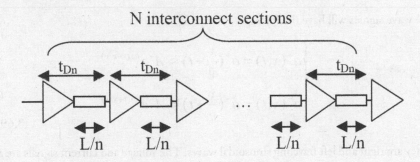

Fig.3.34 General VLSI interconnect repeating circuit

By finding out the delay time of each repeated interconnect segment from the delay graph, the total delta time of the repeated VLSI interconnect can then be calculated based on the graphical technique as shown in figure 3.35.

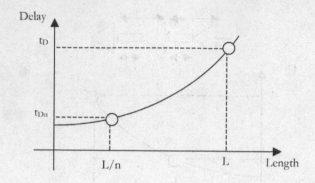

Fig. 3.35. Graphical solution of repeated VLSI interconnect delay time

A3.3 STEADY STATE FREQUENCY DOMAIN MODEL OF LOSSLESS TRANSMISSION LINE

Considering the steady state input signal of the form

$$\begin{cases} V(x,t) = V(x)e^{j\omega t} \\ I(x,t) = I(x)e^{j\omega t} \end{cases}$$
(3.68)

the wave signals will have the forms as

$$\begin{cases} a^{+}(x,t) = a^{+}(\dfrac{x}{v}-t) = A^{+}e^{j\omega(t-x/v)} \\[2mm] a^{-}(x,t) = a^{-}(\dfrac{x}{v}+t) = A^{-}e^{j\omega(t+x/v)} \end{cases}$$
(3.69)

They are right and left traveling sinusoidal waves. The voltage and current signals are as

$$\begin{cases} V(x,t) = \sqrt{Z_o}\,(a^{+}+a^{-}) = \sqrt{Z_o}\,e^{j\omega t}\{A^{+}e^{-jkx}+A^{-}e^{jkx}\} \\[2mm] I(x,t) = \dfrac{1}{\sqrt{Z_o}}(a^{+}-a^{-}) = \dfrac{1}{\sqrt{Z_o}}e^{j\omega t}\{A^{+}e^{-jkx}-A^{-}e^{jkx}\} \end{cases}$$
(3.70)

where

$$k \equiv \frac{\omega}{v}$$
(3.71)

For the transmission line shown below, the effective impedance for the steady state input can be derived as

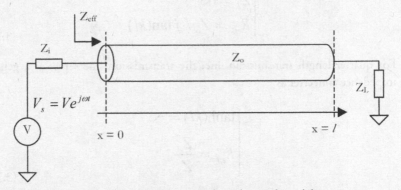

Fig. 3.36. Transmission line electrical model

$$\begin{cases} R_{eff} \equiv \dfrac{V(0,t)}{I(0,t)} = Z_o \dfrac{A^+ + A^-}{A^+ - A^-} \\[3mm] \dfrac{V(l,t)}{I(l,t)} = Z_L = Z_o \dfrac{A^+ e^{jkl} + A^- e^{-jkl}}{A^+ e^{jkl} - A^- e^{-jkl}} \end{cases}$$

$$(3.72)$$

We have

$$R_{eff} = Z_o \frac{Z_L + jZ_o \tan(kl)}{jZ_L \tan(kl) + Z_o}$$

$$(3.73)$$

It can be seen that the equivalent input impedance of the transmission line depends on the transmission line impedance, the transmission line electrical length, and the load impedance.

- For matched load, the equivalent impedance is a pure resistor as

$$\begin{cases} Z_L = Z_o \\ R_{eff} = Z_o \end{cases}$$

$$(3.74)$$

- For transmission line with far end shorted to ground, the equivalent impedance is inductor-like:

$$\begin{cases} Z_L = 0 \\ R_{eff} = jZ_o \tanh(kl) \end{cases}$$

$$(3.75)$$

- For transmission line with far end open, the equivalent impedance is capacitor-like:

$$\begin{cases} Z_L = \infty \\ R_{eff} = Z_o \, / \, j \tan(kl) \end{cases}$$

(3.76)

- For quarter length transmission line, the transmission line serves as a general impedance converter as

$$\begin{cases} \tanh(kl) = \infty \\ R_{eff} = \dfrac{Z_o^2}{Z_L} \end{cases}$$

(3.77)

A3.4 REFLECTION FROM TRANSMISSION LINE DISCONTINUITY

Reflection occurs when there is a transmission line discontinuity as shown in figure 3.37.

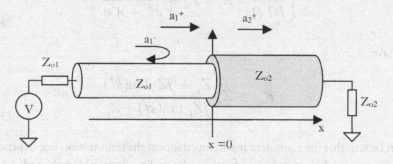

Fig. 3.37. Reflection at transmission line discontinuity

Assuming there is a traveling wave from line 1 to 2 and the load and source are matched. We have that

$$\begin{cases} a_1^+(x,t) = a_1^+(\dfrac{x}{v_1} - t) \\ a_1^-(x,t) = a_1^-(\dfrac{x}{v_1} - t) \\ a_2^+(x,t) = a_1^+(\dfrac{x}{v_2} - t) \end{cases}$$

(3.78)

and

$$\begin{cases} V_1(x,t) = \sqrt{Z_{o1}}\,(a_1^+(\frac{x}{v_1}-t) + a_1^-(\frac{x}{v_1}+t)) \\[2mm] I_1(x,t) = \frac{1}{\sqrt{Z_{o1}}}(a_1^+(\frac{x}{v_1}-t) - a_1^-(\frac{x}{v_1}+t)) \\[2mm] V_2(x,t) = \sqrt{Z_{o2}} \cdot a_2^+(\frac{x}{v_2}-t) \\[2mm] I_2(x,t) = \frac{1}{\sqrt{Z_{o2}}} \cdot a_2^+(\frac{x}{v_2}-t) \end{cases} \tag{3.79}$$

At x = 0, we have that

$$\begin{cases} V_1(0,t) = V_2(0,t) = \sqrt{Z_{o1}}\,(a_1^+(-t) + a_1^-(t)) = \sqrt{Z_{o2}} \cdot a_2^+(-t) \\[2mm] I_1(0,t) = I_2(0,t) = \frac{1}{\sqrt{Z_{o1}}}(a_1^+(-t) - a_1^-(t)) = \frac{1}{\sqrt{Z_{o2}}} \cdot a_2^+(-t) \end{cases} \tag{3.80}$$

We have that

$$\begin{cases} \rho \equiv \dfrac{a_1^-}{a_1^+} = \dfrac{Z_{o2}-Z_{o1}}{Z_{o2}+Z_{o1}} \\[4mm] T_{12} \equiv \dfrac{a_2^+}{a_1^+} = \dfrac{2}{\sqrt{Z_{o2}/Z_{o1}} + \sqrt{Z_{o1}/Z_{o2}}} \end{cases} \tag{3.81}$$

It can be seen that whenever there is an impedance discontinuity, there will be reflection effect in the signal transmission.

As the summary, it can be seen that for a perfectly terminated transmission line, the delay of signal is directly proportional to the length of the transmission line and inversely proportional to the velocity of the signal. However, if there is an impedance discontinuity or poor termination, the transmission line will show significant signal delay degradation or delay uncertainty mainly resulting from the reflection and ringing effects.

References:

[1] M. Afghahi, et al, "Performance of Synchronous and Asynchronous Schemes for VLSI Systems," IEEE Trans. Comput., vol.41, no. 7, pp.858-872, July. 1992.

[2] F. Minami, et al, "Clock Tree Synthesis Based on RC Delay Balancing," IEEE CICC 1992.

[3] R. S. Tsay, "An Exact Zero-Skew Clock Routing Algorithm," IEEE Trans. Computer-Aided Design. Vol. 12, no. 2, Feb. 1993.

[4] H. J. Song, "Delay Penalties and Delay Optimization for Timing Critical VLSI Circuits," un-published work, 1996.

[5] Y. I. Ismail, "On-Chip Inductance Cons and Pros," IEEE Trans. VLSI systems, vol.10, no. 6, pp.685-694, Dec. 2002.

CHAPTER 4
VLSI CIRCUIT DELAY
VARIATION EFFECTS

Electrical signals in VLSI high-speed I/O circuits suffer from various delay variation effects. These delay variation effects are among the key sources of failures in VLSI high-speed I/O circuits.

In time-domain, delay variation effects can be classified into two families, including the time-dependent and time-independent delay variations. Time-independent delay variations are mainly resulted from VLSI device mismatches, PVT, and application condition (such as channel trace length and loading conditions) variations. Time-independent delay variations usually result in delay skew effects in the VLSI high-speed I/O circuits. The time-dependent delay variations, on the other hand, are usually caused by various noise effects that result in jitter of the signals.

VLSI circuit delay variations can be modeled using three key mechanisms, including the time constant modulation, the voltage noise to jitter conversion, and the intersymbol interference (ISI) effects. Time constant modulation effects are usually resulted from the dependency of VLSI circuit time constant on the PVT and application conditions. The voltage noise to jitter conversion effects are resulted from the circuit voltage noise induced delay variations, including the device thermal noise, power supply noise, substrate noise, cross talk, reflection, and ringing-induced delay variations. Voltage noise to jitter conversion effects can be modeled using the voltage noise to jitter transfer functions. The ISI-induced delay variation effects as result of the pattern or duty-cycle dependent signal delay variations that are mainly due to frequency-dependent response of VLSI circuits, such as the bandwidth limitation effects.

4.1 BASIC CIRCUIT DELAY VARIATION MODELS

VLSI RC circuit delay variation effects can be a model using figure 4.1 where a step signal travels through a simple RC network.

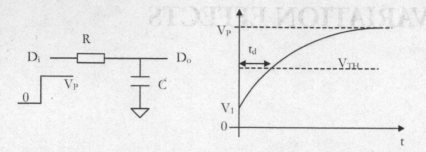

Fig. 4.1. Basic VLSI circuit delay uncertainty model

The output signal waveform of such RC circuit for step input signal can be solved analytically as

$$V_o(t) = V_P - (V_P - V_1)e^{-\frac{t}{RC}} + V_N(t) \tag{4.1}$$

where V_P and V_1 are the peak voltage swing of the step input signal and the initial voltage of the output signal respectively. V_N is the equivalent voltage noise floor contributed from voltage noise sources and dc offset. The delay time t_d with respect to a given voltage threshold V_{TH} is given as

$$t_d = (RC)\ln(\frac{V_P - V_1}{V_P + (V_N - V_{TH})}) = (RC)\ln(\frac{1 - V_1/V_P}{1 + \frac{V_N - V_{TH}}{V_P}}) \tag{4.2}$$

It can be seen from equation 4.2 that there are three major circuit nonideal effects that contribute to the circuit delay time variations:

- Circuit RC time constant variation. Such type of variation is resulted from the PVT variation, and it shows as delay skew effect. This delay variation term may also include duty-cycle distortion effects due to the rise and fall times mismatch and the jitter effects due to the variation of the time constant in time.

- The $(V_N - V_{TH})/V_P$ term represents the noise to jitter conversion effects in the VLSI circuit. It can be seen that either the voltage noise or the threshold variation can result in the delay variation effect. Such delay time uncertainty effect can also introduce the skew in the circuit, if there is a static offset in either the signal or threshold circuits.

- The V_1/V_p term represents the ISI effect that is resulted from the voltage initial condition variation impacted due to the settling of the signal beyond its bit time. Such effect is usually resulted from the bandwidth limitation of the VLSI circuits or the high-speed I/O channel.

4.1.1 TIME CONSTANT MODULATION EFFECTS

The delay time of digital signal passing through a CMOS inverter circuit can usually be approximately modeled using the effective time constant of the circuit as shown in figure 4.2.

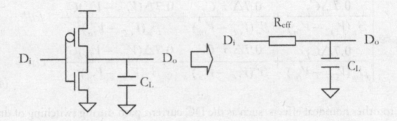

Fig. 4.2. RC delay model of VLSI inverter circuit

The load capacitor in such model includes both the load capacitance and the effective parasitic output capacitance of the inverter. For simplicity, we will ignore the parasitic capacitance of the inverter circuit in the following discussion. For circuit with long interconnect route, the effective load capacitance should also include the parasitic capacitance of the interconnect route.

The effective driver impedance based on the linear device model can be approximately expressed as

$$R_{eff} \approx \begin{cases} \dfrac{1}{\beta_n(V_{CC}-V_{Tn})} & fall \\[4mm] \dfrac{1}{\beta_p(V_{CC}-|V_{Tp}|)} & rise \end{cases}$$

(4.3)

Under this model, the delay time of the digital signal can be approximately expressed as

$$t_d \approx \begin{cases} 0.7 \cdot \dfrac{C_L}{\beta_n(V_{CC}-V_{Tn})} & fall \\[4mm] 0.7 \cdot \dfrac{C_L}{\beta_p(V_{CC}-|V_{Tp}|)} & rise \end{cases}$$

(4.4)

It is important to see the following:

- Delay time of the circuit is directly proportional to the load capacitance.
- Delay time is approximately inversely proportional to the driver strength. However, since the load capacitance also contains the parasitic capacitance of the driver device, there is a diminishing effect in reducing the delay time by increasing the driving devices.
- Delay time is approximately inversely proportional to the difference of the supply voltage and the device threshold voltage.
- The dependency of the delay time on the PVT and load condition implies that there will be a delay time uncertainty if there are PVT and load variations.

$$\Delta t_d \approx \begin{cases} \dfrac{0.7\Delta C_L}{\beta_n(V_{CC}-V_{Tn})} + \dfrac{0.7\Delta\beta_n C_L}{\beta_n^2(V_{CC}-V_{Tn})} + \dfrac{0.7\Delta(V_{CC}-V_{Tn})C_L}{\beta_n(V_{CC}-V_{Tn})^2} & fall \\[4mm] \dfrac{0.7\Delta C_L}{\beta_p(V_{CC}-V_{Tp})} + \dfrac{0.7\Delta\beta_p C_L}{\beta_p^2(V_{CC}-V_{Tp})} + \dfrac{0.7\Delta(V_{CC}-V_{Tp})C_L}{\beta_p(V_{CC}-V_{Tp})^2} & rise \end{cases} \tag{4.5}$$

Due to other nonideal effects, such as the DC current path during switching of driver devices between linear and saturation, a more accurate expression for inverter delay time can be expressed as

$$t_D \approx \frac{C_L}{b(V_{CC}-V_T)^a} \tag{4.6}$$

The parameter α is usually around 1-2. V_T represents the threshold voltages of the PMOS and NMOS devices, for the rising and falling transitions respectively. Such delay model represents general dependency of the delay time on the load capacitance and the driver sizes. The general delay time uncertainty can then be expressed as

$$\Delta t_d \approx \frac{0.7\Delta C_L}{\beta(V_{CC}-V_T)^\alpha} + \frac{0.7\Delta\beta C_L}{\beta^2(V_{CC}-V_T)^\alpha} + \frac{0.7\alpha\Delta(V_{CC}-V_T)C_L}{\beta(V_{CC}-V_T)^{\alpha+1}} \tag{4.7}$$

For the VLSI CMOS current mode logic (CML) differential buffer circuit structure shown in figure 4.3, the delay time of the circuit is approximately determined by the effective RC constant of the buffer output as

$$t_D \approx 0.7RC \tag{4.8}$$

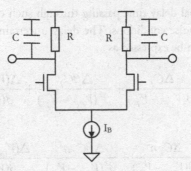

Fig. 4.3. VLSI current model log (CML) buffer circuit structure

It is interesting to see that the delay time of a CML buffer circuit is approximately independent of the power supply voltage, implying high power supply rejection ratio. The major delay uncertainty resources of such CML circuits are the PVT variations of the resistance and capacitance:

$$\Delta t_D \approx 0.7\Delta RC + 0.7R\Delta C \tag{4.9}$$

For VLSI pass gate circuit structures shown in figure 4.4, the delay time of the electrical signal passing through the single and multistage pass gate circuits can be approximately expressed respectively as

$$t_D \propto \frac{C_L}{\beta(V_{CC} - V_T)} \tag{4.10}$$

$$t_D \propto \frac{C_L n^2}{\beta(V_{CC} - V_T)} \tag{4.11}$$

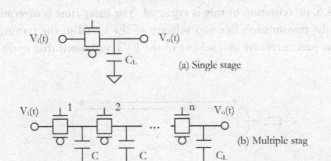

Fig. 4.4. VLSI pass gate circuit structures

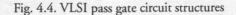

It can be seen that signal delay time passing through such circuits is also sensitive to the various PVT and noise conditions. The delay uncertainty due to PVT and load condition variations can be expressed as

$$\Delta t_d \propto \frac{\Delta C_L}{\beta(V_{CC} - V_T)} + \frac{\Delta \beta C_L}{\beta^2(V_{CC} - V_T)} + \frac{\Delta(V_{CC} - V_T)C_L}{\beta(V_{CC} - V_T)^2} \tag{4.12}$$

$$\Delta t_d \propto \frac{\Delta C_L n^2}{\beta(V_{CC} - V_T)} + \frac{\Delta \beta C_L n^2}{\beta^2(V_{CC} - V_T)} + \frac{\Delta(V_{CC} - V_T)C_L n^2}{\beta(V_{CC} - V_T)^2} \tag{4.13}$$

For complex RC network as shown in figure 4.5, the time constant modulations are mainly contributed from the skew of the R and C values under the PVT conditions.

$$\frac{\Delta t_D}{t_D} \approx \frac{\Delta R}{R} + \frac{\Delta C}{C} \tag{4.14}$$

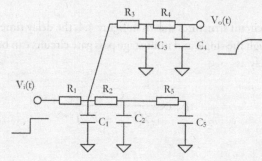

Fig. 4.5. Complex RC network

For ideal VLSI transmission line circuit with proper impedance termination as shown in figure 4.6, no reflection or ring is expected. The delay time is determined by the length of the transmission line and velocity of the signal in the transmission line. All of these parameters are also subject to the PVT variations that result in skew of the signal.

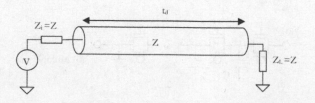

Fig. 4.6. Transmission line channel with proper impedance

$$\frac{\Delta t_D}{t_D} \approx \frac{\Delta L}{2L} + \frac{\Delta C}{2C}$$

(4.15)

4.1.2 NOISE-TO-JITTER CONVERSION EFFECTS

The signal amplification and swing limitation effects of the typical VLSI digital circuits can be modeled as an A/D conversion operation as shown in figure 4.7, where the swing and slew rate degraded input signal in the I/O circuit is amplified and voltage limited to full-swing (digital) signal by the digital gate.

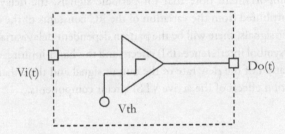

$Vi(t)$ $Do(t)$ Vth

Fig. 4.7. The A/D conversion-based delay time uncertainty model

Under such model, the voltage variation of the VLSI circuit component can be transformed into the delay variation as shown in figure 4.8 as result of the A/D conversion operation.

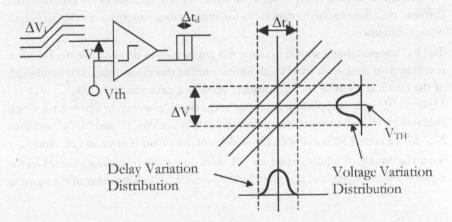

ΔV_i Δt_d Δt_d

V Vth ΔV V_{TH}

Delay Variation Distribution   Voltage Variation Distribution

Fig. 4.8. Conversion of voltage variation to delay variation

Based on figure 4.8, the induced delay variation from the voltage variations in the input signal voltage and the threshold can be expressed as

$$\Delta t_d \equiv J_N(s) \cdot \Delta V = \frac{1}{SlewRate} \Delta(V_i - V_{TH})$$

(4.16)

The voltage to delay variation transfer function is then given as

$$J_N(s) = \frac{1}{SlewRate}$$

(4.17)

- It is important to note that both input signal voltage and threshold voltage variations contribute to the voltage to delay variation conversion effects, and the voltage to delay variation transfer function is inversely proportional to the input signal slope.

- It is also important to note that for periodic signals, the delay variations are mainly contributed from the variation of the RC constants of the circuits; for the nonperiodic signals, there will be the pattern dependent delay variations as a result of the intersymbol interference (ISI) effects; and this band-limiting effect will cause the degradation of the slew rate of the output signal and the enhanced voltage to delay variation effects of the active VLSI circuit components.

4.1.3 THE ISI EFFECTS

Random data pattern passing through the bandwidth-limited VLSI circuit may also introduce delay time variation. Such effect is known as the intersymbol interference (ISI) effects, which is caused by the data transition settling time beyond one data bit (or symbol) time. As a result of data transition settling differences for different data patterns, the delay time of the data varies for different data patterns due to initial output voltage differences.

The RC low-pass circuit shown in figure 4.9 can be used to demonstrate the ISI effect resulting from the circuit bandwidth limitation. For this circuit, the -3dB bandwidth of the circuit is given by its RC constant. Assume a data pattern of 000 . . . 011 . . . 1100 . . . 0111 . . . is applied to the input to the circuit as shown in figure 4.10, which contains N_0 ($N_0 \gg 1$ to ensure the data signal is fully settled) "0" and N_1 "1" and then N_2 "0," the output 50% swing crossover delay of the circuit is given as t_0, t_1, and t_2.

Since the data signal is fully settled at t = 0, delay time t_0 can be calculated based on the signal waveform $V_{oo}(t)$, and it is given by the nominal 50% swing delay of the signal as

$$V_{oo}(t) = (1 - e^{-\frac{t}{RC}})$$

(4.18)

$$t_0 = RC \ln(2) = 0.7RC$$

(4.19)

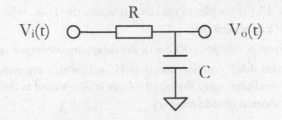

Fig. 4.9. Bandwidth-limited RC circuit channel

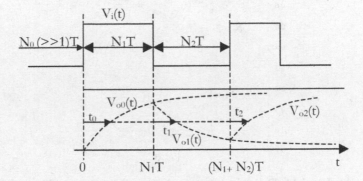

Fig. 4.10. ISI effects due to bandwidth limitation

Delay time t_1 can be calculated based on the signal waveform $V_{o1}(t)$, and it is given by the 50% crossover delay of the signal as

$$V_{o1}(t) = (1 - e^{-\frac{N_1 T}{RC}})e^{-\frac{t - N_1 T}{RC}} \tag{4.20}$$

$$t_1 = RC \ln[2(1 - e^{-\frac{N_1 T}{RC}})] \tag{4.21}$$

Delay time t_2 can be next calculated based on the signal waveform $V_{o2}(t)$, and it is given by the 50% crossover delay of the signal as

$$V_{o2}(t) = [(1 - e^{-\frac{N_1 T}{RC}})e^{-\frac{N_2 T}{RC}} - 1]e^{-\frac{t - N_1 T - N_2 T}{RC}} + 1 \tag{4.22}$$

$$t_2 = RC \ln[2[1 - (1 - e^{-\frac{N_1 T}{RC}})e^{-\frac{N_2 T}{RC}}]] \tag{4.23}$$

It can be seen that t_1 and t_2 are pattern dependent (function of N_1 and N_2).

Shown in figure 4.11 is the plot of the t1 value versus the T/RC ratio and the pattern parameter N_1. It can be seen that

- for low frequency signals (T/RC >>1), the delays are converged to 0.7RC, and
- the worst-case delay variation occurs at $N_1 = 1$, which represents the so-called lone-pulse condition, when the initial voltage is fully settled to the rail and current bit has the shortest time to ramp up.

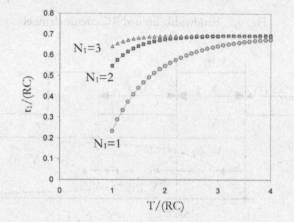

Fig. 4.11. Delay time t_1 versus the data frequency and N_2

Since the lone-pulse has shorter time duration and the center of the lone-pulse eye is pushed out with respect to other bits as shown in figure 4.12.

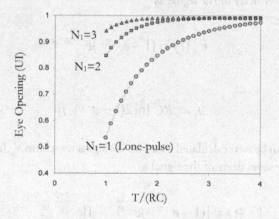

Fig. 4.12. Eye opening versus the data frequency and N_1

Shown in figure 4.13 is the delay time t_2 versus the data rate and the pattern parameter N_1 N_2. It can also be seen that

- for low data frequency (T/RC>>1), delay time t_2 is also approached the normal RC delay of 0.7RC, and

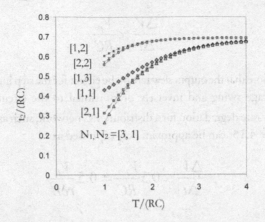

Fig. 4.13. Delay time t_2 versus the data frequency, N_1 and N_2

• delay time t_2 depends on both N_1 and N_2, and the worst-case delay variation is also related to the lone-pulse condition.

In addition to the random data pattern, VLSI clock signals will also suffer from the ISI-like effects when the duty-cycle distortions (DCD) or jitters are presented. Such effect can result in DCD and jitter amplification effects, which will be discussed in details in a later chapter.

4.2 VLSI CIRCUIT SLEW RATE MODEL

Based on the discussion, signal slew rate plays an important role in the voltage variation to delay variation conversion effect. In VLSI circuits slew rate degradation are typically caused by the bandwidth limitation (or filtering effect) in the VLSI circuit components.

4.2.1 VLSI PASSIVE CIRCUITS

Passive RC circuit structures, such as the on-chip interconnect and the data transmission channel, may degrade the slew rate of the electrical signals. For the lumped RC circuit structure shown in figure 4.14, the slew rate for a step input can be approximately expressed as

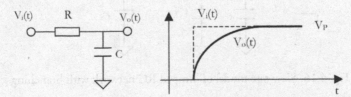

Fig. 4.14. Slew rate model of lumped RC circuits

$$\frac{\Delta V}{\Delta t} \propto \frac{V_P}{RC}$$

$$(4.24)$$

It is important to note that the output slew rate of the circuit for the step input is proportional to the input voltage swing and inversely proportional to the circuit's RC constant. Similarly, the slew rate degradation for a distribute RC network, such as the interconnect as shown in figure 4.15, can be approximately expressed as

$$\frac{\Delta V}{\Delta t} \propto 0.5 \frac{V_{PP}}{RC} = 0.5 \frac{V_P}{rcL^2}$$

$$(4.25)$$

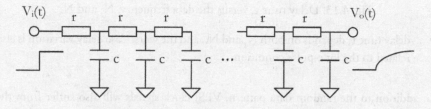

Fig. 4.15. Slew rate model of distributed RC circuits

where r and c in the circuit are the unit length resistance and capacitance, and L the length of interconnect. Similar to the lumped RC circuit, the slew rate is proportional to the signal swing and inversely proportional to the RC products and square of the interconnect length.

For lumped passive RC network with branches as shown in figure 4.16, the slew rate degradation can be approximately modeled as

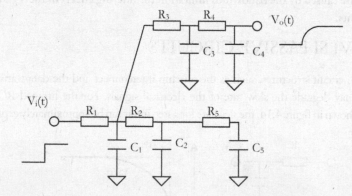

Fig. 4.16. Slew rate model of lumped RC network with branching

$$\frac{\Delta V}{\Delta t} \propto \frac{V_P}{[R_1(C_1 + C_2 + C_3 + C_4 + C_5) + R_3(C_3 + C_4) + R_4 C_4]} \tag{4.26}$$

It can be seen that similar to other RC circuit structures, output slew rate is proportional to the signal swing and inversely proportional to the RC products, implying the solution for slew rate improvement using higher signal swing and circuit bandwidth.

VLSI pass gate circuit structures as shown in figure 4.17 can be modeled using equivalent pass gate resistance.

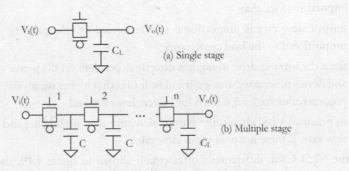

(a) Single stage

(b) Multiple stage

Fig. 4.17. Slew rate model of VLSI pass gate circuit structures

The output slew rates of such circuits and can be approximately expressed based on the equivalent RC circuit as

$$\frac{\Delta V}{\Delta t} \propto \frac{V_P}{C} \beta (V_{CC} - V_T) \tag{4.27}$$

$$\frac{\Delta V}{\Delta t} \propto \frac{V_P}{Cn^2} \beta (V_{CC} - V_T) \tag{4.28}$$

It can be seen that the slew rate of the output is proportional to the input signal swing. Such a fact implies that a slightly large signal may help to reduce the jitter injection effects.

4.2.2 VLSI ACTIVE CIRCUITS

For the VLSI active devices, such as the CMOS inverter circuit, the slew rate of the output signal can be approximately expressed as

$$\frac{\Delta V}{\Delta t} \approx \frac{I}{C_L} \approx \frac{\beta (V_{CC} - (V_{TN} + |V_{TP}|))}{C_L} \tag{4.29}$$

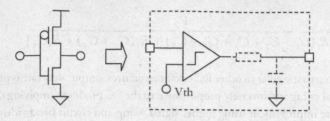

Fig. 4.18. Slew rate model of VLSI inverter circuit

It is important to see that

- output slew rate is proportional to the device drive capability and inversely proportional to the load capacitance;

- since the inverter drive strength is directly dependent on the power supply voltage and device process technology, it can be found that higher supply voltage and faster process technology are desired for better slew rate; and

- in practical VLSI circuit, the output slew rate may slightly depend on the input slew rate, if there is a poor input slew rate.

For the VLSI CML differential buffer circuit shown in figure 4.19, the slew rate is proportional to the bias current and inversely proportional to the load capacitance as

$$\frac{\Delta V}{\Delta t} \approx \frac{V_{PP}}{RC} = \frac{RI_B}{RC} = \frac{I_B}{C} \qquad (4.30)$$

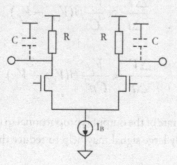

Fig. 4.19. Slew rate modeling of differential buffer circuit

Note that the output slew rate of a differential CML buffer is proportional to the bias current, which implies the trade-off of the bias current level and power dissipation with the timing uncertainty in the differential CMOS-based VLSI I/O circuit design. Similar to other circuit structures, minimized capacitance loading is also preferred for better slew rate performance.

4.3 VLSI CIRCUIT THRESHOLD MODELS

The threshold voltage of VLSI inverter can typically be defined as the input voltage value where the output voltage changes the state.

For the CMOS inverter circuit shown in figure 4.20, the threshold voltage of the CMOS inverter can be approximately derived based on the CMOS transistor I-V model as

$$V_{TH} \approx \frac{\sqrt{\beta_P}(V_{CC} - |V_{TP}|) + \sqrt{\beta_N}(V_{Tn})}{(\sqrt{\beta_N} + \sqrt{\beta_P})}$$

(4.31)

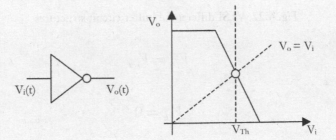

Fig. 4.20. Threshold voltage of inverter threshold voltage

It is important to see that the threshold voltage of the CMOS inverter is the function of the P/N device ratio, the P/N MOS threshold values, and the power supply voltage. All of them are strong functions of the device PVT conditions. As a result, delay variation can be introduced from the device PVT variations.

$$\Delta V_{TH} \approx \Delta[\frac{\sqrt{\beta_P}(V_{CC} - |V_{TP}|) + \sqrt{\beta_N}(V_{Tn})}{(\sqrt{\beta_N} + \sqrt{\beta_P})}]$$

(4.32)

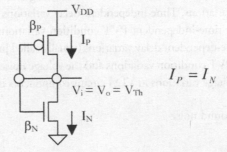

Fig. 4.21. Modeling of CMOS inverter threshold voltage

For the VLSI CML buffer circuit structures shown in figure 4.22, the threshold of the circuits can be expressed respectively as

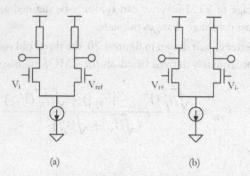

Fig. 4.22. VLSI differential buffer circuit structures

$$V_{Th} = V_{ref}$$

(4.33)

$$V_{Th} = 0$$

(4.34)

It can be seen that their VLSI circuit structures provide high accuracy voltage threshold independent of the PVT variations:

$$\Delta V_{Th} = \Delta V_{ref}$$

(4.35)

$$\Delta V_{Th} = 0$$

(4.36)

4.4 VLSI CIRCUIT NOISE AND JITTER MODELS

Practical VLSI circuit components suffer from various voltage variation effects that may result in delay time variations. Time-independent delay variations (usually called skew) can be caused by the time-independent PVT condition variations of the VLSI circuit components. The time-dependent delay variations (usually called jitter) can be caused by the time-dependent PVT condition variations and the voltage noise effects. Major source of time-dependent voltage variations in VLSI circuit components include the following:

- power supply/ground noise
- thermal noise
- cross-coupling/cross talk noise
- reflections/ringing effect-induced noise

These voltage noise introduced delay variations (jitter) may significantly impact the performance of the VLSI circuits that can be modeled through certain voltage noise-to-jitter injection mechanisms.

4.4.1 POWER SUPPLY VOLTAGE NOISES

Shown in figure 4.23 is a simplified power supply/ground noise model based on the di/dt and IR noise coupled to the device from all other switching devices with shared power supply.

The di/dt and IR noises on the Vcc and Vss can be induced by the changes of the supply current (or called Icc(t)), and they can be expressed as

$$\Delta V_{cc} = (\Delta V_{cc})_{dI/dt} + (\Delta V_{cc})_{IR} = L\frac{dI_{cc}}{dt} + \Delta I_{cc}R \tag{4.37}$$

$$\Delta V_{ss} = (\Delta V_{ss})_{dI/dt} + (\Delta Vss)_{IR} = L\frac{dI_{ss}}{dt} + \Delta I_{ss}R \tag{4.38}$$

Power supply/ground noise may introduce jitter in VLSI circuits through the dependency of the delay of the circuit block on the supply voltages.

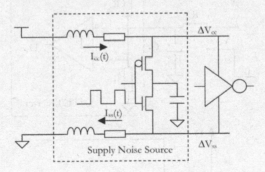

Fig. 4.23. Power supply noise source model

Shown in figure 4.24 is the conceptual model for the supply noise-based jitter injection for the passive RC circuits. For the step voltage signal input, the output voltage of the RC circuit can be expressed as

$$V_o(t) = V_{cc}(1 - e^{-t/RC}) \tag{4.39}$$

The delay time of the circuit with respect to the A/D conversion threshold voltage V_{th} = kV_{cc} is given as

$$kV_{cc} = V_{cc}(1 - e^{-t_d/RC})$$

(4.40)

We have that

$$t_d = RC\ln(1 - \frac{V_{th}}{V_{cc}}) = RC\ln(1 - k)$$

(4.41)

It is important to note that for such passive RC circuit structure delay is independent of the supply voltage if the threshold of the circuit tracks the voltage swing of the signal. As a result, this circuit can reject the supply noise. It is important to point out that the power supply noise rejection capability of the passive RC circuits is mainly due to the tracking-based logic threshold proportional to the V_{cc}. This can be better explained with the passive RC circuit structure shown in figure 4.24, where a fixed logic threshold is applied.

In this case, the Vcc dependent delay time of the circuit is given as

$$t_d = RC\ln(1 - \frac{V_{th}}{V_{cc}})$$

(4.42)

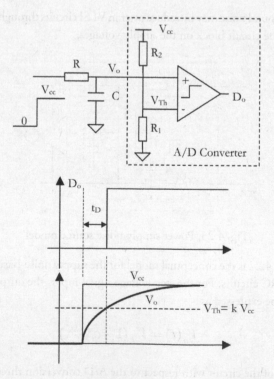

Fig. 4.24. Passive RC network power supply jitter model with tracked threshold

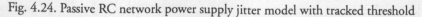

Consequently, timing jitter will be injected through supply noise-induced delay variation.

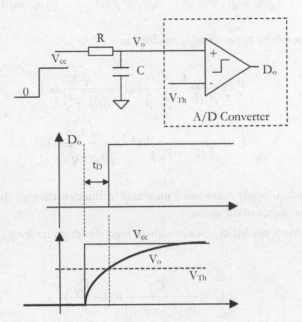

Fig. 4.25. Passive RC network power supply jitter model with fixed threshold

For VLSI inverter circuit shown in figure 4.26, similar jitter injection model can be generated.

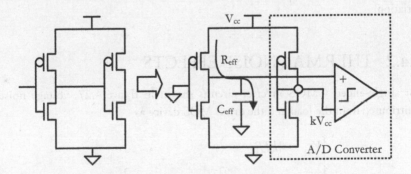

Fig. 4.26. CMOS inverter power supply jitter model

For this circuit, we have that

$$R_{eff} \approx \frac{1}{\beta_P(V_{cc} - |V_{TP}|)}$$

(4.43)

$$V_{Th} = \frac{\sqrt{\beta_P}}{(\sqrt{\beta_N} + \sqrt{\beta_P})} V_{cc} + \frac{\sqrt{\beta_N}(V_{Tn}) - \sqrt{\beta_P}(|V_{TP}|)}{(\sqrt{\beta_N} + \sqrt{\beta_P})} \approx \frac{\sqrt{\beta_P}}{(\sqrt{\beta_N} + \sqrt{\beta_P})} V_{cc} \tag{4.44}$$

The delay time can be approximately modeled as

$$t_d^p \approx \frac{C}{\beta_P(V_{cc} - |V_{TP}|)} \ln(1 - \frac{\sqrt{\beta_P}}{\sqrt{\beta_P} + \sqrt{\beta_N}})$$

$$t_d^n \approx \frac{C}{\beta_P(V_{cc} - V_{Tn})} \ln(1 - \frac{\sqrt{\beta_N}}{\sqrt{\beta_P} + \sqrt{\beta_N}}) \tag{4.45}$$

It can be seen that supply noise-based jitter may be injected through the variation of the effective resistance of the devices.

On the other hand, for MOS inverter without logic threshold tracking (fixed V_{th}), we have that

$$t_d^p \approx \frac{C}{\beta_P(V_{cc} - |V_{TP}|)} \ln(1 - \frac{V_{ref}}{V_{cc}})$$

$$t_d^n \approx \frac{C}{\beta_P(V_{cc} - V_{Tn})} \ln(1 - \frac{V_{ref}}{V_{cc}}) \tag{4.46}$$

It implies that additional jitter may be injected through the threshold reference variation.

4.4.2 THERMAL NOISE EFFECTS

For single-ended CMOS inverter circuits shown in figure 4.27, voltage noise is contributed from the load and the driver MOS device as

$$\begin{cases} V_{N1}^2 \approx 4kTR_{eq} \\ V_{N2}^2 \approx I_N^2 R_{eq}^2 = [\frac{2}{3} 4kTg_m]R_{eq}^2 = [\frac{2}{3} A]4kTR_{eq} \end{cases} \tag{4.47}$$

$$V_{Noise}^2 \approx (V_{N1}^2 + V_{N2}^2) \cdot \frac{1}{RC_L} \approx \frac{4kT}{C_L}(1 + \frac{3}{2} A) \tag{4.48}$$

For signal with peak swing V_{PP} and effective gain a_v, the thermal voltage noise-induced jitter in the CMOS inverter can be modeled as

$$\frac{\Delta t_{RMS}}{t_d} \approx \sqrt{\frac{2kT}{C_L}} \sqrt{1 + \frac{2}{3} a_V} \frac{1}{V_{PP}}$$

(4.49)

or

$$\Delta t_{RMS} \propto \frac{\sqrt{T C_L}}{V_{PP}}$$

(4.50)

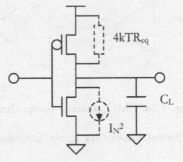

Fig. 4.27. Thermal noise-induced jitter in CMOS inverter

It can be seen that the RMS jitter is proportional to the square root of the load capacitance and inversely proportional to the signal voltage swing.

In practical VLSI applications, an inverter chain of increasing drive capability is commonly used for driving large capacitive load. For the buffer chain with n buffers of same fan-out A as shown in figure 4.28, the random jitter can be expressed as

$$t_{RMS}^2 \approx \sum (t_d \sqrt{\frac{2kT}{C_L}} \sqrt{1 + \frac{2}{3} a_V} \frac{1}{V_{PP}})^2 = t_1^2 (\frac{C_o}{C_g} + A) \sum_{k=0}^{n-1} \frac{1}{A^k}$$

$$= t_1^2 \frac{(A + \frac{C_o}{C_g})(A - A^{-(n-1)})}{A-1} \approx t_1^2 \frac{(A + \frac{C_o}{C_g})A}{A-1}$$

(4.51)

where

$$t_1^2 \equiv (\frac{\sqrt{2kT(1+\frac{2}{3}a_V)}}{V_{PP}} \frac{t_o}{C_{o1}})^2 C_{g1}$$

(4.52)

It can be seen that the random jitter is inversely proportional to the peak voltage of the signal as

$$t_{RMS} \propto \frac{\sqrt{T}}{V_{PP}}$$

(4.53)

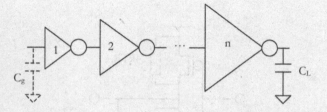

Fig. 4.28. CMOS inverter buffer chain

It can be seen that the optimal Fan-out for minimal random jitter is approximately given as

$$A_{min} \approx 1 + \sqrt{1 + \frac{C_o}{C_g}} \approx 2.4 \sim 3$$

(4.54)

Similarly, for VLSI differential buffer circuit structure shown in figure 4.29, the thermal voltage noise can be modeled as

$$V_{Noise}^2 = \frac{2kT}{C_L}(1 + \frac{3}{2}A)$$

(4.55)

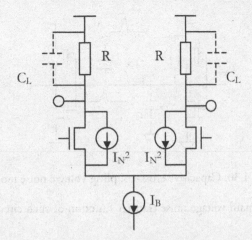

Fig. 4.29. Thermal voltage noise in VLSI differentia buffer circuit

The random jitter due to thermal noise is given as

$$\left(\frac{t_{RMS}}{t_d}\right)^2 = \left(\frac{1}{\ln(2)}\right)^2 \frac{\frac{2kT}{C_L}(1+\frac{3}{2}A)}{V_{PP}^2} = \left(\frac{1}{\ln(2)}\right)^2 \frac{2kT(1+\frac{3}{2}A)}{C_L R I_B V_{PP}} \qquad (4.56)$$

or as

$$t_{RMS} \propto \frac{R\sqrt{TC_L}}{V_{PP}} = \frac{\sqrt{TC_L}}{I_B} \qquad (4.57)$$

It is important to note that for both single-ended and differential VLSI buffer circuit structures, thermal noise-induced jitter is inversely proportional to the swing of the signals. In addition, the random jitter is proportional to the square root of the loading capacitance.

4.4.3 CROSS-COUPLING NOISES

Signal cross-coupling effects may occur at the power delivery network, substrate, and interconnects.

Shown in figure 4.30 is the capacitive cross-coupling voltage noise model, where the victim node of cross-coupling noise injection is driven by a buffer with equivalent output resistance R, the capacitance on the victim node as C, and the cross-coupling capacitance as C_x.

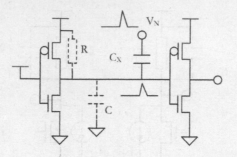

Fig. 4.30. Capacitive cross-coupling voltage noise model

The frequency domain voltage noise transfer function of such circuit can be derived as

$$\frac{\Delta V}{V_N} = \frac{sC_X}{s(C_X + C) + 1/R}$$

(4.58)

It can be seen that it represents a high-pass frequency response as shown in figure 4.31. It can also be seen that reduction in the cross-capacitance to load capacitance ratio as well as low driver impedance will help on cross-coupling voltage noise injection.

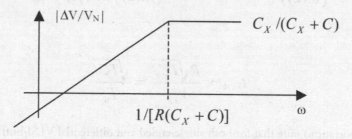

Fig. 4.31. Cross-coupling noise transfer function

The jitter transfer function for step noise input is approximately modeled as

$$\frac{\Delta t}{V_N} \approx \frac{RC_X}{V_{pp}}$$

(4.59)

It can be seen that capacitive coupling noise is proportional to the cross-coupling capacitance and the driver resistance, and inversely proportional to the signal swing.

Cross-coupling noise may also be generated through the mute inductance of signal interconnect (either on-chip or off-chip) as the circuit configuration shown in figure 4.32.

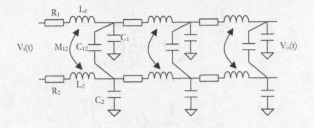

Fig. 4.32. Capacitive and inductor cross talk voltage noise model

Based on the closed form LRC delay model given in the literatures, voltage noise and the delay uncertainty will be induced from the neighboring switching interconnects via the mute inductance.

$$t_d = 1.047 \cdot \sqrt{\sum_k C_k L_{ik}} \cdot e^{-\varsigma_i/0.85} + 0.695 \cdot \sum_k C_k R_{ik} \tag{4.60}$$

Another commonly seen cross-coupling noise in VLSI circuits is the substrate coupling due mainly to the manufacture limitation that all circuits are fabricated on the same substrate. In order to avoid design issues, such as the latch-up issue, low-resistance substrates are typically required that results in strongly resistive and capacitive noise coupling form the common integrated circuit substrates.

4.4.4 RINGING AND REFLECTION EFFECTS

Ringing effects are usually caused by underdamped LC circuits. For high-speed I/O circuits, ringing effect may induce significant timing uncertainty that degrades the circuit performances. Shown in figure 4.33 is a circuit model for the lumped LC circuit.

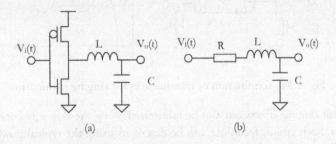

Fig. 4.33 Ringing effects model of VLSI LC circuits

The s-domain signal transfer function of the circuit can be expressed as

$$\frac{V_o}{V_i} = \frac{1}{(\frac{s}{\omega_o})^2 + \frac{1}{Q}(\frac{s}{\omega_o}) + 1}$$

(4.61)

$$\omega_o = \frac{1}{\sqrt{LC}} , Q = \frac{1}{R}\sqrt{\frac{L}{C}}$$

(4.62)

For Q > 0.7, there is a peaking effect in the transfer function and results in certain level of ringing in the data transmission. This ringing effect can be eliminated through the termination of the LC circuit. The critically damped (or matched) R for the circuit can be derived as

$$R = \frac{1}{0.7}\sqrt{\frac{L}{C}} \approx \sqrt{\frac{L}{C}}$$

(4.63)

Ringing effects can also be seen in the improperly terminated transmission line load in I/O circuits as shown in figure 4.34. Elimination of this ringing effect can be achieved through termination of the transmission line at the driver or load side with the impedance equal to the impedance of the transmission line as

$$R_L = Z_o = \sqrt{\frac{L}{C}}$$

(4.64)

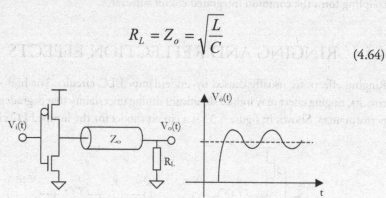

Fig. 4.34. Termination of transmission for ringing elimination

VLSI circuit ringing effects can also be minimized using the slew rate control circuit techniques. Such circuit technique can be described using the typical power spectral density of a digital signal shown in figure 4.35. It can be seen that the energy of the signal is typically confined within a frequency range known as the knee frequency.

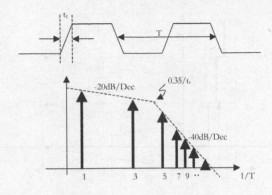

Fig. 4.35. Power spectral density of digital signal

$$f_{Knee} \approx 0.35 / t_r \tag{4.65}$$

Based on the above fact, slew rate control can be used to effectively eliminate the high-frequency contents that are within the resonant frequency of the LC resonant tank and therefore eliminating the ringing. However, such technique is limited to low-speed I/O circuit applications where the useful signal spectral is significantly lower than the resonant frequency of the LC circuits. For high-frequency applications where the resonant frequency of the LC circuit is close to the useful signal frequency, it is difficulty or even impossible for eliminating the resonant frequency content without impacting the useful signal.

Reflection occurs for the improperly terminated transmission line channel. Reflection may induce significant voltage noises and delay variations. It has been observed that channel-related delay variations are among the dominant sources of delay variations of the VLSI high-speed I/O circuits. Shown in figure 4.36 is a reflection diagram of a high-speed I/O channel with high source impedance (e.g., $Z_i = 3Z_o$) and open load ($Z_L \gg Z_o$), where the near and far end transmission and reflection coefficients are given as $t_i = 0.25$, $\rho_I = 0.5$, $\rho_L = 1$.

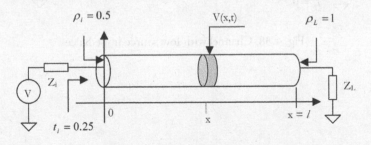

Fig. 4.36. Channel with higher source impedance

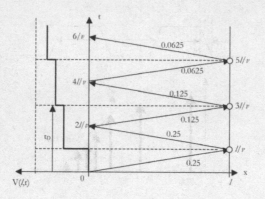

Fig. 4.37. Reflection diagram of channel with higher source impedance

It can be seen that in such case, there is a significantly higher delay time for a signal to pass the 50% swing at the receiver end, mainly due to higher source impedance that results in lower current to charge and discharge the channel.

For open load channel with lower source impedance (e.g., $Z_i = Z_o/3$) with the near and far end transmission and reflection coefficients as $t_i = 0.75$, $\rho_1 = -0.5$, $\rho_L = 1$, the reflection diagram of the channel under such termination condition is shown in figure 4.38. It can be seen that there is significant overshoot and ringing effects for signal to propagate to the receiver, mainly resulting from the higher source current for the channel. Such overshoot and ringing effect may result in significant delay uncertainty and cross talk effects.

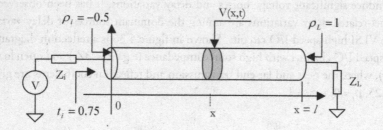

Fig. 4.38. Channel with low source impedance

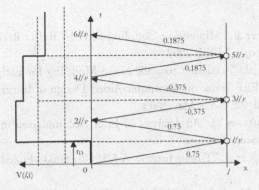

Fig. 4.39. Reflection diagram of channel with lower source impedance

Shown in figure 4.40 is a channel with low source and load impedance (e.g., $Z_i = Z_L = Z_o/3$) with the near and far end transmission and reflection coefficients given as $t_i = 0.75$, $\rho_I = -0.5$, $\rho_L = -0.5$.

As shown in the reflection diagram in figure 4.41, there is significantly longer propagate delay mainly resulting from the mismatch of the channel impedance to the source and load impedances.

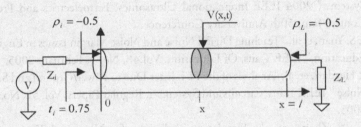

Fig. 4.40. Channel with low source and load impedance

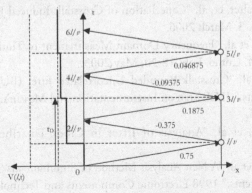

Fig. 4.41. Reflection diagram of channel with lower source and load impedance

References:

[1] M-F Hsiao et al, "Minimizing Coupling Jitter by Buffer Resizing for Coupled Clock Networks," IEEE 2003.

[2] Li Ding, et al, "Accurate Crosstalk Noise Modeling for Early Signal Integrity Analysis," IEEE Trans. on Computer-Aided Design of Integrated Circuits and Systems, Vol. 22, No.5, May 2003.

[3] T. Shimamura, et al, "An Analysis of Jitter Accumulation in a Chain of PLL Timing Recovery Circuits,"

[4] C. W. Zhang, et al, "Timing Jitter in A 1.35Ghz Single-Ended Ring Oscillator," 2002 IEEE.

[5] J. Lee, et al, "Analysis and Modeling of Bang-Bang Clock and Date Recovery Circuits," IEEE JSSC Vol.39, No. Sept. 2003.

[6] A. V. Mezhiba, et al," Scaling Trends of On-Chip Power Distribution Noise," IEEE

[7] J. Zhang, et, al," Crosstalk Noise Model for Shielded Interconnects in VLSI-based Circuits,"

[8] C. W. Zhang, et, al, "Simulation of Timing Jitter in Ring Oscillators," IEEE 2003.

[9] A. M. Fahim, "Jitter Analysis of Digital Frequency Dividers in Communication Systems," 2004 IEEE International Ultrasonics, Ferroelectrics and Frequency Control Joint 50th Anniversary Conference.

[10] J. S. Yuan, et al, "Teaching Digital Noise and Noise Margin Issues in Engineering Education," IEEE Trans. Of Education. Vol.48, No. 1, February 2005.

[11] P. Heydari, et, al, "Analysis of the PLL Jitter Due to Power/Ground and Substrate Noise," IEEE Trans. Circuits and Systems-I: Regular Papers, Vol. 52, No.6, June, 2005.

[12] C. W. Zhang, et, al, "Simulation Technique for Noise and Timing Jitter in Electronic Oscillators," IEE Proc.—Circuits Devices Syst. Vol. 151, No. 2, April 2004.

[13] J. F. Buckwalter, et, al, "Cancellation of Crosstalk-Induced Jitter," IEEE JSSC, Vol. 41, No. 3, March 2006.

[14] L. Schiano, et al, "Frequency Domain Measurement of Timing Jitter in ATE," IMTC 2004, Como, Italy. 18-20, May 2003.

[15] A. Kuo, et al, "Crosstalk Bounded Uncorrelated Jitter (BUJ) for High-Speed Interconnects," IEEE Tran on Instrumentation and Measurement, Vol. 54, N0. 5, Oct 2005.

[16] R. Darapu, et al, "Analysis of Jitter in Clock Distribution Networks," IEEE2004.

[17] H. R. Cha, et al, "A New Analysis Method of Simultaneous Switching Noise in CMSO Systems," 1998 Electronic Components and Technology Conference.

[18] M. S. Laurent, et al, "Impact of Power-Supply Noise on Timing in High-Frequency Microprocessor," IEEE Tran. On Advanced Packaging, vol. 27, no. 1, Feb. 2004.

[19] Y. I. Ismail, "On-Chip Inductance Cons and Pros," IEEE Tran. On VLSI Systems, vol. 10, no. 6, Dec. 2002.

[20] L. D. Huang, "Global Wire Bus Configuration with Minimum Delay Uncertainty," IEEE DATE 2003.

[21] K. Moiseev, "An Effective Technique for Simultaneous Interconnect Channel Delay and Noise Reduction in Nonometer VLSI Design," IEEE 2006.

[22] A. Roy, et al, "Impact of Inductance on the Figure of Merit to Optimize Global Interconnect," IEEE 2006.

[23] B. Mesgarzadeh, et al, "Jitter Characteristic in Resonant Clock Distribution," IEEE 2006.

[24] M. Hashimoto, et al, "Interconnect Structure for High-Speed Long Distance Signal Transmission," IEEE 2002.

[25] H. R. Cha, et al, "A New Analytic Model of Simultaneous Switching Noise in CMOS Systems," 1998 Electronic Components and Technology Conference.

[26] P. Heydari, "Characterizing the Effects of the PLL Jitter Due to Substrate Noise in Discreet-time Delta-Sigma Modulators," IEEE Tran Circuits and Systems-I: Regular Papers, vol. 52, no. 6, June 2005.

[27] J. Buckwalter, et al, "Data-Dependent Jitter and Crosstalk-Induced Bounded Uncorrelated Jitter in Copper Interconnects," 2004 IEEE MTT-S Digest.

CHAPTER 5
STATISTICAL VLSI CIRCUIT DELAY VARIATION MODELS

Skew and jitter are the dc and ac delay deviations of the data or clock signal zero-crossings with respect to the ideal time reference clock.

Practical high-speed I/O circuits usually suffer from various delay variation effects including both the delay skew and the jitter effects. VLSI high-speed I/O circuit delay variations are typically caused by the PVT variations, the device mismatches, the voltage and phase noises, the reflection and ringing, and the channel bandwidth limitation effects. Delay variations may significantly degrade the performances of the high-speed I/O circuits. Delay variation minimizations of the circuits are among the most important and most challenging design tasks of the VLSI high-speed I/O circuits.

Statistical delay variation modeling methods are widely used in VLSI high-speed I/O circuit developments for identifying the sources of circuit delay variations in circuit debug, for predicting the circuit and system performance, such as the bit error rate (BER), for analyzing the jitter margin of the circuit, and for predicting the yield of the implemented VLSI high-speed I/O circuits and systems at targeted data rates.

The jitter signature and jitter decomposition techniques based on the statistical delay variation analysis methods provide very effective ways for post-silicon debug and product validation in the practical high-speed I/O circuit design and development.

5.1 DELAY VARIATIONS IN VLSI HIGH-SPEED I/O CIRCUITS

At the microscopic level, a successful (or error-free) data transmission of the VLSI I/O circuit is determined by the bounded tracking phase error based on I/O timing constraint equation developed in previous chapter as

$$
\begin{cases}
-E_{max} < E \equiv \delta_{DP} + \delta_{TX} - \delta_{T/2} - \delta_{RX} < E_{max} \\
E_{max} \equiv \dfrac{T - (T_{setup} + T_{hold})}{2}
\end{cases}
\tag{5.1}
$$

Practical VLSI high-speed I/O circuits suffer from various delay variation effects that limit their achievable performances, such as the bit error rate (BER). Strong dependency of delay time to PVT and application conditions can usually cause the high-speed I/O circuit to deviate from the desired operation condition. In worst cases, such I/O circuit may not even meet the phase tracking error criteria given in equation 5.1 and cause the data transmission to fail.

The microscopic timing variation effects in high-speed I/O circuit can be modeled using the time-domain I/O SFG model as shown in figure 5.1 as

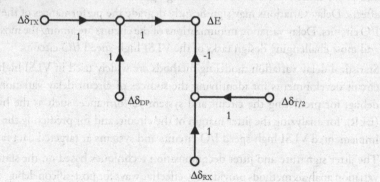

Fig. 5.1. The I/O delay variation SFG model

5.1.1 TIME-INDEPENDENT DELAY VARIATION EFFECTS

Practical VLSI high-speed I/O circuits are usually designed to be "centered" at zero tracking phase error condition under the nominal operation to maximize the timing noise margin. Such design condition was described by so-called maximum timing margin (MxTM) condition as introduced in previous chapter, with the average across all delay variation effects (both time dependent and time independent):

$$\overline{E} \equiv \overline{(\delta_{DP} + \delta_{TX} - \delta_{T/2} - \delta_{RX})} = 0 \qquad (5.2)$$

Due to PVT variations, device mismatch effects and application conditions, the tracking phase error E will vary from unit to unit in practical VLSI I/O circuits.

For the Gaussian-type variation distribution of tracking phase error E (as result of various manufacture nonidealities) as

$$\rho(E) = A \exp(-\frac{(E - \overline{E})^2}{2\overline{E}^2}) \qquad (5.3)$$

the yield loss of the VLSI high-speed I/O circuit can be caused by the time-independent timing uncertainties, such as PVT variation, device mismatch, PCB skew, etc. VLSI high-speed I/O circuit yield can be approximately modeled by the probability of tracking phase error within the maximum available timing window under the constraint of the jitter (time-dependent delay variations) condition as

$$Yield \approx \int_{-E^*_{max}}^{+E^*_{max}} \rho(E)dE = \int_{-E^*_{max}}^{+E^*_{max}} A \exp(-\frac{E^2}{2\overline{E}^2})dE \qquad (5.4)$$

where the modified maximum delay time budget $E_{max}{}^*$ is given as the original maximum high-speed I/O delay time budget E_{max} minus the delay time budget loss due to the jitter effects (represented by the total jitter (TJ) of the high-speed I/O circuit with respect to a specified bit error rate):

$$E^*_{max} \equiv \frac{T - (T_{setup} + T_{hold})}{2} - \frac{TJ}{2} \qquad (5.5)$$

5.1.2 TIME-DEPENDENT TIMING VARIATION EFFECTS

Practical VLSI high-speed I/O circuits also suffer from various jitter effects. Consequently, the delay parameters in the I/O timing constraint equation will vary from time to time during the data transmission operations. For the VLSI high-speed I/O circuits under the MxTM operation condition, the microscopic (bit level) tracking phase error E is basically determined by the delay variations in the basic I/O timing constraint equation as shown below:

$$\begin{cases} \overline{E} \equiv \overline{(\delta_{DP} + \delta_{TX} - \delta_{T/2} - \delta_{RX})} = 0 \\ E = \Delta E \equiv \Delta\delta_{PD} + \Delta\delta_{TX} - \Delta\delta_{T/2} - \Delta\delta_{RX} \end{cases} \tag{5.6}$$

In the above model, $\Delta\delta_{Tx}$ and $\Delta\delta_{Rx}$ represent the delay deviations in the transmitter and the receiver clock references away from its desired delay values respectively. $\Delta\delta_{DP}$ represents the delay deviation in the transmitter synchronization circuit clock-to-out delay and channel delay respectively. $\Delta\delta_{T/2}$ represents the deviation in the eye-centering phase shift. At the macroscopic level, each delay deviation term may vary from bit to bit in data stream in time due to jitter effects. In the high-speed I/O circuit operation, the timing constraint in the basic I/O timing constraint equation must be met for each bit for error-free data transmission. Data transmission error can typically be validated through the dynamic and static verifications. In the timing domain dynamic verification methods, the timing equation is checked for each bit data transmission for large numbers of data bit ($>10^6$) either using computer simulation or from post-silicon testing. This verification type provides direct measurement of the I/O circuit performance.

However, since dynamical verification is a very time-consuming process, it is usually impractical to directly simulate or measure large amounts of data bit beyond 10^6 level. In addition, such amount of data typically is not even sufficient for verifying the performance specification beyond 10E-12 BER as usually required in modern VLSI high-speed I/O circuits. As a result, static verifications based on statistical analysis techniques are widely used in various high-speed I/O circuit for performance verifications.

For commonly used non-return-to-zero (NRZ) data coding scheme, bit error can only occur if there is a 1-to-0 or 1-to-0 data transition. As a result, the bit error rate (BER) with respect to the NRZ data stream can be approximately modeled as the product of the probability of data transition and the probability of the tracking errors that are higher than the maximum allowed time margin as

$$BER \approx P_{Transition} \cdot P(\Delta E > E_{max}) \tag{5.7}$$

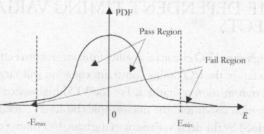

Fig. 5.2. The probability density function of tracking phase error

Equation 5.7 can usually be expressed as an *I/O link timing budget equation* as

$$\begin{cases} 2 \cdot \Delta E + (T_{Setup} + T_{Hold}) \le T = UI \\ \Delta E \equiv \Delta(\delta_{DP} + \delta_{TX} - \dfrac{T}{2} - \delta_{RX}) \end{cases} \tag{5.8}$$

In this I/O link timing budget equation, the unit interval (UI = T) represents the overall available timing budget for the I/O link at the given data rate (f = 1/T). The $(T_{Setup} + T_{Hold})$ term represents the timing penalty due to the metastability of the receiver sampler. The $2\Delta E$ term represents the peak-to-peak delay variation (jitter and skew) of the I/O circuits for given transmitted data bits (e.g., 10^{12} bits for BER of 10E-12). Since jitter and skew effects are usually uncorrelated, the I/O link timing equation can also be approximately written as

$$\begin{cases} 2 \cdot \Delta E_{Jitter} + 2 \cdot \Delta E_{Skew} + (T_{Setup} + T_{Hold}) \le UI \\ \Delta E \equiv \Delta(\delta_{DP} + \delta_{TX} - \dfrac{T}{2} - \delta_{RX}) \end{cases} \tag{5.9}$$

5.2 JITTER MODELING

Jitter is defined as a short-term variation of the signals from their ideal location in time. Jitters within VLSI high-speed I/O circuit are resulted from various circuit nonidealities such as ringing, reflection, EMI, ground bounce, switching power supply noise, thermal noise, cross talk effects, etc.

5.2.1 JITTER TERMINOLOGIES

Jitter in clock signals can usually be characterized using their absolute jitter, period jitter, and cycle-to-cycle jitter.

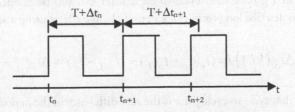

Fig. 5.3. Clock with transition time jitter

- Absolute jitter is the absolute timing difference of the clock signal related to a reference clock. Suppose $\{t_n\}$ is the transition time points of a clock signal of nominal period T as shown in figure 5.3. The sequence $\{t_n - nT\}$ defines the absolute jitter of the clock signal:

$$(\Delta t_{abs})_n = t_n - nT = \sum_{i=1}^{n} \Delta t_i$$

(5.10)

where $\{\Delta t_i\}$ is the difference of each clock period with respect to the ideal clock period. Absolute jitter is also called the timing interview error (TIE), the tracking jitter, the aperture jitter, or the aperture uncertainty. Absolute jitter is usually used to describe the tracking error or long-term jitter between two clocks with one as golden clock.

- Period jitter is defined as the variation of the clock period. Period jitter is defined by the time sequence:

$$\Delta t_n = t_{n+1} - t_n - T$$

(5.11)

Period jitter may lead to the reduction of the time margin between two clocked digital sequential circuits in a pimple. In general case, period jitter can also be defined over k clock periods to describe long-term jitter effects using a sequence as

$$(\Delta t(kT))_n = t_{n+k} - t_n - nT = \sum_{i=1}^{k} \Delta t_i$$

(5.12)

- Cycle-to-cycle jitter is the difference in clock period between the adjacent clock cycles. Cycle-to-cycle jitter is defined by the time sequence:

$$(\Delta t_{cc})_n = (t_{n+2} - t_{n+1}) - (t_{n+1} - t_n) = \Delta t_{n+2} - \Delta t_n$$

(5.13)

Cycle-to-cycle jitter describes the local change in clock period, from one clock cycle to an adjacent cycle. Cycle-to-cycle jitter is also called adjacent period jitter. Cycle-to-cycle jitter may impact performance of multiple synchronous clock systems. In a general case, cycle-to-cycle jitter can also be defined over k clock periods to describe long-term clock period change effects using a sequence as

$$(\Delta t_{cc}(kT))_n = (t_{n+2k} - t_{n+k}) - (t_{n+k} - t_n) = \Delta t_{n+2k} - \Delta t_n$$

(5.14)

It can be seen that cycle-to-cycle jitter is the first difference of the period jitter, and the absolute jitter is the integration of the period jitter.

For a 2.5Ghz clock example shown in figure 5.4, these jitters can be demonstrated using Table. 5.1.

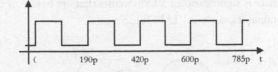

Fig. 5.4. Nonideal 2.5Ghz clock signal

Table 5.1. Jitter representation of nonideal clock signal

t (ps)	0	200	400	600	800
Rise (ps)	0	190	420	600	780
AJ (ps)	0	-10	20	0	-15
PJ (ps)	-	-10	30	-20	-15
CCJ (ps)	-	-40	-50	5	-

In practical high-speed I/O circuits, each type of jitter in the above definitions may impact the I/O circuit performance in different ways. For example, in the common clock I/O circuits, the period jitter is the major concern since the data transmitted is captured at the next clock edge as shown in figure 5.5.

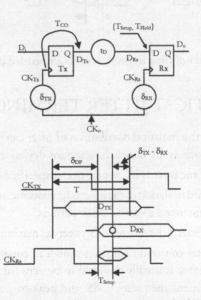

Fig. 5.5. Impact of period jitter to common clock I/O circuit

While for the embedded clock I/O circuits, the absolute jitter is important since the recovered clock, which averages the input data delays for many periods, is used for the data capture. The absolute jitter between the transmitter clock and the recovered clock determines the performance of the high-speed I/O circuits.

Cycle-to-cycle jitter is significant for VLSI circuits that are based on the cycle-to-cycle clock period parameters, such as VLSI DLL circuits.

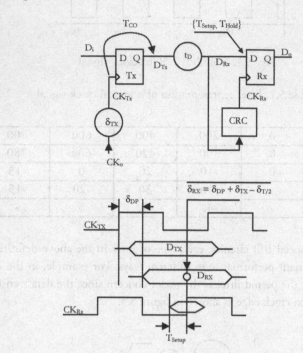

Fig. 5.6. Impact of time interval error to embedded clock I/O circuit

5.2.2 STATISTICAL JITTER TERMINOLOGIES

For each type of jitter, the statistical distribution of jitter can be described based on its statistical parameters such as the mean, the standard deviation, and the peak values as

- the mean value of the jitter is its average over a specified clock periods,
- the RMS or standard deviation of the jitter is its roots mean square value of variation from its mean value over a specified clock periods,
- The peak-to-peak jitter is the difference between its maximum and minimum values.

For the basic I/O timing constraint equation shown in equation 5.1, the mean value of the tracking phase error is usually designed to be zero for optimal timing margin. For different types of jitters, the mean, RMS, and peak-to-peak jitters may have their certain meanings. For example, for the deterministic jitter, the peak-to-peak jitter may represent actual value. While for the random jitter, the peak-to-peak jitter is typically unbounded, and the RMS value is more meaningful.

5.2.3 PHASE AND VOLTAGE NOISE JITTER MODEL

The phase noise of a clock signal can be modeled by a phase-modulated signal as

$$x(t) = A\cos(\omega_o t + \theta(t)) \tag{5.15}$$

where A and ω_o are the amplitude and angular frequency of the carrier. The phase noise of such clock signals can also be expressed as absolute time jitter as

$$\Delta t_{Jitter} = \theta(t)/\omega_o \tag{5.16}$$

For small phase noise, equation 5.15 can be further expressed as

$$x(t) \approx A\cos(\omega_o t) - \theta(t)A\sin(\omega_o t) = x_s(t) + x_n(t) \tag{5.17}$$

where $x_s(t)$ and $x_n(t)$ are the ideal signal and the noise components of the jittering signal respectively. The relationship among the phase noise, time jitter, and the voltage noise can, in general, be expressed as

$$\Delta t_{Jitter} \approx \theta(t)/\omega_o = x_n(t)/SlewRate \tag{5.18}$$

5.2.4 POWER SPECTRUM DENSITY OF PHASE NOISE

Jittering clock can be expressed as a phase-modulated signal for small phase noise. If the clock is measured in the frequency domain, the baseband noise is translated by the ideal clock frequency as shown in equation 5.17. The RMS power of the clock can be calculated as

$$\overline{x(t)^2} = \lim_{T \to \infty} \frac{1}{T} \int_{-T/2}^{T/2} x(t)^2 dt \approx \overline{x_s(t)^2} + \overline{x_n(t)^2} = \frac{A^2}{2} + \frac{A^2}{2}\overline{\theta(t)^2} \tag{5.19}$$

The total RMS phase noise power is the integration of the power spectral density as

$$\overline{\theta(t)^2} = \lim_{T \to \infty} \frac{1}{T} \int_{-T/2}^{T/2} \theta(t)^2 dt \equiv \int_{-\infty}^{\infty} s_\theta(f) df \tag{5.20}$$

The total power spectral density of the signal consists of the power density of the ideal signal and the phase noise component as

123

$$\overline{x(t)^2} \approx \frac{A^2}{2} + \frac{A^2}{2} \int_{-\infty}^{\infty} S_\theta(f) df$$

(5.21)

On the other hand, single sideband (SSB) phase noise $L(f_m)$ is defined as the ratio of noise power in $1\,Hz$ bandwidth at the offset f_m from the clock-to-clock power. It can be seen that the SSB phase noise is directly related to the phase noise spectral density as

$$L(f_m) = \frac{1}{2} \frac{(A^2/2) S_\theta(f-f_o)}{(A^2/2)} = S_\theta(f_m)/2$$

(5.22)

5.2.5 JITTER ESTIMATION BASED ON PHASE NOISE

Based on equation 5.20 and 5.22, we have that

$$\overline{\theta(t)^2} = 2 \int_0^{\infty} L(f) df$$

(5.23)

The RMS voltage noise can be estimated from the clock peak value and the integral of the SSB phase noise as

$$(x_n)_{RMS} = \sqrt{\overline{x_s(t)^2} \cdot 2 \int_0^{\infty} L(f) df} = \sqrt{\overline{x_s(t)^2}} \cdot \sqrt{2 \int_0^{\infty} L(f) df}$$

(5.24)

The RMS absolute jitter can then be estimated as

$$\sqrt{\overline{\Delta t_{abs}^2}} \approx \sqrt{\overline{\theta(t)^2}} / \omega_o = \sqrt{2 \int_0^{\infty} L(f) df} / \omega_o$$

(5.25)

On the other hand, the general k period jitter is given as

$$(\Delta t_n)(kT) \equiv t_{n+k} - t_n - kT = [t_{n+k} - (k+n)T] - [t_n - nT]$$
$$\equiv \Delta t_{abs}(k+n) - \Delta t_{abs}(n)$$

(5.26)

The frequency domain expression of the two jitter types can be derived using z-transform as

$$(\Delta t(kT))(z) = (z^k - 1)\Delta t_{abs}(z) \tag{5.27}$$

$$(\Delta t(kT))(f) = (e^{2\pi fkT} - 1)\Delta t_{abs}(f) \tag{5.28}$$

As a result, the k-period RMS period jitter can be approximately expressed as

$$\sqrt{\overline{\Delta t^2(kT)}} \approx \sqrt{2 \times 2^2 \int_0^\infty L(f)\sin^2(\pi kTf)df \, / \, \omega_o} \tag{5.29}$$

Similarly, the RMS jitter for the k-period cycle-to-cycle jitter is given as

$$\sqrt{\overline{\Delta t_{cc}^2(kT)}} \approx \sqrt{2 \times 2^4 \int_0^\infty L(f)\sin^4(\pi kTf)df \, / \, \omega_o} \tag{5.30}$$

For the typical phase noise including the white noise and the flicker noise, we have that

$$S_\theta(f) = f_o^2 \left[\frac{c_w}{f^2} + \frac{c_f}{f^3}\right] \tag{5.31}$$

The RMS value of the k-period cycle-to-cycle jitter can be derived as

$$\sqrt{(\Delta t_{cc}^2(kT))} = \sqrt{2c_w|kT| + 8\ln(2)c_f(kT)^2} \tag{5.32}$$

It can be seen that this expression provides a way to extract the C_w and C_f parameters from the cycle-to-cycle jitter measurements.

5.2.6 TIME-DOMAIN JITTER MODEL

For random jitter contributed from white noise, the period jitter at each clock period is totally uncorrelated, and the RMS value of the absolute jitter of N clock periods is given as

$$\sqrt{(\overline{\Delta t_{abs}^2})_N} = \sqrt{(\sum_{i=1}^N \Delta t_i)^2} = \sqrt{N\overline{\Delta t^2}} = \sqrt{N}\sqrt{\overline{\Delta t^2}} \propto \sqrt{N} \tag{5.33}$$

where $\sqrt{\overline{\Delta t^2}}$ is the RMS value of the period jitter. It can be seen that absolute jitter for the white noise is unbounded and will grow at a rate of $\sqrt{N}$, depending on duration of the jitter measurement.

Similarly, for the white noise-induced cycle-to-cycle jitter, the RMS jitter can be expressed in terms of the RMS value of the period jitter as

$$\sqrt{(\Delta t_{cc}^2)} = \sqrt{(\Delta t_{n+1} - \Delta t_n)^2} = \sqrt{2\overline{\Delta t^2}} = \sqrt{2}\sqrt{\overline{\Delta t^2}}$$

(5.34)

5.2.7 THE EYE DIAGRAM

The eye diagram is commonly used to specify the signal integrity, such as the voltage noise and jitter of the signal at specific circuit nodes. An eye diagram is a composite of all the bit periods of captured bits superposed on each other related to a bit reference clock. The open area within the eye diagram is referred to as eye opening. In practical applications, an eye mask can typically be constructed to represent the minimal signal integrity requirement at the specific point in the circuit. If the waveform is repeatable, a sampling scope can be used to build an eye diagram from individual samples taken at random delays on many waveforms. Eye diagrams can be either monochrome or color coded to indicate the density of waveform samples at any given point of the display.

There are several common methods of eye diagram generation in high-speed I/O circuit jitter testing, including the oscilloscope-based method with a long-persistence display mode, the clock recovery-based method using in either software or hardware, or an external clock-recovery circuit.

The eye diagram can often be specified by the probability density function (PDF) or the cumulative density function (CDF). PDF is the probability that a random variable will be within a given range. PDF has unit area. CDF is the probability that a random variable will be less than a given value. It is the integral of PDF, which starts at 0 and ends at 1.

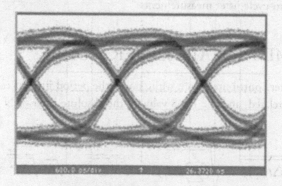

Fig. 5.7. The eye diagram example

5.3 JITTER SIGNATURE AND DECOMPOSITION

Jitter in high-speed I/O circuit can be decomposed into basic jitter types that have well-understood behaviors. Jitter decomposition can be used to extrapolate system performance to cases that would be difficult or time-consuming to measure directly. Jitter decomposition techniques are commonly used to model the underlying physical effects in I/O system to provide insight into the precise cause or causes of excessive jitter. Jitter decomposition techniques allow correlating measured jitter behavioral to specified jitter signatures.

Based on the jitter decomposition method, the total jitter in high-speed I/O circuits can be decomposed into two basic jitter families, including the deterministic jitter (DJ) and the random jitter (RJ), where the deterministic jitters are defined as repeatable and predictable timing variations. Deterministic jitter is bounded, and the bounds can usually be observed or predicted with high confidence based on a reasonably low number of measurements.

The deterministic jitters can be further partitioned into several jitter types, including

- periodic jitter (PJ);
- data-dependent jitter (DDJ), including the intersymbol interference (ISI) and the duty-cycle distortion (DCD);
- bounded, uncorrelated jitter (BUJ).

The random jitter, on the other hand, is usually assumed to have a Gaussian distribution for the jitter modeling based on the facts that the primary source of random noise in many electrical circuits is thermal noise, which have a Gaussian distribution, and the fact that the composite effect of many uncorrelated noise sources, no matter what the distributions of the individual sources, approaches a Gaussian distribution according to the central limit theorem. It is important to note that for a Gaussian distribution, the peak value is inherently unbounded.

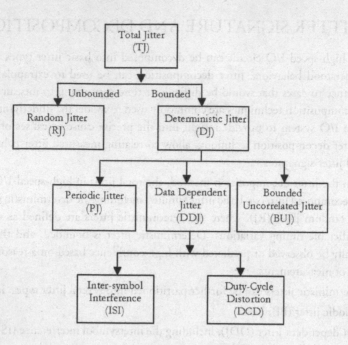

Fig. 5.8. The decomposition of jitters

5.3.1 PERIODIC JITTER (PJ)

Periodic jitter (PJ) is the jitter that repeats in a cyclic fashion. Periodic jitter is also called sinusoidal jitter since any periodic waveform can be decomposed into a Fourier series of harmonically related sinusoids. By convention, periodic jitter is uncorrelated with any periodically repeating patterns in a data stream.

Periodic jitter is typically caused by external deterministic noise sources coupling into a system. An example of PJ in high-speed I/O circuit is VRM noise due to switching power supply noise or a strong local RF carrier injection. PJ may also be caused by an unstable clock-recovery PLL.

Single-tone periodic jitter can be modeled in voltage domain as

$$V(t) = V_o \cos(\omega t + \phi_m \cos(\omega_m t))$$

(5.35)

where the phase error (or jitter) is given as

$$\phi(t) = \phi_m \cos(\omega_m t)$$

(5.36)

The probability density function of the single-tone periodic jitter can be derived as

$$\rho[\phi(t)] \propto \frac{1}{\sqrt{\phi_m^2 - \phi(t)^2}}$$

$$(5.37)$$

The typical signature of the single-tone PJ shows high density at the peak jitter and low density in the middle as shown in figure 5.9. Measurement can be used to determine the jitter frequency and the phase, for example, from the Fourier transformation with peak detection on edge data acquired using a fast, real-time oscilloscope. For multiple PJ sources case, if the frequencies of these PJ components are not harmonically related to each other, it is possible to estimate the total PJ PDF by convolving the PDFs of individual PJ components.

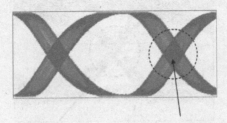

(a) Eye diagram

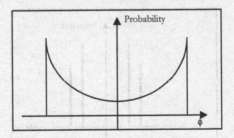

(b) Probability density function

Fig. 5.9. Single-tone periodic jitter signature

5.3.2 INTERSYMBOL INTERFERENCE (ISI)

The data-dependent jitter (DDJ) is the timing jitter that is correlated with the bit sequence in a data stream. One type of the DDJ is the intersymbol interference (ISI) that is caused by the band-limited channel, where waveform cannot reach a full high or low state unless there are several bits in a row of the same polarity. ISI is predictable, and it depends on the transmitted data pattern where the timing of each edge of the transmitted signal depends on the data bit pattern preceding this edge. ISI effect can also be explained using the channel's frequency response and the differences in frequency contents associated with different data patterns where fast-changing data patterns behave as high-frequency signals and slow-changing data patterns behave as low-frequency signals. Because of the signal channel's frequency selectivity, different bit patterns propagate at difference speeds through the channel. This difference in propagation speed causes bits swear into adjacent bits, resulting in ISI.

The typical signature of ISI jitter consists of sole impulses as shown in figure 5.10, which are resulted from the limited number of data patterns. To calculate the total ISI, the probability of each bit pattern and the magnitude of its ISI-induced jitter need to be evaluated.

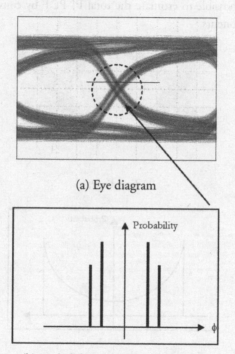

(a) Eye diagram

(b) Probability distribution function

Fig. 5.10. DDJ jitter signature

The ISI can be modeled using the PDF with isolated peaks as

$$\rho[\phi] = \sum_{i=1}^{N} P_i \delta(\phi - \phi_i)$$

$$(5.38)$$

5.3.3 DUTY-CYCLE DISTORTION (DCD)

Duty-cycle dependent jitters (DCD) are the other type of data-dependent jitter (DDJ) that are associated with timing differences in the rising or falling edges of the signals. DCD in VLSI circuits are usually caused by

- rising/falling time mismatches and
- decision threshold error.

The typical signature of DCD jitter is two equal magnitude peaks with the distance related to the duty-cycle error or threshold offset as shown in figure 5.11.

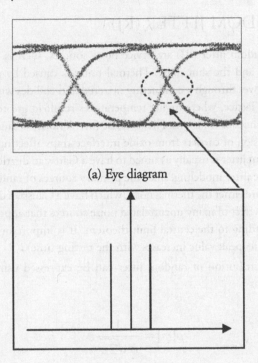

(a) Eye diagram

(b) Probability distribution function

Fig. 5.11. DCD jitter signature

The PDF of the DCD jitter can be mathematically expressed using the isolated jitter peaks as

$$\rho[\phi] = \frac{1}{2}\left[\delta(\phi + \phi_o/2) + \delta(\phi - \phi_o/2)\right]$$

(5.39)

5.3.4   BOUNDED UNCORRELATED JITTER (BUJ)

Bounded uncorrelated jitters (BUJ) in VLSI high-speed I/O circuits are typically resulted from cross talk effect, e.g., noise coupling from adjacent data lines or from on-chip digital random switching devices. BUJ is bounded due to the fact that it has a finite range of coupling effect on signal transitions, and it is uncorrelated because there is no correlation between signal transitions of adjacent lines or on-chip switching circuits. The exact model of BUJ depends on the data pattern, coupling signal, and coupling mechanism.

5.3.5   RANDOM JITTER (RJ)

The causes of random jitter (RJ) are device noise sources, such as the thermal noise, the flicker noise, and the shot noise. Thermal noise is caused by electron scattering when electron moves through a conducting material and collides with silicon atoms or impurities in the lattice, where higher temperatures result in greater atom vibrations and increased chances of collisions. Flicker noise (or $1/f$ noise) is caused by the random capture and emission of carriers from oxide interface traps affecting carrier density in transistor. Random jitter is usually assumed to have a Gaussian distribution as shown in figure 5.12 for the jitter modeling, since the primary sources of random noise in many electrical circuits are either the thermal noise, which have a Gaussian distribution, or due to the composite effect of many uncorrelated noise sources that approach the Gaussian distribution according to the central limit theorem. It is important to note that RJ is unbounded (i.e., its peak value increases with the testing time).

The Gaussian distribution of random jitter can be expressed using the probability density function as

$$\rho[\phi] = \frac{1}{\sigma\sqrt{2\pi}} e^{-\left(\frac{\phi^2}{2\sigma^2}\right)}$$

$$(5.40)$$

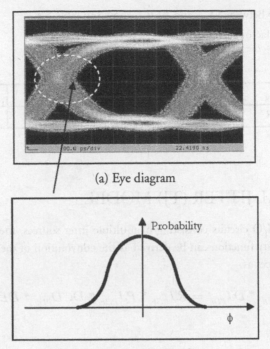

(a) Eye diagram

(b) Probability distribution function

Fig. 5.12. Signature of random jitter

It has been observed that Gaussian-based RJ depends on the number of samples. The relationship between the RMS and peak-to-peak jitter as shown can be used to compute the peak-to-peak jitter for targeted BER as

$$Jitter_{peak-b-peak} = 2Q_{BER} \cdot Jitter_{RMS} \tag{5.41}$$

where parameter Q_{BER} is a function of the BER that is defined as

$$\frac{1}{2} erfc(\frac{Q_{BER}}{\sqrt{2}}) \equiv \frac{1}{\sqrt{\pi}} \int\limits_{\left(\frac{Q_{BER}}{\sqrt{2}}\right)}^{\infty} e^{-t^2} dt = BER \tag{5.42}$$

The relationship between Q_{BER} and BER is shown in table 5.1.

Table 5.1. Q_{BER} factor versus BER

Q_{BER}	6.4	6.7	7.0	7.3	7.6
BER	10^{-10}	10^{-11}	10^{-12}	10^{-13}	10^{-14}

5.4 TOTAL JITTER (TJ) MODEL

For high-speed I/O circuits consisting of multiple jitter sources, the total jitter (TJ) probability density function can be derived as the convolution of the individual PDF of the jitter sources as

$$TJ_{PDF} = RJ_{PDF} * DJ_{PDF} = RJ_{PDF} * PJ_{PDF} * DCD_{PDF} * DDJ_{PDF} \quad (5.43)$$

5.4.1 THE DUAL-DIRAC JITTER MODEL

One of the commonly used behavioral TJ models is the dual-Dirac (or double Dirac) model. The dual-Dirac model can be used to quickly estimate total jitter in VLSI high-speed I/O circuit that is associated with low bit error rate. Under the dual-Dirac model, the PDF of the total jitter is behaviorally modeled as the convolution of a Gaussian-based RJ and a twin-peak DJ (called dual-Dirac distribution) as shown in figure 5.13 as

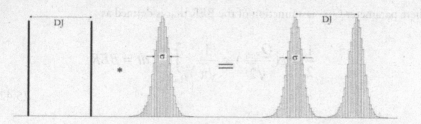

Fig. 5.13. The dual-Dirac TJ model

$$TJ_{PDF} = RJ_{PDF} * DJ_{PDF} = \frac{1}{\sigma\sqrt{2\pi}} e^{-\left(\frac{\phi^2}{2\sigma^2}\right)} * \frac{1}{2}[\delta(\phi+\phi_L) + \delta(\phi-\phi_R)]$$

$$(5.44)$$

where

$$DJ_{\delta\delta} = \phi_R + \phi_L \tag{5.45}$$

Under this model, total jitter (TJ) for a given BER can be expressed using DJ and the scaled RJ as

$$TJ(BER) \approx 2Q_{BER} \cdot \sigma + (\phi_R + \phi_L) = 2Q_{BER} \cdot \sigma + DJ_{\delta\delta} \tag{5.46}$$

5.4.2 THE BATHTUB CURVE AND THE BER

With the help of the dual-Dirac model and recall the basic I/O timing constraint equation that

$$|E| < E_{Max} \tag{5.47}$$

In addition, for high-speed I/O circuit operated under the MxTM condition, the average tracking phase error is zero:

$$\overline{E} = \overline{\delta_{DP} + \delta_{TX} - \delta_{RX} - \frac{T}{2}} = 0 \tag{5.48}$$

The high-speed I/O tracking phase error E is mainly contributed by the total jitter of the equivalent datapath delay parameter as

$$|E| = \Delta(\delta_{DP} + \delta_{TX} - \delta_{RX} - \frac{T}{2}) = TJ \tag{5.49}$$

The eye-opening (or delay margin T_{Eye}) as shown in figure 5.14 for the high-speed I/O data transmission with respect to targeted BER transmission is then given as

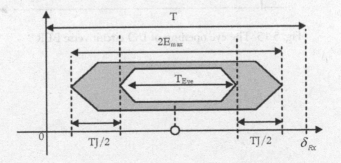

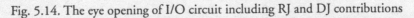

Fig. 5.14. The eye opening of I/O circuit including RJ and DJ contributions

$$T_{Eye} = 2E_{Max} - 2Q_{BER} \cdot \sigma - DJ_{\delta\delta} = 2E_{Max} - 2Q_{BER} \cdot \sigma - (\phi_R + \phi_L)$$
$$= (2E_{Max} - \phi_L - \sigma \cdot Q_{BER}) - (Q_{BER} \cdot \sigma + \phi_R) \tag{5.50}$$

Equation 5.50 can be rewritten in the form as

$$\begin{cases} T_{Eye}(Q_{BER}) = T_L(Q_{BER}) - T_R(Q_{BER}) \\ T_L(x) \equiv (2E_{Max} - \phi_L - \sigma \cdot x) \\ T_R(x) \equiv (\sigma \cdot x + \phi_L) \end{cases} \tag{5.51}$$

Such a function can be plotted versus the Q_{BER} factor as shown in figure 5.15.

It is important to note that the eye openings and the theoretically unbounded peak value of Gaussian probability distribution together leads to a useful way to predict for the BER of the link under a specific jitter condition. This method is based on the thought that there will be an inevitable eye diagram closure for signal containing some Gaussian jitter if we accumulate samples for a long-enough time. A plot called a bathtub plot can be used to estimate the maximum data bits before the eye is completely closed (i.e., data transmission failure).

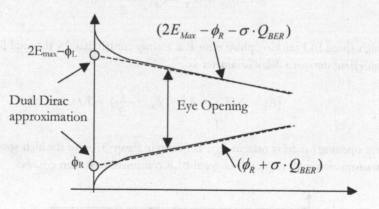

Fig. 5.15. The eye opening of I/O circuit verse BER

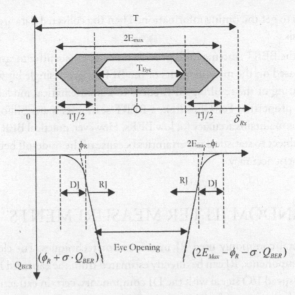

Fig. 5.16. The bathtub curve

5.5 JITTER MEASUREMENT TECHNIQUES

There has been several proposed jitter measurement techniques associated with different measurement equipments, such as time interval analyzers (TIAs), oscilloscopes, and BER testers (BERTs).

One of the commonly used jitter measurement techniques is based on the real-time oscilloscopes for capturing large number of data points in one pass and reconstructing the waveform for display or eye diagram construction by interpolation. This method offers jitter component extraction based on the extracted transition event with respect to a golden time reference. The extracted jitter then can be used for other jitter analysis such as spectral analysis. The oscilloscope-based method provides the easiest way to measure jitter by plotting eye diagrams on an oscilloscope and applying histogram function to zero-crossing. This procedure yields values for the horizontal or vertical eye openings as well as for the total jitter. However, such jitter measurement method does not provide much insight into the jitter properties.

The second jitter measurement method is based on the sampling scope. The sampling scope method is based on the bookkeeping of sampling positions to provide the improvement in jitter measurement. However, such method may lead to accuracy problems at low-error rates due to undersampling operations of the sampling scope since not every data bit is analyzed under this technique.

The other jitter measurement method based on TIA is implemented with or without golden time reference based on collecting many single-shot edge-to-edge time measurements to get the timing information. Then the collected data are used for jitter spectral analysis.

Alternatively, the BERT scan methods provide more adequate jitter measurement. Such methods are based on the measurements of the BER of every single bit within the data eye and the fitting of the resulting BER curve to a mathematical model to obtain the required jitter properties. For Gbps link, a BERT scan tests many billions of bits each second and thus maintains accuracy for low BERs. However, practical BER measurements are typically subject to statistical uncertainties because of the trade-off between test time and measurement accuracy.

5.5.1 RANDOM JITTER MEASUREMENTS

There are several commonly used RJ measurement techniques. For clocklike signals without DJ components, RJ can be directly estimated from the captured histogram. For a general high-speed I/O signal with the DJ components, certain extraction techniques need to be used to separate the RJ from the DJ components.

One of the RJ measurement methods is based on the curve-fitting algorithm. The curve-fitting technique is based on the principle that the tails of the jitter histograms are populated with Gaussian random jitter even in the presence of DJs. Curve-fitting algorithm is used to find the best fitting to the tail regions of the histogram. The RJ estimation can then be estimated by the standard deviation of the fitted Gaussian distribution. However, such curve-fitting techniques suffer from major limitations that require large number of samplers for fitting the tails and that non-Gaussian RJ cannot be effectively extracted.

The second RJ measurement method is based on the jitter spectral analysis that performs the Fourier transformation of the captured jitter signals. Because RJ is stochastic, it exhibits on the spectral graph as small amplitude white noise floor across all frequencies. The RMS value of the noise floor is approximately the RMS value of the RJ.

RJ in the high-speed I/O circuits can also be measured indirectly based on the slope of the measured bathtub using the BERT measurements. However, this RJ measurement method tends to overestimate the RJ value. In addition, only a Gaussian-type RJ can be measured by such technique.

5.5.2 DETERMINISTIC JITTER MEASUREMENTS

Different DJ components in high-speed I/O signals can be extracted using different jitter extraction methods. DCD in a VLSI high-speed I/O circuit can usually be measured by transmitting clocklike data patterns and measuring the periods of logic high and logic low. Such clocklike data patterns can be used to eliminate the ISI effects in such measurement process. On the other hand, the RJ contents in such method can be eliminated by averaging large number of measurements.

Single-tone PJ in the circuits can be measured based on the jitter histogram of the clocklike data patterns. The histogram of such measurement using oscilloscope or TIA contains both the RJ and PJ components. The PJ components can be estimated by the separation of the peaks.

ISI jitter can be measured by transmitting data patterns containing both long and short bit run lengths. In this case, the ideal timing event for the ith edge in the pattern related to a reference edge would occur at nT, while the actual timing event may contain deviations, which can be expressed as $nT+x_i$, where x_i is the deviation of the ith edge. The measurement of x_i values contains both the RJ and PJ components that can be eliminated by averaging. The measured distribution of the averaged x_i can be used to extract the ISI PDF. The limitation of this method is that this method needs a repeating pattern and that low frequency PJ might not be averaged out. Such components may show up in the measurement results and cause reduced accuracy.

Spectral analysis method provides another method for measuring the signal DJ contents. In such method, PJ components usually show as fixed frequency peaks in the spectral graph. PJ magnitude can be computed by an inverse Fourier transform of the isolated peak from the spectral graph. Both DCD and ISI are pattern dependent and will show in the spectral graph at multiple of f_o/N, where f_o is the bit rate and N is the data pattern length. A proposed method is to first perform an inverse transform of the combined components, then to construct a histogram for each of the rising and falling edges. The difference between the mean values of the two histograms can be used to extract DCD, while the difference between peak-to-peak values of the histogram can be used to extract ISI.

Spectral analysis can also be used to separate the RJ from TJ if the RJ has the Gaussian distribution. Such method is also easy to separate the PJ contents. However, BUJ and DDJ separation is very difficult.

APPENDIX 5

A5.1 RELATIONSHIP BETWEEN PEAK AND RMS VALUES OF RANDOM JITTER

For Gaussian-based RJ, the PDF is modeled as

$$\rho[\phi] = \frac{1}{\sigma\sqrt{2\pi}} e^{-\left(\frac{\phi^2}{2\sigma^2}\right)} \tag{5.52}$$

The peak-to-peak jitter ϕ_m can be modeled as the probability of all of N-1 sampler of N samples contained within ϕ_m given as

$$N-1 = N \int_{-\phi_m/2}^{\phi_m/2} \frac{1}{\sigma\sqrt{2\pi}} e^{-\left(\frac{\phi^2}{2\sigma^2}\right)} d\phi \tag{5.53}$$

$$\Rightarrow N-1 = N \frac{2}{\sqrt{\pi}} \int_{0}^{\phi_m/2} e^{-\left(\frac{\phi^2}{2\sigma^2}\right)} d\left(\frac{\phi}{\sqrt{2\sigma}}\right) \tag{5.54}$$

$$\Rightarrow N-1 = N \frac{2}{\sqrt{\pi}} \int_{0}^{\left(\frac{\phi_m}{2\sqrt{2\sigma}}\right)} e^{-x^2} dx \tag{5.55}$$

$$\Rightarrow 1-1/N = \frac{2}{\sqrt{\pi}} \int_{0}^{\left(\frac{\phi_m}{2\sqrt{2\sigma}}\right)} e^{-x^2} dx \tag{5.56}$$

By definition, the BER is given as 1/N and RMS jitter is σ, we have that

$$\Rightarrow 1-BER = \frac{2}{\sqrt{\pi}} \int_{0}^{\left(\frac{\phi_m}{2\sqrt{2\sigma}}\right)} e^{-x^2} dx \tag{5.57}$$

$$\Rightarrow BER = 1 - \frac{2}{\sqrt{\pi}} \int_{0}^{\left(\frac{\phi_m}{2\sqrt{2}\sigma}\right)} e^{-x^2} dx \equiv erfc(\frac{\phi_m}{2\sqrt{2}\sigma}) \qquad (5.58)$$

For simplicity, we may introduce as scalar parameter as Q_{BER}, which can be used to linearize the relationship between the peak-to-peak and RMS jitter of Gaussian-based RJ as

$$erfc(\frac{Q_{BER}}{\sqrt{2}}) \equiv BER \qquad (5.59)$$

Finally, the relationship between the peak-to-peak and RMS jitter can be given as

$$\phi_m \equiv 2Q_{BER} \cdot \sigma \qquad (5.60)$$

References:

[1] M. Li, et al, "Transfer Function for the Reference Clock Jitter in a Serial Link: Theory and Applications," ITC International Test Conference, 2005.

[2] P. K. Hanumolu, et al," Jitter in High-Speed Serial and Parallel Links," ISCAS 2005.

[3] J. Buckwalter, et, al, "Predicting Data-Dependent Jitter," IEEE Tran on Circuits and Systems, Vol. 51, No. 9, Sept. 2005.

[4] C. Pease, et al, "Practical Measurement of Timing Jitter Contributed by a Clock-and Data Recovery Circuit," IEEE Tran on Circuits and Systems, Vol. 52, No. 1, January 2005.

[5] A. Kuo, et al, "Jitter Models and Measurement Methods for High-Speed Serial Interconnects," ITC International Test Conference, 2005.

[6] C. Madden, et al, "System Level Deterministic and Random Jitter Measurement and Extraction for Multi-gigahertz Memory Buses," ???

[7] A. Kuo, et al, "Crosstalk Bounded Uncorrelated Jitter (BUJ) for High-Speed Interconnects," IEEE Tran on Instrumentation and Measurement, Vol. 54, N0. 5, Oct 2005.

[8] B. Analui, et al," Data-Dependent Jitter in Serial Communications," IEEE Tran on Microwave Theory and Techniques Vol. 53. No. 11, Nov. 2005.

[9] K. K. Kim, et, al, "On the Modeling and Analysis of Jitter in ATE Using Matlab," Proc. 2005 20th IEEE International Symposium on Defect and Fault Tolerance in VLSI System.

[10] J. Buckwalter, et al, "Data-Dependent Jitter and Crosstalk-Induced Bounded Uncorrelated Jitter in Copper Interconnects," 2004 IEEE MTT-S Digest.

[11] G. Balamurugan, et al, "Modeling and Mitigation of Jitter in Multi-Gbps Source-Synchronous I/O Links," Proc. of 21st International Conference on Computer Design (ICCD'03).

[12] R. Stephens, et al, "Analyzing Jitter and High Data Rates," IEEE Optical Communication, Feb. 2005.

[13] C-K. Ong, et al, "Random Jitter Extraction Technique in a Multi-Gigahertz Signal," Proc. of Design Automation and Test in Europe Conference and Exhibition (DATE'04).

[14] N. Ou, et al, "Jitter Models for the Design and Test of Gbps-Speed Serial Interconnects," IEEE Design & Test of Computers, 2005.

[15] S. Sunter, et al, "On-Chip Digital Jitter Measurement, from Megahertz to Gigahertz," IEEE Design & Test of Computers, 2005.

[16] M-L Liu, et al, "A New Probability Density Function Deconveolution Method to Remove the Effect of Timing Jitter," IEEE 6th CAS Symp. On Emerging Technologies: Mobile and Wireless Comm. Shanghai, China, May 31-June 2, 2005.

[17] M Amemiya, et, al, "Jitter Accumulation for Periodic Pattern Signals," IEEE Tran. On Communications, Vol. COM-34, No. 5, May 1986.

[18] Y Cai, et al, "Jitter Testing for Multi-Gigabit Backplane SerDes—Techniques to Decompose and Combine Various Types of Jitter," ITC International Test Conference, 2002.

[19] U-K. Moon, et al, "Spectral Analysis of Time-Domain Phase Jitter Measurements," IEEE Tran on Circuits and Systems-II: Analog Signal Processing. Vol 49, N0.5, May 2002.

[20] M Shimanouchi, "An Approach to Consistent Jitter Modeling for Various Jitter Aspects and Measurement Methods," ITC International Test Conference, 2001.

[21] M. P. Li, et al, "A New Method for Jitter Decomposition through Its Distribution Tail Fitting," ITC International Test Conference, 1995.

[22] G. F. Hughes, et al, "Bit Errors due to Channel Modulation of Media Jitter," IEEE Tran on Magnetics. Vol 34, N0.5, Sept 1998.

[23] J. Chin, et al, "Phase Jitter = Timing Jitter?" IEEE Communication Letters, Vol. 2, No. 2, Feb 1998.

[24] K. Murakami, "Jitter in Synchronous Residual Time Stamp," IEEE Tran. On Communications, Vol. 44, No. 6, June, 1996.

[25] D. Hong, et al, "Bit-Error-Rate Estimation for High-Speed Serial Links," IEEE Transactions on Circuits and Systems Vol.53, No.12, Dec. 2006.

[26] P. K. Hanumolu, et al, "Analysis of PLL Clock Jitter in High-speed Serial Links," IEEE Transactions on Circuits and Systems Vol.50, No.11, November 2003.

[27] O. Oliaei, "Extraction of Timing Jitter From Phase Noise," IEEE2003.

[28] D. C. Lee, "Analysis of Clock Jitter in Phase-Locked Loops," IEEE Transactions on Circuits and Systems Vol.49, No.11, November 2002.

[29] G. Balamurugan, et al, "Modeling and Mitigation of Jitter in High-speed Source-Synchronous Inter-Chip Communication," IEEE 2003.

[30] A. Fakhfakh, et al, "Study and Behavioral Simulation of Phase Noise and Jitter in Oscillators," IEEE 2001.

[31] R. Drapu, et al, "Analysis of Jitter in Clock Distribution Networks," IEEE 2004.

[32] L. Schiano, et al, "Frequency Domain Measurement of Timing Jitter in ATE," IMTC 2004.

[33] C. W. Zhang, et al, "Timing Jitter in A 1.35-Ghz Single-Ended Ring Oscillator," IEEE 2002.

[34] C. W. Zhang, et al, "Simulation Technique for Noise and Timing Jitter in Electronic Oscillator," IEE Proc.—Circuits Devices Syst. Vol.151, No.2 April 2004.

[35] A. M. Fahim, "Jitter Analysis of Digital Frequency Dividers in Communication Systems," 2004 IEEE International Ultrasonic, Ferroelectrics and Frequency Control Joint 50th Anniversary Conference.

[36] D. Elad, et al, "Short Term Oscillator Frequency Stability as a Function of Its Phase Noise Distribution," IEEE 2000.

[37] M. Mansuri, et al, "Jitter Optimization Based on Phase-Locked Loop Design Parameters," IEEE JSSC, Vol. 37, N0. 11, November 2002.

[38] P. S. Lin, et al, "Analysis and Control of Timing Jitter in Digital Logic Arising from Noise Voltage Sources," IEEE 1993.

[39] D. A. Howe, et al, "Clock Jitter Estimation based on PM Noise Measurement," 2003 IEEE International Frequency Control Symposium and PDA Exhibition Jointly with the 17th European Frequency and Time Forum.

[40] W. F. Egan, "Phase Noise Modeling in Frequency Dividers," 45th Annual Symposium on Frequency Control.

[41] K. Vichienchom, et al, "Analysis of Phase Noise due to Bang-Bang Phase Detector in PLL-Based Clock and Data Recovery Circuits," IEEE 2003.

[42] P. Heydari, "Analysis of DLL Jitter Due to Substrate Noise," IEEE 2002.

[43] M. Hatamian, "Time Delay Estimation in Presence of Jitter," IEEE 1985.

[44] W. F. Walls, et al, "Computation of Time-Domain Frequency Stability and Jitter from PM noise Measurement," 2001 IEEE International Frequency Control Symposium and PDA Exhibition.

[45] M. Li, et al, "Transfer Function for the Reference Clock Jitter in A Serial Link: Theory and Applications," ITC International Test Conference 2004.

[46] J. J. Huang, et al, "An Infrastructure IP for On-Chip Clock Jitter Measurement," ICCD 2004.

[47] K. H. Cheng, et al, "BIST for Clock Jitter Measurement," IEEE 2003.

[48] C. Pease, et al, "Practical Measurement of Timing Jitter Contributed by a Clock-and-Data Recovery Circuit," IEEE Tran. Circuits and Systems—I: Regular Papers, Vol.52, No. 1, January 2005.

[49] A. Kuo, et al, "Jitter Models and Measurement Methods for High-Speed Serial Interconnects," ITC International Test Conference 2004.

[50] C. Madden, et al, "System Level Deterministic and Random Jitter Measurement and Extraction for Multi-Gighertz Memory Buses," IEEE 2004.

[51] J. Buckwalter, et al, "Predicting Data-Dependent Jitter," IEEE Tran. Circuits and Systems—II: Express Briefs, Vol. 51, No. 9, Sept. 2004.

[52] A. Kuo, et al, "Crosstalk Bounded Uncorrelated Jitter (BUJ) for High-Speed Interconnects," IEEE Tran. Instrumentation and Measurement, Vol. 54, N0. 5, Oct. 2005.

[53] B. Analui, et al, "Data-Dependent Jitter in Serial Communication," IEEE Tran. Microwave Theory and Techniques, Vol. 53, No. 11, Nov. 2005.

[54] K. K. Kim, et al, "On the Modeling and Analysis of Jitter in ATE Using Matlab," IEEE DFT 2005.

[55] T. Xia, et al, "Time-to-Voltage Converter for On-Chip Jitter Measurement," IEEE Tran. Instrumentation and Measurement, Vol. 52, No. 6, Oct. 2003.

[56] P. Stephens, "Analyzing Jitter for High Data Rates," IEEE Optical Communication, 2004.

[57] J. Walker, et al, "Modeling of the Synchronization Process Jitter Spectrum with Input Jitter," IEEE Tran. Communication, Vol. 47, No. 2, Feb. 1999.

[58] J. Buckwalter, et al, "Data-Dependent Jitter and Crosstalk-Induced Bounded Uncorrelated Jitter in Copper Interconnects," 2004 IEEE MTT-S Digest.

[59] M. Shinagawa, et al, "Jitter Analysis of High-Speed Sampling Systems," IEEE JSSC, vol.25, no. 1, Feb. 1990.

[60] C. Fulton, et al, "Delay Jitter First-order Statistical Function in High-Speed Multi-Media Networks," IEEE 1996.

[61] M. J. E, Lee, et al, "Jitter Transfer Characteristics of Delay-Locked Loops—Theories and Design Techniques," IEEE JSSC, vol.38, no. 4, April. 2003.

[62] B. Kim, et al, "PLL/DLL System Noise Analysis for Low Jitter Clock Synthesizer Design,"

[63] R. C. H. Van de Beek, et al, "Low Jitter Clock Multiplication: A Comparison Between PLLS and DLLs," IEEE Tran. Circuits and Systems: II: Analog and Digital Signal Processing, vol. 49, no. 8, August 2002.

[64] P. K. Hanumolu, et al, "Jitter in High-Speed Serial and Parallel Links," IEEE ISCAS 2004.

[65] M. Takahashi, et al, "VCO Jitter Simulation and Its Comparison with Measurement," IEEE 1999.

[66] T. J. Yamaguchi, et al, "Extraction of Peak-to-Peak and RMS Sinusoidal Jitter Using an Analytical Signal Method," IEEE 2001.

[67] C. K. Ong, et al, "Random Jitter Extraction Technique in Multi-Gighertz Signal," Proceedings of the IEEE Design, Automation and Test in Europe Conference and Exhibition, 2004.

[68] S. Saad, et al, "Analysis of Accumulated Timing-Jitter in the Time-domain," IEEE Tran. Instrument and Measurement, vol. 47, no. 1, Feb. 1998.

CHAPTER 6
VLSI DELAY TIME AND PHASE CONTROL CIRCUITS

Electrically controllable delay circuits serve as important VLSI time-domain signal processing circuit elements for VLSI high-speed I/O circuit implementations. These circuits are widely used for compensating the delay uncertainties that are resulted from various circuit nonidealities in the manufacture and operation of VLSI high-speed I/O circuits. There are three widely used VLSI delay control circuit families, including the voltage-controlled delay line (VCDL) circuits, the voltage-controlled oscillator (VCO) circuits, and the phase interpolator (PI) circuits.

VLSI voltage-controlled delay line (VCDL) and voltage-controlled oscillator (VCO) circuits offer time-domain signal processing capability, which allows electrically controllable delay time of the signals. The control signals of VCDL and VCO circuits are typically analog voltages signals. However, they can also be in other analog (such as current) or digital signal forms. VLSI phase interpolation (PI) circuits, on the other hand, usually serve as the digital to delay converters that are based on phase weighting of multiple phase clock inputs.

VLSI voltage-controlled delay line, voltage-controlled oscillator, and phase interpolator circuits are usually used in the phase control loops in VLSI high-speed I/O circuits implemented in the forms of delay-locked loop (DLL) circuits, phase-locked loop (PLL) circuits, and data recover circuits (DRC) for adaptive compensations of circuit or system delay time uncertainties.

6.1 VLSI VOLTAGE-CONTROLLED DELAY LINE (VCDL) CIRCUITS

Most VLSI voltage-controlled delay circuit elements are based on the electrically controlled circuit RC time constant as shown in the conceptual circuit in figure 6.1.

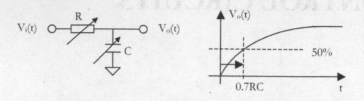

Fig. 6.1. RC time constant-based VCDL circuit structure

In VLSI VCDL circuits, the delay of the VCDL can be electrically controlled through their effective time constants using the voltage-controlled resistor (VCR) elements, the discrete resistor banks, the varactor, or the capacitor banks. Delay time of the RC-based VCDL elements can also be controlled based on the voltage threshold value adjustment as shown in figure 6.2.

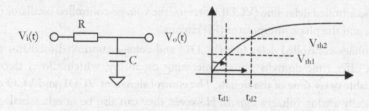

Fig. 6.2. Voltage threshold-based VCDL circuit structure

In addition, the control of VCDL delay time can also be realized in full digital fashion using a signal multiplexer circuit as shown in figure 6.3.

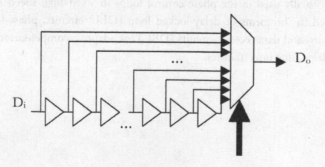

Fig. 6.3. Digital VCDL circuit structure

6.1.1 VCDL SIGNAL PROCESSING CIRCUIT MODEL

The time-domain signal processing representations of the VCDL circuit are shown in figure 6.4.

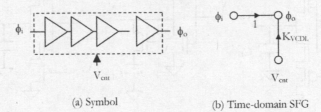

(a) Symbol   (b) Time-domain SFG

Fig. 6.4. VLSI VCDL circuit representations

Such circuit can also be mathematically modeled in time-domain control equation as

$$\phi_o = \phi_i + K_{VCDL} V_{cnt} \tag{6.1}$$

where K_{VCDL} is defined as the gain of the VCDL circuits. Note that the control signal of VCDL can, in general, have many forms, including the voltage, current, digital, analog, or mixed signal forms.

6.1.2 BASIC VLSI VCDL CIRCUIT STRUCTURES

The delay time of a VLSI VCDL based on CMOS inverter as shown in figure 6.5 can be approximately expressed as

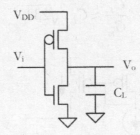

Fig. 6.5. Basic voltage controllable delay circuit

$$t_d \approx \frac{0.7 C_L}{K_p (\frac{W}{L})(V_{DD} - V_T)} \tag{6.2}$$

It can be seen that the delay time of such circuit can be electrically controlled based on three basic control schemes including the effective supply voltage, the effective loading capacitance, and the effective device size (strength) control as shown in figure 6.6.

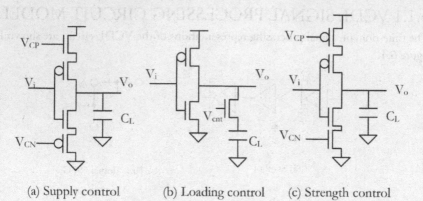

(a) Supply control   (b) Loading control   (c) Strength control

Fig. 6.6. Basic control schemes of VCDL circuits

The delay time of the VLSI VCDL circuit can also be controlled based on the effective transconductance of the circuit as shown in figure 6.7. Such circuit implementation is attractive because of its good noise performance in terms of both lower switching noise and higher noise rejection ability. For such circuit, the effective (small signal) Gm of the circuit is approximately given as

$$G_m \approx \sqrt{2\beta I_C} \qquad (6.3)$$

The time constant of such circuit can be approximately expressed as

$$\frac{C_L}{G_m} \approx C_L / \sqrt{2\beta I_C} \qquad (6.4)$$

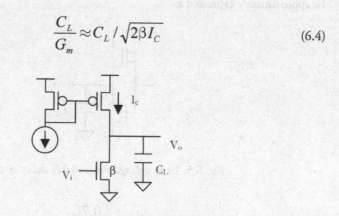

Fig. 6.7. Time constant-based VLSI VCDL circuit

Shown in figure 6.8, figure 6.9, and figure 6.10 are several typical VLSI VCDL circuit implementations that are developed based on various delay control circuit techniques, including current steeling, switched capacitor bank, digital multiplexing, etc.

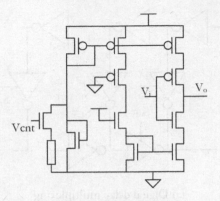

(a) Analog current steeling

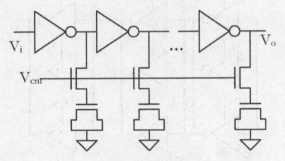

(b) Analog switching cap

Fig. 6.8. VLSI Analog VCDL circuit structures

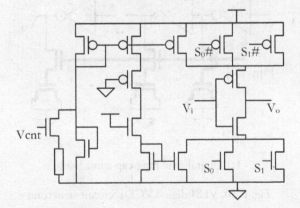

Fig. 6.9. VLSI Analog/Digital current steeling VCDL circuit structure

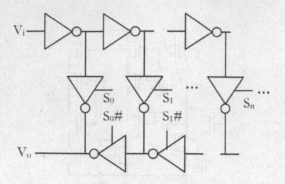

(a) Digital delay multiplexing

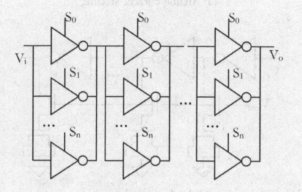

(b) Digital dealy selection

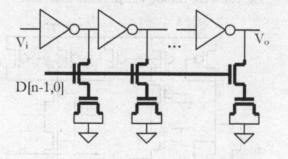

(c) Digital switching capacitor bank

Fig. 6.10. VLSI digital VCDL circuit structures

6.1.3 VLSI DIFFERENTIAL VCDL CIRCUIT STRUCTURES

VLSI differential VCDL circuits are widely used in VLSI high-speed I/O circuits for better noise immunity and lower switch noise. Shown in figure 6.11 is a widely used VLSI differential VCDL circuit structure that is based on the differential CML buffer circuit structure employing the symmetrical voltage-controlled resistor (VCR) load structure.

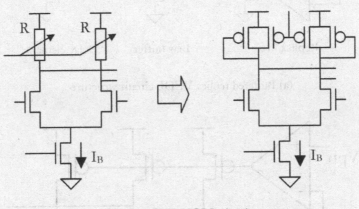

Fig. 6.11. VLSI CML VCDL circuit structure

Such VCDL circuits are typically used with the replica bias circuit shown in figure 6.12 to maintain proper operation point supporting large signal swing and wide delay control range.

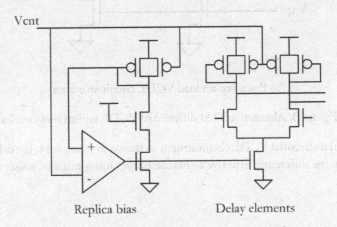

Replica bias   Delay elements

Fig. 6.12. VLSI differential VCDL circuit with replica bias

Shown in figure 6.13 are alternative replica bias circuit implementations for the VCR-based VLSI differential VCDL circuits.

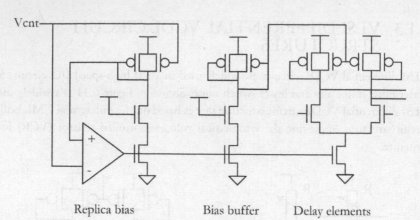

Vcnt

Replica bias   Bias buffer   Delay elements

(a) Buffered replica VCDL circuit structure

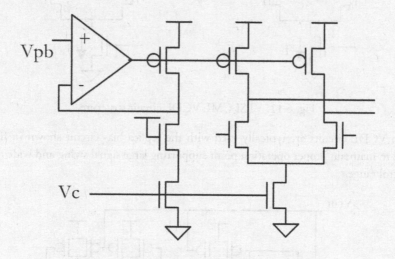

Vpb

Vc

(b) Pmos replica load VCDL circuit structure

Fig. 6.13. Alternative VLSI differential VCDL replica bias circuits

In the VLSI differential VCDL circuit structures shown in figure 6.14, the delay tuning is based on the differential effective resistance tuning using negative resistance circuit technique.

6.1.3 VLSI Differential VCDL Circuit Structures

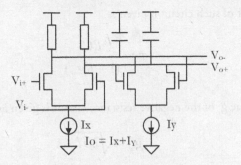

(a) Negative resistance tuning VCDL circuit

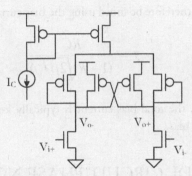

(b) Alternative negative resistance tuning VCDL circuit

Fig. 6.14. VLSI differential VCDL based on negative resistance tuning

The operation of such VCDL circuit can be modeled using the equivalent circuit shown in figure 6.15.

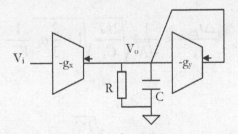

Fig. 6.15 Equivalent circuit mode of differential VCDL

The transfer function of such circuit can be expressed as

$$\frac{V_o}{V_i} = \frac{g_x}{(\frac{1}{R} - g_y) + sC}$$

(6.5)

The time constant of such circuit is given as

$$\tau = \frac{RC}{(1 - Rg_y)}$$

(6.6)

The effect G_m value g_y of the negative resistance circuit is given as

$$g_y \approx \sqrt{2\beta I_y}$$

(6.7)

where β is the MOS device size of the negative resistor circuit.

The time constant can therefore be tuned using the bias current I_y as

$$\tau = \frac{RC}{(1 - R\sqrt{2\beta I_y})}$$

(6.8)

In such VCDL circuit, the total bias current is typically kept constant for constant common mode voltage bias.

6.1.4 VLSI VCDL CIRCUIT PHASE NOISE MODELS

Since most VLSI VCDL circuits are based on the delay buffer circuit structures, the open loop noise characteristics of the VCDL circuits are similar to the VLSI buffer circuit structure. For the first-order approximation, the random phase noise of the N-stage VCDL circuit can be approximately expressed as the roots means square of random phase noise of each buffer stage:

$$\frac{\Delta t_{RMS}}{t_d} \approx \frac{1}{\sqrt{N}} \sqrt{\frac{2kT}{C_L}} \sqrt{1 + \frac{2}{3} a_V} \frac{1}{V_{PP}}$$

(6.9)

or

$$\Delta t_{RMS} \propto \frac{\sqrt{TC_L}}{V_{PP}}$$

(6.10)

In the equation, N is the number of VCDL stage and t_d is the total delay of the delay line. C_L and a_V are the single stage effect load capacitance and the buffer gain respectively. It has been assumed in the above expression that delay of each stage is totally uncorrelated. The expression will be slightly different if high order effects, such as the interaction between delay stages, are included.

For differential VLSI VCDL circuits, a delay uncertainty due to thermal noise is given in literature as

$$\frac{\Delta t_{RMS}}{t_d} \approx \sqrt{\frac{2kT}{C_L}} \frac{\xi}{V_{GS} - V_T}$$

(6.11)

in which ξ is a design-dependent parameter, V_{GS}-V_T is the overdrive voltage of the delay cell.

For N-stage VCDL circuit, the random phase noise at the output of the circuit can be approximately expressed using the phase noise of the single stage delay line as

$$\left(\Delta t_{RMS}\right)_{N-stage} \approx \sqrt{N}\left(\Delta t_{RMS}\right)_{Cell}$$

(6.12)

6.1.5 VCDL TIME UNCERTAINTY DUE TO MISMATCH

Because of random device mismatch effects, the delay of different cells in the VLSI VCDL circuit will vary and result in timing jitter or static phase error in high-speed I/O circuit applications.

Assume that the random mismatch effect of the k-th delay cell is expressed using the delay time of the cell $(t_d)_k$ and the random mismatch parameter e_k as

$$(t_d)_k = t_d(1 + e_k)$$

(6.13)

where t_d is the mean delay of each delay cell.

For a "free" running or uncontrolled VCDL with N replica delay stages and assume the delay uncertainty of the delay stages due to random mismatch are totally uncorrelated, the total delay time uncertainty of the VCDL is approximately given as

$$\sum_{k=1}^{N}(t_d)_k = t_d(1 + \sum_{k=1}^{N}e_k)$$

(6.14)

The total delay time uncertainty of m delay cells is given as

$$\left(\Delta t_{RMS}\right)_{VCDL}^2 = \overline{\left(\sum_{k=1}^{N}(t_d)_k - N \cdot t_d\right)^2} = \overline{(t_d\sum_{k=1}^{N}e_k)^2} = N\left(\Delta t_{RMS}\right)_{Cell}^2$$

(6.15)

We have that

$$\left(\Delta t_{RMS}\right)_{VCDL} \approx \sqrt{N}\left(\Delta t_{RMS}\right)_{Cell} \tag{6.16}$$

It can be seen that the highest delay uncertainty is located at the final stage of the uncontrolled VCDL, and the RMS uncertainty value is proportional to the square root of the delay stages in the VCDL.

For a VCDL within the delay-locked loop, the sum of the delay uncertainty is zero, assuming the ideal phase detector. We have that

$$\left(t_d\right)_k = \frac{T(1+e_k)}{N + \sum\limits_{k=1}^{N} e_k} \tag{6.17}$$

where T is the total mean delay of the M-stage VCDL circuit. The total delay of m VCDL stage under a given constraint is given as

$$\sum\limits_{k=1}^{m}\left(t_d\right)_k = \sum\limits_{k=1}^{m}\frac{T(1+e_k)}{N + \sum\limits_{k=1}^{N} e_k} = T\frac{m}{N}\frac{(1+\frac{1}{m}\sum\limits_{k=1}^{m} e_k)}{1+\frac{1}{N}\sum\limits_{k=1}^{N} e_k} = m \cdot t_d \frac{(1+\frac{1}{m}\sum\limits_{k=1}^{m} e_k)}{1+\frac{1}{N}\sum\limits_{k=1}^{N} e_k} \tag{6.18}$$

The RMS uncertainty value at the m-th VCDL stage within an N-stage DLL is given as

$$\left(\Delta t_{RMS}\right)^2_{m-stage} = [m \cdot t_d(\frac{(1+\frac{1}{m}\sum\limits_{k=1}^{m} e_k)}{1+\frac{1}{N}\sum\limits_{k=1}^{N} e_k} - 1)]^2 \tag{6.19}$$

For small mismatch variation, we can ignore the high order mismatch terms. The RMS delay uncertainty can be derived as

$$\left(\Delta t_{RMS}\right)^2_{m-stage} \approx \overline{\left(m \cdot t_d\right)^2 \left(\frac{1}{m}\sum_{k=1}^{m} e_k - \frac{1}{N}\sum_{k=1}^{N} e_k\right)^2}$$

$$= \overline{\left(m \cdot t_d\right)^2 \left[\left(\frac{1}{m}\sum_{k=1}^{m} e_k\right)^2 + \left(\frac{1}{N}\sum_{k=1}^{N} e_k\right)^2 - 2\frac{1}{m}\sum_{k=1}^{m} e_k \cdot \frac{1}{N}\sum_{k=1}^{N} e_k\right.}$$

$$= \overline{\left(m \cdot t_d\right)^2 \left[\left(\frac{1}{N}\sum_{k=1}^{N} e_k\right)^2 - \left(\frac{1}{m}\sum_{k=1}^{m} e_k\right)^2\right]}$$

$$= \left(m \cdot t_d\right)^2 \left[\frac{1}{m} - \frac{1}{N}\right]\overline{\left(e_k\right)^2} \tag{6.20}$$

Such delay uncertainty can be expressed in terms of the delay uncertainty of the delay cell, the delay line stage N, and the tap location m as

$$\left(\Delta t_{RMS}\right)^2_{m-th\_stage} \approx \frac{m(N-m)}{N}\left(\Delta t_{RMS}\right)^2_{cell} \tag{6.21}$$

It is interesting to see that within an ideal DLL, delay uncertainties at both ends of the VCDL circuit are the lowest. Such behavior is expected because of the delay feedback control loop of DLL. It can also be seen that the highest delay uncertainty is at the middle (i.e., m=N/2) of the VCDL with the RMS value given as

$$\left(\Delta t_{RMS}\right)_{N/2} \approx \frac{\sqrt{N}}{2}\left(\Delta t_{RMS}\right)_{cell} \tag{6.22}$$

6.2 VOLTAGE-CONTROLLED OSCILLATOR (VCO)

There are two commonly used VLSI voltage-controlled oscillator (VCO) circuit families in VLSI high-speed I/O circuits including the ring oscillator VCO and LC oscillator VCO circuits. In the ring oscillator VCO circuits shown in figure 6.16, multiple VCDL circuit stages with odd number of signal inversions are connected in a ring configuration. Such circuit provides self-oscillation operation when the loop gain is larger than unity, and there is a 180-degree loop phase shift at the VCO oscillation frequency.

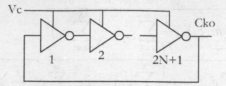

(a) Basic VCO circuit

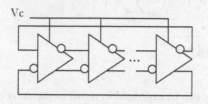

(b) Differential VCO circuit

Fig. 6.16. VSLI Ring VCO circuit structures

The LC oscillator VCO circuit as shown in figure 6.17 is based on an LC-resonant tank with a differential Gm circuit that provides the positive feedback for the VCO operation. The closed-loop Gm circuit serves as a negative resistance to compensate for the loss in the inductor and capacitor resonant circuit. Such a circuit can also offer self-oscillation for high-quality clock reference generation.

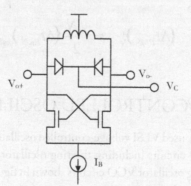

Fig. 6.17. VLSI LC VCO circuit structure

6.2.1 VCO SIGNAL PROCESSING CIRCUIT MODEL

The time-domain signal processing representations of the VLSI VCO are shown in figure 6.18.

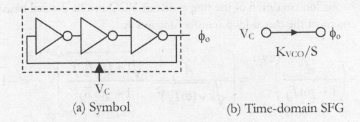

(a) Symbol   (b) Time-domain SFG

Fig. 6.18. VLSI VCO circuit representations

Such VCO circuit can be modeled mathematically using a time-domain equation as

$$\phi_o = \frac{K_{VCO}}{s} V_C \qquad (6.23)$$

where K_{VCO} is defined as the VCO gain. The control signals of VLSI VCO circuits can also have many forms, including the voltage, current, digital, analog, or mixed signal forms. Note that VCO circuit serves as an integrator for the time-domain signal.

6.2.2 VLSI RING OSCILLATOR VCO CIRCUIT MODEL

The frequency response of the voltage signal of a single VCDL stage in the VCO can typically be modeled as a single-pole circuit with the transfer function as

$$H(s) = \frac{A}{1 + sT_d} \qquad (6.24)$$

Such a circuit stage can be characterized by the dc gain and the equivalent delay time T_d of the circuit. The closed-loop transfer function of a (2N+1)-stage VLSI VCO circuit can be expressed as

$$F(s) = \frac{\left(\dfrac{A}{1 + sT_d}\right)^{2N+1}}{1 + \left(\dfrac{A}{1 + sT_d}\right)^{2N+1}} \qquad (6.25)$$

The self-oscillation condition of the ring oscillator VCO can be derived based on the stability criteria of the closed-loop transfer function as

$$\left(\frac{A}{1+j\omega T_d}\right)^{2N+1} = \left(\frac{A}{\sqrt{1+(\omega T_d)^2}} \cdot \frac{\sqrt{1+(\omega T_d)^2}}{1+j\omega T_d}\right)^{2N+1} = -1 \tag{6.26}$$

Such equation yields the self-oscillation conditions as

$$\begin{cases} \left|\dfrac{A}{1+(\omega T_d)^2}\right| = 1 \\ \omega T_d \approx \dfrac{\pi}{2N+1} \end{cases} \tag{6.27}$$

The VCO gain can be derived based on the equation 6.25 as

$$K_{VCO} \approx \frac{\partial}{\partial V_C}\left(\frac{\pi}{(2N+1)T_d}\right) \tag{6.28}$$

Shown in figure 6.19 is a typically used VLSI ring oscillator VCO circuit implementation that employs the MOS symmetrical active VCR load with replica biasing circuit.

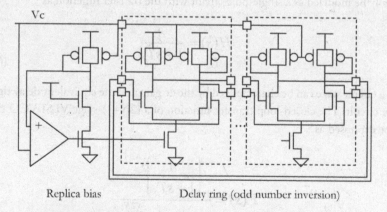

Fig. 6.19. Replica biasing symmetrical load VCO circuit structure

The VLSI PMOS VCR circuits shown in figure 6.20 serve as the core circuit for voltage-controlled RC time constant tuning for the VCDL circuit.

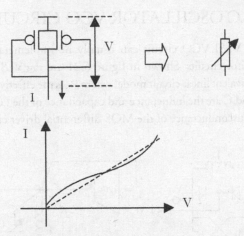

Fig. 6.20. VLSI symmetrical load VCR for delay elements

The effective conductance and the delay time of such CML driver circuit under the bias current I_B condition can be derived respectively as

$$I_B = 2\frac{\beta}{2}(V_{cc} - V - |V_T|)^2$$

(6.29)

$$G = \partial(I_B)/\partial V = 2\beta(V_{cc} - V - |V_T|) = \sqrt{2\beta I_B}$$

(6.30)

$$t_d \propto \frac{C}{G} = \frac{C}{2\beta(V_{cc} - V - |V_T|)} = \frac{C}{\sqrt{2\beta I_B}}$$

(6.31)

The VCO operation frequency and the VCO gain under the bias current I_B condition can further be derived respectively as

$$f \propto \frac{g_m}{C} = \frac{2\beta}{C}(V_{cc} - V - |V_T|) = \frac{\sqrt{2\beta I_B}}{C}$$

(6.32)

$$K_{VCO} \equiv \frac{\partial f}{\partial V_C} \propto \frac{2\beta}{C} \propto (I_B)^0$$

(6.33)

Note that the VCO frequency is determined by the square root of the VCO control current that offer linear VCO tuning with VCO gain approximately independent of the control voltage. In addition, the VCO gain can be adjusted based on the MOS device sizing.

6.2.3 VLSI LC OSCILLATOR VCO CIRCUIT MODEL

High-performance VLSI VCO circuit can usually be implemented using on-chip inductor and capacitor circuits. Shown in figure 6.21 are the VLSI LC VCO circuit structure and its equivalent linear circuit model where R_p is the effective parallel resistance of the LC tank, L and C are the inductance and capacitance of the LC tank circuit. Gm is the equivalent transconductance of the MOS differential driver circuit.

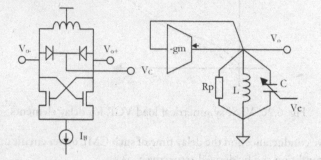

Fig. 6.21. VLSI LC VCO and equivalent linear circuit model

The s-domain transfer function of the circuit can be expressed as

$$V_o(s)(\frac{1}{R_P} - G_m + \frac{1}{Ls} + sC) = 0$$

(6.34)

The self-oscillation conditions are given as

$$\begin{cases} \dfrac{1}{R_P} \leq G_m \\ \omega_o = \sqrt{\dfrac{1}{LC}} \end{cases}$$

(6.35)

It can be seen that the oscillation frequency is fully determined by the passive L and C parameters. Consequently, such type of VCO can provide extremely lower phase noise clock signal.

6.2.4 VLSI POLYPHASE VCO CIRCUIT STRUCTURES

Polyphase VCO circuits are commonly used to generate polyphase clocks, where the clocks have identical clock frequency and their phases are equally spaced within the clock period. Such polyphase clocks have found important applications in high-speed I/O data recovery circuits. In addition, VLSI PI circuits typically rely on polyphase clocks for their operations. Shown in figure 6.22 is a polyphase ring oscillator VCO circuit structure.

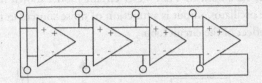

Fig. 6.22. VLSI ring oscillator polyphase VCO circuit structures

Such circuit can be extended using multiple rings as shown in figure 6.23 to generate high phase spacing resolution.

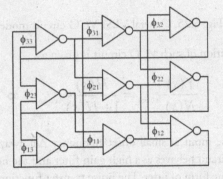

Fig. 6.23. VLSI multiple ring oscillator polyphase VCO circuit structures

Polyphase VCO can also be implemented using coupled LC oscillator as shown in figure 6.24.

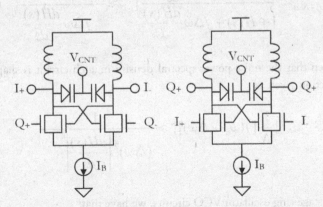

Fig. 6.24. VLSI coupled-LC oscillator-based I/Q clock generation circuit

6.2.5 VLSI VCO PHASE NOISE MODELS

Considering a general s-domain VLSI VCO circuit model shown in figure 6.25, the phase noise at the oscillator output is a function of noise sources in the VCDL circuit and the filtering effect of the circuit loop.

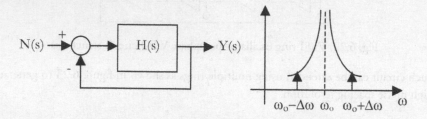

Fig. 6.25. General VLSI VCO circuit model

The oscillation condition of such VCO circuit is given as

$$\frac{Y(s)}{N(s)}\Big|_{s=j\omega_o} = \frac{H(s)}{1+H(s)}\Big|_{s=j\omega_o} = \infty$$

(6.36)

If there is small noise input at small offset frequency $\Delta\omega$ away from the oscillation frequency ω_o, such circuit behaves as a high gain filter and the noise can appear at the oscillator output in the form of jitter. The noise transfer function at the $\Delta\omega$ frequency offset can therefore be derived as

$$\frac{Y(s)}{N(s)}\Big|_{s=j(\omega_o+\Delta\omega)} \approx \frac{H(s)+j\Delta\omega\dfrac{dH(s)}{ds}}{1+H(s)+j\Delta\omega\dfrac{dH(s)}{ds}}\Big|_{j\omega_o} \approx -\frac{1}{j\Delta\omega\dfrac{dH(s)}{ds}}\Big|_{j\omega_o}$$

(6.37)

It can be seen that the noise power spectral density in such circuit is shaped by the transfer function given as

$$\Big|\frac{Y}{N}[j(\omega_o+\Delta\omega)]\Big|^2 \approx \frac{1}{(\Delta\omega)^2\left|\dfrac{dH(s)}{ds}\right|^2_{s=j\omega_o}}$$

(6.38)

For (2N+1) stage ring oscillator VCO circuits, we have that

$$\frac{dH(s)}{ds}\bigg|_{s=j\omega_o} = \frac{d\left(\frac{A}{1+sT_d}\right)^{2N+1}}{ds}\bigg|_{s=j\omega_o} = -\frac{(2N+1)T_d}{A}\left(\frac{A}{1+sT_d}\right)^{2N+2}_{s=j\omega_o} \quad (6.39)$$

And that

$$\left|\frac{Y}{N}[j(\omega_o+\Delta\omega)]\right|^2 \approx \frac{1}{(\Delta\omega)^2\left|\frac{dH(s)}{ds}\right|^2} = (\frac{A}{\pi})^2(\frac{\omega_o}{\Delta\omega})^2 \quad (6.40)$$

It can be seen that the thermal noise for each delay stage is approximately given as

$$|N|^2 \approx 4R_{eq}kT(1+\frac{3}{2}A) \quad (6.41)$$

The ring oscillator VCO power density per unit bandwidth at $\Delta\omega$ frequency offset due to the thermal noise can be further derived as

$$|Y[j(\omega_o+\Delta\omega)]|^2 \approx (\frac{A}{\pi})^2 4R_{eq}kT(1+\frac{3}{2}A)(\frac{\omega_o}{\Delta\omega})^2 \propto R_{eq}kT(\frac{\omega_o}{\Delta\omega})^2 \quad (6.42)$$

For the LC oscillator VCO circuits shown in figure 6.26, the s-domain noise transfer function of the VCO is approximately given as

$$\frac{Y(s)}{N(s)} = \frac{-sL \cdot g_m}{LCs^2 + (1/R_p - g_m)Ls + 1} \quad (6.43)$$

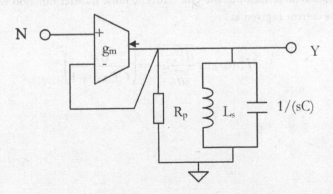

Fig. 6.26. Conceptual LC-oscillator circuit structure

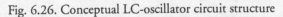

The oscillation conditions are given as

$$\begin{cases} g_m = 1/R_p \\ \omega_o = \sqrt{\dfrac{1}{LC}} \end{cases}$$

(6.44)

Note that the LC circuit will oscillate at frequency equals to the resonant frequency of the LC tank, independent of active component parameters such as the Gm of the biasing circuit. Phase noises of the LC VCO are contributed from the noise resistances in the circuit as shown in figure 6.27.

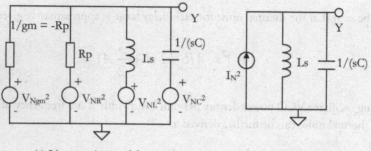

(a) Linear noise model   (b) Equivalent circuit

Fig. 6.27. LC VCO phase noise model

The equivalent noise current is given as

$$I_N^2 \equiv \overline{\frac{di_N}{df}}^2 = 4kT\left(F_{Gm}|G_m| + \frac{1}{R_P} + \frac{R_L}{(L\omega)^2} + R_C(C\omega)^2\right)$$

(6.45)

F_{Gm} is the access noise factor of the Gm cell. The noise transfer function with respect to the noise current is given as

$$H_N^2(\omega) \equiv \frac{\overline{dV_{oN}}^2}{\overline{di_N}^2} = \left(\frac{L\omega}{CL\omega^2 + 1}\right)^2$$

(6.46)

At the $\Delta\omega$ offset from the resonant frequency ωo we have that

$$H_N^2(\omega_o + \Delta\omega) \equiv \frac{\overline{dV_{oN}}^2}{\overline{di_N}^2} = \left| \frac{1}{CL(\omega_o + \Delta\omega)^2 + 1} \right|^2 \approx \frac{\omega_o L}{4\left(\dfrac{\Delta\omega}{\omega_o}\right)^2}$$

(6.47)

The total output noise voltage is given as

$$\frac{\overline{dV_N}^2}{df} \approx 4kT(F_{Gm}|G_m| + \frac{1}{R_P} + \frac{R_L}{(L\omega_o)^2} + R_C(C\omega_o)^2)\frac{\omega_o L}{4\left(\dfrac{\Delta\omega}{\omega_o}\right)^2}$$

(6.48)

By introducing the effective resistance and the quality factor of the LC tank as

$$\begin{cases} R_{eff} \equiv R_C + R_L + \dfrac{1}{R_P(\omega_o C)^2} \\[3mm] Q \equiv \dfrac{1}{R_{eff}}\sqrt{\dfrac{L}{C}} \end{cases}$$

(6.49)

we have the voltage noise density as

$$\frac{\overline{dV_N}^2}{df} \approx kT(F_{Gm} + 1)R_{eff}\left(\frac{\omega_o}{\Delta\omega}\right)^2$$

(6.50)

The single sideband phase noise density as

$$PN \equiv \frac{\displaystyle\int_{\Delta\omega-1/2}^{\Delta\omega+1/2} \overline{dV_N}^2}{carrier\_power} \approx \frac{kT(F_{Gm}+1)R_{eff}\left(\dfrac{\omega_o}{\Delta\omega}\right)^2}{V_A^2/2}$$

(6.51)

where V_A is the signal magnitude. In practical cases, it can be seen that due to the nonlinearity of a VCO circuit, a typical phase noise curve will include three areas as

$$PN = 10\log[\frac{2FkT}{Ps}(1 + \left(\frac{\omega_o}{2Q\Delta\omega}\right)^2)(1 + \frac{\Delta\omega_{1/f^3}}{\Delta\omega})]$$

(6.52)

6.3   VLSI PHASE INTERPOLATION (PI) CIRCUITS

There are two basic VLSI phase interpolation schemes that provide perfectly linear phase interpolation operations, including the sinusoidal phase interpolation and the triangular (or linear) phase interpolations. The sinusoidal interpolation algorithm is based on the weighted summation of two single-tone sinusoidal signals of same frequency and 90 degrees phase difference (known as I/Q signals). The linear phase interpolation, on the other hand, is based on the weighted summation of two or more triangular (linear ramping) signals.

Sinusoidal phase interpolation methods usually offer advantage of implementation simplicity in signal precondition that rely on single-tone signals, which can be generated using simple lowpass or bandpass filtering operations by eliminating the harmonic contents of the input I/O clock references. Sinusoidal phase interpolation typically suffers from the constraint in VLSI circuit implementation since it usually requires nonlinear weighting coefficients for perfectly linear phase interpolation.

The triangular phase interpolation method, on the other hand, offers the advantage of highly VLSI circuit implementation simplicity in the weighting coefficient generations. However, triangular phase interpolation typically suffers from design complexity in VLSI signal precondition circuit implementations for highly linear triangular signal generation for perfectly linear phase interpolation.

Practical VLSI phase interpolations are usually based on the combination of the two phase interpolation algorithms that employs the single tonelike preconditioning using band-limiting circuits and the linear weighting coefficients. Such an approximate approach provides a workable solution for the trade-off between design simplicity and circuit performance.

6.3.1   SINUSOIDAL PHASE INTERPOLATION

Phase interpolation based on I/Q sinusoidal clock signals can be realized by weighted addition of two sinusoidal signals of same frequency with $\pi/2$ (i.e., 90°) phase difference as shown in figure 6.28.

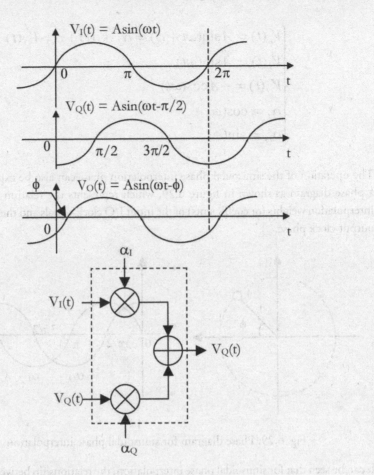

Fig. 6.28. Phase interpolation based on I/Q sinusoidal signals

The sinusoidal phase interpolation operation can be mathematically expressed by the phase interpolation equation as

$$V_o(t) = A\sin(\omega t - \phi) = A[\cos(\phi)\sin(\omega t) - \sin(\phi)\cos(\omega t)] \qquad (6.53)$$

This equation implies that sinusoidal signal output of arbitrary phase shift can be realized by the weighted combination of I/Q signal sinusoidal clock input as

$$\begin{cases} V_o(t) = A\sin(\omega t + \phi) = \alpha_I \cdot V_I(t) + \alpha_Q \cdot V_Q(t) \\ V_I(t) = A\sin(\omega t) \\ V_Q(t) = -A\cos(\omega t) \\ \alpha_I = \cos(\phi) \\ \alpha_Q = \sin(\phi) \end{cases}$$

(6.54)

The operation of the sinusoidal phase interpolation phase can also be expressed using a phase diagram as shown in figure 6.29, which represents the relation between the interpolation weights (or coefficients) of the input I/Q clock signals and the interpolated output clock phase.

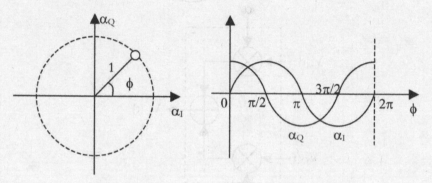

Fig. 6.29. Phase diagram for sinusoidal phase interpolation

It can be seen that for sinusoidal phase interpolation, the relationship between the phase and the phase interpolation and the interpolation weighting coefficient is nonlinear. In addition, we can also see that the sum of the two weighting coefficients is not constant:

$$\begin{cases} \phi = \tan^{-1}(\dfrac{\alpha_Q}{\alpha_I}) \\ \alpha_I^2 + \alpha_Q^2 = 1 \end{cases}$$

(6.55)

Such nonlinear properties have major drawbacks in VLSI circuit implementation, which may significantly increase the implement complexity for the interpolation circuit implementation.

6.3.2 TRIANGULAR PHASE INTERPOLATION

The nonlinear phase relation and the nonconstant weighting coefficient sum in the sinusoidal phase interpolation can be resolved using the triangular phase interpolation technique as shown in figure 6.30. The linear relation between the interpolation phase and the interpolation weight coefficients and the constant sum of the weighting coefficients offers by the triangular-based phase interpolation is very suitable for the VLSI circuit implementation.

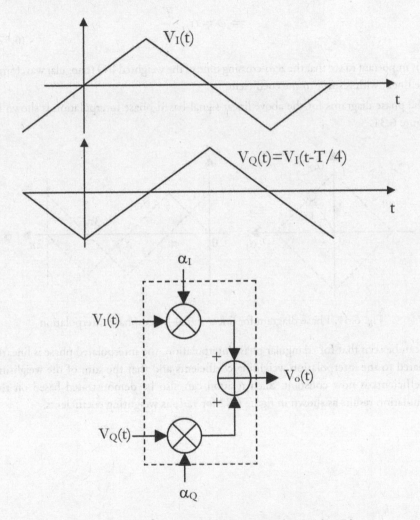

Fig. 6.30. Linear signal-based phase interpolation algorithm

The operation of the triangular phase interpolation can be expressed by the following equation:

$$\begin{cases} \alpha_I \cdot V_m(\frac{\phi}{T}) + \alpha_Q \cdot V_m(\frac{\phi - T/4}{T}) = 0 \\ \alpha_I + \alpha_Q = 1 \end{cases}$$

(6.56)

$$\Rightarrow \phi = \alpha_Q \cdot \frac{T}{4}$$

(6.57)

It is important to see that the zero-crossing time of the weighted I/Q triangular waveforms are linear with respect to the coefficient.

The phase diagrams for the above linear signal-based phase interpolation is shown in figure 6.31.

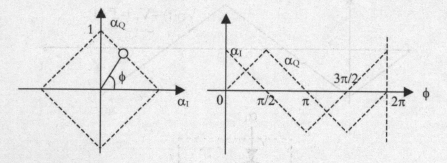

Fig. 6.31. Phase diagram for linear signal-based phase interpolation

It can be seen that for triangular phase interpolation, the interpolated phase is linearly related to the interpolation weighting coefficients and that the sum of the weighting coefficients is now constant. Such relation can also be demonstrated based on the simulation results as shown in figure 6.32 for various weighting coefficients.

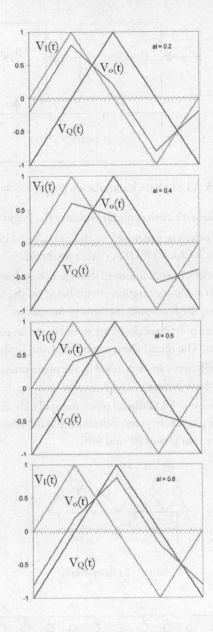

Fig. 6.32. Simulation waveform for linear-based phase interpolation

6.3.3 VLSI PI CIRCUIT IMPLEMENTATIONS

A typical VLSI phase interpolator circuit usually consists of four major building components as shown in figure 6.33, including the polyphase clock generator, the conditioner, the mixer, and the control circuit.

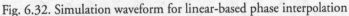

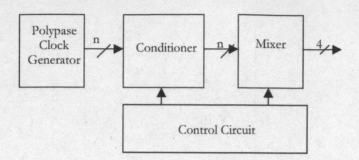

Fig. 6.33. VLSI phase interpolation circuit block diagram

A VLSI phase interpolation circuit typically includes the following functions:

- The polyphase clocks are generated in the polyphase clock generation circuit, employing either a VCO, a VCDL, or other methods.

- The polyphase clocks are preconditioned through either a lowpass or a bandpass filter or through a linear ramps generation circuit based on the sinusoidal or triangular method that shapes the harmonic contents in the clocks.

- The preconditioned polyphase clocks are mixed based on programmable weighted addition operation. The mixed clock is amplified and limited for output.

- The weighting coefficients are controlled by the phase interpolation control circuit for output clock phase tuning.

Shown in figure 6.34 are two VLSI digital phase interpolation circuit implementations that are based on direct phase interpolator circuit structure, where one additional phase is interpolated from the two phases ($0°$ and $90°$).

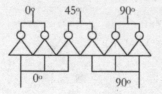

(a) 1-to-2 interpolation

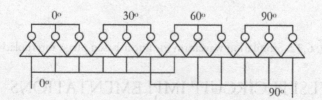

(b) 1-to-3 interpolation

Fig. 6.34. VLSI direct phase interpolation circuits

Shown in figure 6.35 is a VLSI R-ring phase interpolation circuit structure, where the signal conditioning is realized using a simple capacitor. The I/Q phase mixing coefficients in this phase interpolation circuit structure are implemented through the weighted addition of the conditioned I/Q clock phase signal based on the ratio of the resistors as shown in figure 6.36:

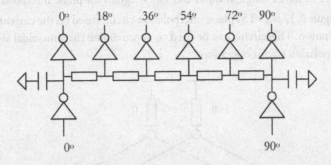

Fig. 6.35. VLSI R-ring 1-to-5 phase interpolation circuit

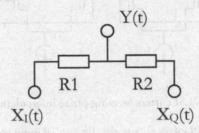

Fig. 6.36. VLSI resistor ratio phase interpolation circuit

Based on the weighted summation of two clock signals as

$$Y(t)(\frac{1}{R_1}+\frac{1}{R_2}) = X_I(t)\frac{1}{R_1}+X_Q(t)\frac{1}{R_2} \tag{6.58}$$

We have that

$$\Rightarrow \begin{cases} Y(t) = \alpha_I \cdot X_I(t) + \alpha_Q \cdot X_Q(t) \\ \alpha_I \equiv \dfrac{R_2}{R_1+R_2} \\ \alpha_Q \equiv \dfrac{R_1}{R_1+R_2} \end{cases} \tag{6.59}$$

It can be seen that

$$\alpha_I + \alpha_Q = 1 \tag{6.60}$$

Therefore, such PI circuit belongs to the linear coefficients type phase interpolation, and it is preferred to have triangular input I/Q clock signals for phase interpolation.

Shown in figure 6.37 is a VLSI phase interpolation circuit based on the current I/Q clock mixing operation. This circuit can be used to approximate the sinusoidal signal-based phase interpolation circuit operations.

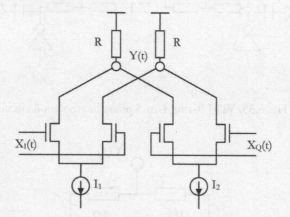

Fig. 6.37. VLSI current steeling phase interpolation circuits

For this current steeling PI mixer circuit, the differential input and output signal relation can be expressed as

$$\begin{cases} Y(t) = R \cdot g_{m1} \cdot X_I(t) + R \cdot g_{m2} \cdot X_Q(t) \\ g_{m1} = \sqrt{2\beta I_1} \\ g_{m2} = \sqrt{2\beta I_2} \end{cases} \tag{6.61}$$

by letting

$$\begin{cases} Y(t) = A[\alpha_I \cdot X_I(t) + \alpha_Q \cdot X_Q(t)] \\ \alpha_I = \sqrt{I_1 / (I_1 + I_2)} \\ \alpha_Q = \sqrt{I_2 / (I_1 + I_2)} \end{cases} \tag{6.62}$$

we can see that

$$\alpha_I^2 + \alpha_Q^2 = \frac{I_1}{I_1 + I_2} + \frac{I_2}{I_1 + I_2} = 1 \tag{6.63}$$

It implies that such a phase interpolation circuit preferred the sinusoidal I/Q clock signals for phase interpolation.

Differential VLSI circuit shown in figure 6.38 can usually be used to control weight current for phase interpolation.

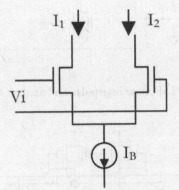

Fig. 6.38. VLSI phase interpolation coefficient generation circuit

For such circuit structure, we have that

$$\begin{cases} I_1 = \dfrac{I_B}{2} + \dfrac{g_m}{2} V_i \\[3mm] I_2 = \dfrac{I_B}{2} - \dfrac{g_m}{2} V_i \end{cases} \tag{6.64}$$

and

$$I_1 + I_2 = I_B \tag{6.65}$$

Shown in figure 6.39, figure 6.40, and figure 6.41 are VLSI phase interpolation mixer circuits based on the symmetrical load differential buffer circuit structure. Such circuits provide regulated signal voltage swing and the bandwidth of the mixer core and preconditioner employing a replica biasing circuit structure.

Above, circuit structures can be used to implement the PI mixer circuit by including the digital control circuit that provides the digitally controlled weights of the PI mixing operations.

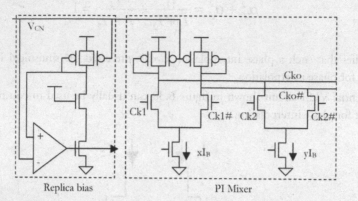

Fig. 6.39. VLSI phase interpolation based on CML circuit

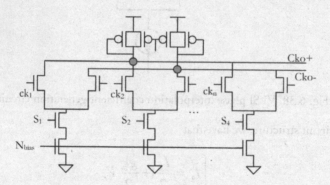

Fig. 6.40. VLSI PI mixer circuit employing the replica biasing circuit

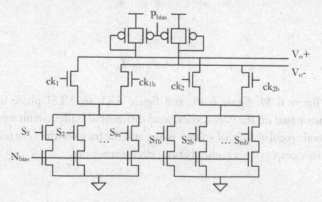

Fig. 6.41. Alternative VLSI PI mixer circuit

6.3.4   VLSI PI CIRCUIT PHASE NOISE MODEL

Similar to VLSI VCDL circuits, VLSI PI circuit structures are similar to the basic delay buffer circuit structures. Therefore, the open loop noise characteristics of PI circuits will be similar to the VLSI buffer circuit structure. For the first-order approximation, the random phase noise of the PI circuit can be approximately expressed as

$$\frac{\Delta t_{RMS}}{t_d} \approx \sqrt{\frac{2kT}{C_L}}\sqrt{1+\frac{2}{3}a_V}\frac{1}{V_{PP}} \qquad (6.66)$$

or

$$\Delta t_{RMS} \propto \frac{\sqrt{TC_L}}{V_{PP}} \qquad (6.67)$$

References:

[1]   F. Yang, et al, "A CMOS Low-Power Multiple 2.5-3.125 Gb/s Serial Link Macrocell for High IO Bandwidth Network ICs," IEEE JSSC, Vol 37, No. 12, Dec. 2002.

[2]   T. Kim, et al, "Phase Interpolator Using Delay Locked Loop," SSMSD 2003.

[3]   L. Yang, et al, "An Arbitrarily Skewable Multiphase Clock Generator Combing Direct Interpolation with Phase Error Averaging," 2003 IEEE.

[4]   J-M Chou, et al, "A 125Mhz 8b Digital-to-Phase Converter," ISSCC 2003.

[5]   R. Kreienkamp, et al, "A 10-Gb/s CMOS Clock and Data Recovery Circuit with an Analog Phase Interpolator," IEEE2003 Custom Integrated Circuits Conference.

[6]   A. Maxim, "A 0.16-2.55Ghz CMOS Active Clock Deskewing PLL Using Analog Interpolator," IEEE JSSC. Vol. 40, No. 1, Jan. 2006.

[7]   J-M. Chou, et al, "Phase Averaging and Interpolation Using Resistor Strings or Resistor Rings for Multi-Phase Clock Generation," IEEE Trans. Circuits & Systems, Vol.53, No. 5, May 2006.

[8]   W. Rhee, et al, "A Semi-Digital Delay-Locked Loop Using an Analog-Based Finite State Machine," IEEE Trans. Circuits & Systems, Vol.51, No.11, Dec. 2004.

[9]   B. W. Garlepp, et al, "A Portable Digital DLL for High-Speed CMOS Interface Circuits," IEEE JSSC Vol.34, No. 5, May 1999.

[10] G. Jovannovic, et al," Delay Locked Loop with Linear Delay Element," Serbia and Montenegro, Nis, September 28-30, 2006.

[12] J. G. Maneatis, "Low-Jitter Process-Independent DLL and PLL Based on Self-Biased Techniques," IEEE JSSC, Vol. 31, No. 11, Nov. 1996.

[13] B. W. Garlepp, et al, "A Portable Digital DLL for High-speed CMOS Interface Circuits," IEEE JSSC, Vol. 34, No. 5, May 1999.

[14] S-S Hwang, et al, "A DLL Based 10-320 Mhz Clock Synchronizer," ISCAS 2000, May 28-31, Geneva, Switzerland.

[15] C. Kim, et al, "A Low-Power Small-Area +/- 7.28ps Jitter 1-Ghz DLL-Base Clock Generator", IEEE JSSC, Vol. 37, No. 11, Nov. 2002.

[16] Y Arai, "A High-Resolution Time Digitizer Utilizing Dual PLL Circuits," IEEE2004.

[17] H. J. Song, "Digital Delay Locked Loop for Adaptive De-skew Clock Generation," US Patent, No. 6275555, 2001.

[18] H. J. Song, "Programmable De-Skew Clock Generation based on Dual Digital Delay Locked Loop Structure," SOCC2001.

[19] G.C. Hsieh, et al, "Phase-Locked Loop Techniques—A Survey," IEEE Transactions on Industrial Electronics, Vol.43, No.6, Dec.1996.

[20] P. Heydari, et al, "Analysis of the PLL Jitter Due to Power/Ground and Substrate Noise," IEEE Transactions on Circuits and Systems—I: Regular Papers, Vol 51, No. 12, Dec. 2004.

[21] T. M. Almeida, et al, "High Performance Analog and Digital PLL Design," IEEE 1999.

[22] V. F. Kroupa, "Noise Properties of PLL Systems," IEEE Transactions on Communications, Vol COM-30, No. 10, Oct. 1982.

[23] A. Maxim, "A 0.16-2.55-Ghz CMOS Active Clock De-Skewing PLL Using Analog Phase Interpolation," IEEE, JSSC, Vol. 40, No.1, Jan. 2006.

[24] M. Inoue, et al, "Over-Sampling PLL for Low-Jitter and Responsive Clock Synchronization," IEEE2006.

[25] T. Toifl, et al, "A 0.94ps-RMS-Jitter 0.016-mm2 2.5Ghz Multiphase Generation PLL with 360o Digitally Programmable Phase Shift for 10-Gb/s Serial Links," IEEE JSSC Vol. 40, No.12, December 2005.

[26] J. F. Bulzacchelli, et al, "A 10-Gb/s 5-Tap DFE/4-Tap FFE Transceiver in 90nm CMOS Technology," IEEE JSSC Vol. 41, No. 12, December 2006.

[27] B. G. Kim, et al, "A 250 Mhz-2Ghz Wide Range Delay-Locked Loop," IEEE JSSC Vol. 40, No.6, June 2005.

[28] T. Matano, et, al, "A 1-Gb/s/pin 512-Mb DDRII SDRAM Using a Digital DLL and a Slew-Rate-Controlled Output Buffer," IEEE JSSC Vol.38, No. 5, May 2003.

[29] R. C. H. Van de Beek, et, al, "Low-Jitter Clock Multiplication: A Comparison Between PLLs and DLLs," IEEE Tran. Circuits and System—II: Analog and Digital Signal Processing, Vol.49, No. 8, Aug. 2002.

[30] T. C. Weigandt, et, al, "Timing Jitter Analysis for High-frequency Low-power CMOS Ring-Oscillator Design," Proc. Int. Symp., Circuits and Systems. London, UK., June 1994.

[31] P. Vancorenland, et, al, "A Wideband IMRR Improved Quadrature Mixer/LO Generator," IEEE.

[32] P. Minami, et, al, "A 1Ghz Portable Digital Delay-Locked Loop with Infinite Phase Capture Range," 2000 IEEE ISSCC.

[33] J. G. Maneatis, et, al, "Precise Delay Time Generation Using Coupled Oscillators," 1993 IEEE ISSCC.

[34] M. Kokubo, et, al, "Spread-Spectrum Clock Generator for Serial ATA using Fractional PLL Controlled by $\Delta\Sigma$ Modulator with Level Shifter" 2005 IEEE ISSCC.

[35] P. Lu, et, al, "A Low-Jitter Frequency Synthesizer with Dynamic Phase Interpolation for High-Speed Ethernet," 2000 IEEE ISSCC.

[36] T. Matsumoto, "High-Resolution On-Chip Propagation Delay Detector for Measuring Within-Chip and Chip-to-Chip Variation," 2004 Symposium on VLSI Circuits Digest of Technique Papers.

[37] M. Bazes, et, al, "An Interpolating Clock Synthesizer," IEEE JSSC, vol. 31, no. 9, Sept. 1996.

[38] Y. J. Jung, et, al, "A Dual-Loop Delay-Locked Loop Using Multiple Voltage-Controlled Delay Lines," IEEE 2001.

[39] M. G. Johnson, et, al, "A Variable Delay Line PLL for CPU-Coprocessor Synchronization," IEEE JSSC, vol. 23, no. 5, Oct. 1988.

[40] A. Ghaffari, et, al, "A Novel Wide-Range Delay Cell for DLLs," ICECE 2006.

[41] M. T. Hsieh, et, al, "Clock and Data Recovery with Adaptive Loop Gain for Spread Spectrum SerDes Applications," IEEE 2005.

[42] J. M. Chou, et, al, "Phase Averaging and Interpolation Using Resistor Strings or Resistor Rings for Multi-Phase Clock Generation," IEEE Tran. Circuits and Systems-I: Regular Papers, vol. 53, No. 5, May 2006.

[43] C. C. Wang, et, al, "Clock-and-Data Recovery Design for LVDS Transceiver Used in LCD Panels," IEEE Tran. Circuits and Systems-II: Express Briefs, vol. 53, No. 11, Nov. 2006.

[44] U. Yodprasit, et, al, "Realization of a Low-Voltage and Low-Power Colpitts Quadrature Oscillator," IEEE Tran. Circuits and Systems-II: Express Briefs, vol. 53, No. 11, Nov. 2006.

CHAPTER 7
VLSI SYNCHRONIZATION CIRCUITS

Delay synchronization is a fundamental delay (phase) manipulation operation in VLSI time-domain signal processing circuits where the delay (phase) of a data signal can be "forced" to a known clock reference. VLSI synchronization circuits are widely used in VLSI high-speed I/O for data signal phase noise filtering and for received data sampling. Static latched (or flip-flop) based synchronization circuits are the most basic and most commonly used VLSI synchronization circuits. VLSI synchronization circuits can also be realized using dynamic latch based on the sample/hold circuit structures.

The performance of VLSI synchronization circuits can usually be expressed using the input setup and hold times, the clock to output delay time, the transparent datapath delay, and the input voltage thresholds. VLSI synchronization circuits employing the basic bistable circuits usually suffer from the metastability constraints that determine the setup hold times, voltage thresholds, and the clock-to-output delay parameters of the circuits.

VLSI synchronization circuits are widely used within high-speed I/O, the transceiver circuits for data transmission synchronization, and for data sampling. Synchronization circuits are also widely used within the VLSI high-speed I/O circuit for the cross clock domain synchronization, frequency division, phase detection, and data rate conversions, such as parallel-in-serial-out (PISO) and serial-in-parallel-out (SIPO) circuit operations.

7.1 MODELING OF SYNCHRONIZATION CIRCUITS

The typical time-domain signal processing operation of the VLSI synchronization circuit is shown in figure 7.1. Under such time-domain signal processing model, the input data stream is synchronized by the synchronization circuit elements such that the phase of the output data signal is "forced" to the phase value of the reference clock as soon as the setup and hold times are met (ignoring clock-to-output delay):

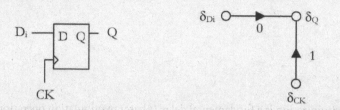

(a) D-FF based synchronization circuit   (b) Time-domain SFG Expression

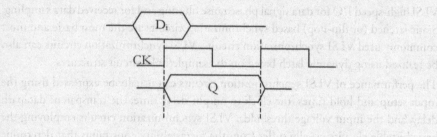

(c) Timing Diagram

Fig. 7.1. VLSI synchronization circuit expressions

The time-domain signal processing operation of the synchronization circuits can be mathematically expressed as

$$\delta_Q = f(\delta_{Di}, \delta_{CK}) = \delta_{CK} \tag{7.1}$$

The VLSI synchronization circuits can usually be implemented using the sample/hold circuit structures that are typically operated in two clock phases. During the sampling clock phase, the input data value is captured by the synchronization circuit. During the hold phase, the captured data signal is held within the synchronization circuits. The hold operation of the VLSI synchronization circuit can be realized using a bistable circuit as shown in figure 7.2.

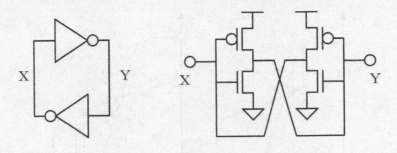

Fig. 7.2. Basic VLSI bistable circuit structure

The large signal response of such bistable circuit can be modeled using a feed-forward (H_F = Y/X) and a feedback (H_B = X/Y) transfer function as shown in figure 7.3.

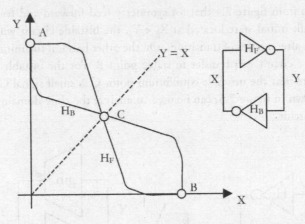

Fig. 7.3. Bitable circuit model of VLSI synchronization circuits

It can be seen that such circuit has three equilibrium points A, B, and C, where point A and B are stable equilibrium points and C point is an unstable equilibrium point.

Shown in figure 7.4 are the large signal state transition models of the bistable circuit that can be used to demonstrate the transition responses of the basic bistable circuit structure.

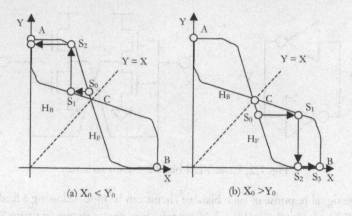

(a) $X_0 < Y_0$ (b) $X_0 > Y_0$

Fig. 7.4. State transition diagram of bistable circuit

It can be seen from figure 7.4 that for symmetry feed-forward and feedback paths with the circuit initial state located at $X_0 < Y_0$, the bistable circuit will transfer to stable point A after a series of transitions. On the other hand, if the initial state starts at $X_0 > Y_0$, the circuit will transfer to stable point B. For the bistable circuit with the initial state near the unstable equilibrium point C, a small signal Gm-C circuit model, as shown in figure 7.5, can be used to analyze the time-domain behavior of the bistable circuit.

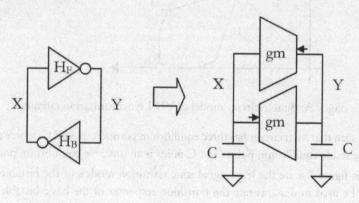

Fig. 7.5. Gm-C based small signal model of bistable circuit

In the above model, the Gm and C are the effective transconductance and the parasitic capacitance of the bistable circuit. The behavior of such circuit near the unstable equilibrium point C can be expressed in differential equation as

$$\begin{cases} \dfrac{dX}{dt} = -\dfrac{g_m}{C} Y \\[2mm] \dfrac{dY}{dt} = -\dfrac{g_m}{C} X \\[2mm] X\big|_{t=0} = +\Delta \\[2mm] Y\big|_{t=0} = -\Delta \end{cases} \tag{7.2}$$

The solution of the equation is given as

$$\begin{cases} X = \Delta e^{\frac{g_m}{C} t} \\[3mm] Y = -\Delta e^{\frac{g_m}{C} t} \end{cases} \tag{7.3}$$

Note that for bistable circuit with initial state near the unstable equilibrium point C, the time constant for the circuit to transfer to the stable points is approximately given as

$$\tau \sim \frac{C}{g_m} \ln\left(\frac{1}{\Delta}\right) \tag{7.4}$$

where Δ is the initial voltage difference deviated from the unstable equilibrium point C. It can be seen that

- the time constant is proportional to the Gm/C parameter and that is directly proportional to the effective bandwidth of the circuit, and
- the time constant is proportional to $\ln(1/\Delta)$, implying there is a metastable region the circuit may be trapped at the unstable equilibrium point C for a significant amount of time and that may cause circuit to fail.

7.1.1 VLSI BISTABLE CIRCUIT STRUCTURES

Shown in figure 7.6 are several commonly used VLSI bistable configurations for realizing the synchronization circuits. These circuits usually include a write port for the data sampling operations.

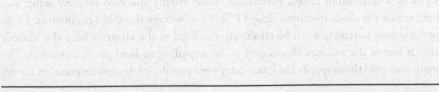

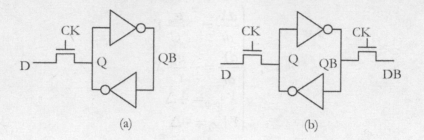

(a)   (b)

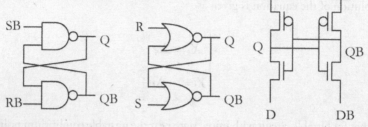

(c)   (d)   (e)

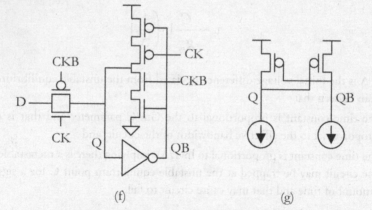

(f)   (g)

Fig. 7.6. Basic VLSI bistable circuit structures

7.1.2 SETUP/HOLD TIMES, TCO, AND VOLTAGE THRESHOLDS

VLSI synchronization circuit performances are usually specified by their setup and hold times, the clock-to-output delay (TCO) and voltage threshold parameters. These performance parameters can be effectively modeled as the effort to keep the bistable circuit out of the metastability region at the sampling to hold phase transition. The setup and hold times specify the input delay time penalty of the synchronization circuit to sample the input data bits. The TCO parameters, on the other hand, are the delay time penalty for the synchronization circuits to drive the load.

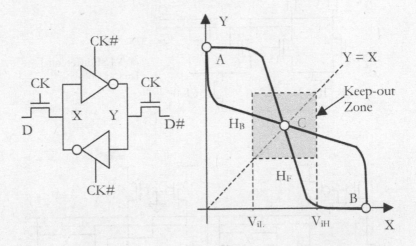

Fig. 7.7. Keep-out zone of VLSI synchronization circuit in hold state

For the synchronization circuit shown in figure 7.7, the metastability effect can be avoided by keeping the internal voltage of the bistable circuit out of the metastability region at the end of the sampling process. For such circuit, the sum of the setup and hold times are directly determined by the internal RC time constant for the circuit to effectively pass the metastability during the transition from one stable state to another during the sampling operation.

The TCO of the synchronization circuit, in turn, is the time for the internal state transition to be presented at the synchronization circuit output. It can be seen that these timing parameters are directly related to the effective time constant of the bistable circuit.

7.1.3 VLSI DIFFERENTIAL SYNCRONIZATION CIRCUITS

Differential synchronization circuits that offer better noise immunity and minimized metastability effects are widely used in VLSI high-speed I/O circuits. Differential synchronization circuits typically have narrow voltage threshold regions, and therefore they can be used to handle low swing input signals in the VLSI high-speed I/O applications. In addition, differential synchronization circuits are also commonly used in VLSI high-speed I/O transmitter to minimize the TCO time variations. Shown in figure 7.8 are two commonly used VLSI differential latch and flip-flop circuit structures.

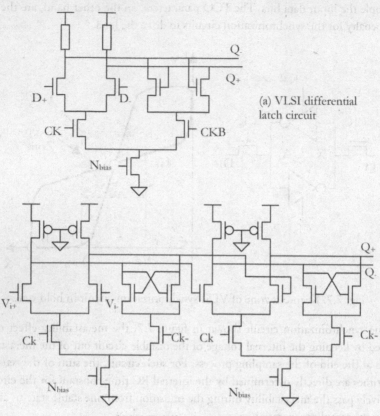

(a) VLSI differential latch circuit

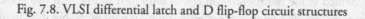

(b) VLSI differential flip-flop circuit

Fig. 7.8. VLSI differential latch and D flip-flop circuit structures

Such circuit structures have regulated tail current offering very low switch noise operations. In addition, their differential data and clock inputs offer high common-mode noise (such as power supply noise) immunity. As a result, such circuit structures provide minimized delay time uncertainty operations.

Shown in figure 7.9 are sense amplifier-based VLSI differential flip-flop circuit structures, where the differential input data signal path offers high noise immunity and input differential pair also offers signal amplification capability for receiving low swing differential voltage signal. A single-ended clock is used in this circuit to minimize the clock loading effect. The sampler consists of two cascaded differential latch circuit structure with the first one for signal amplification and sampling and the second one as data keeper. Such circuit is equivalent to a D flip-flop circuit. During the hold (precharge) phase, the input latch is precharged to Vcc to erase the past data signal information to minimize the hysteresis effects. The output latch keeps the last sampled data bit. During the sampling process, the input differential data is amplified and then sampled into the first latch circuit. The output of the input latch is then used to overwrite the data inside the output latch circuits.

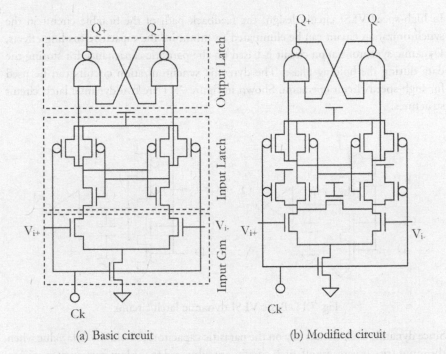

(a) Basic circuit   (b) Modified circuit

Fig. 7.9. Sense amplifier-based differential VLSI sampler circuits

Shown in figure 7.10 are two alternative circuit implementations of the sense amplifier differential sampler circuits.

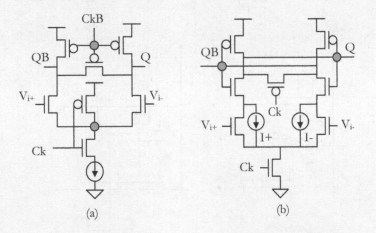

Fig. 7.10. Alternative sense amplifier-based differential sampler circuit structures

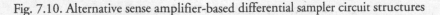

7.2 VLSI DYNAMIC SYNCHRONIZATION CIRCUITS

In high-speed VLSI circuit design, the feedback path of the bistable circuit in the synchronization circuit can be eliminated to minimize the capacitive loading effects. Dynamic synchronization circuit is based on the parasitic capacitance for storing the data during the holding phase. The dynamic synchronization circuits can be used for high-speed circuit operation. Shown in figure 7.11 are two dynamic latch circuit structures.

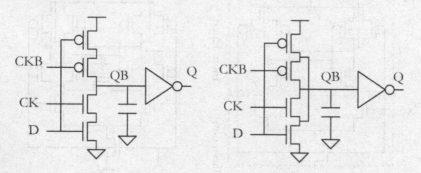

Fig. 7.11. Basic VLSI dynamic latch circuits

Since dynamic latch circuits rely on the parasitic capacitors to hold the data value when the input tristate gates are off, such circuits are subjected to the low frequency operation limitation and noise margin degradation as shown in figure 7.12.

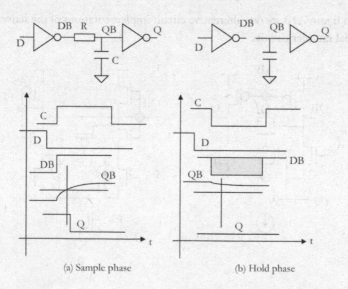

(a) Sample phase   (b) Hold phase

Fig. 7.12. Circuit operation of VLSI dynamic latch circuit

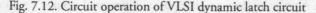

7.2.1 SETUP AND HOLD TIMES OF DYNAMIC LATCH CIRCUITS

The input to output transfer function of the dynamic latch circuit shown in figure 7.13 is similar to a CMOS inverter circuit. Since there is no regeneration circuit, the circuit must rely on the capacitor to store the sampled voltage value. As a result, the setup and hold times of the dynamic latch are determined by the effective time constant of the latch keeper node for switching from one valid state to another during the sampling process.

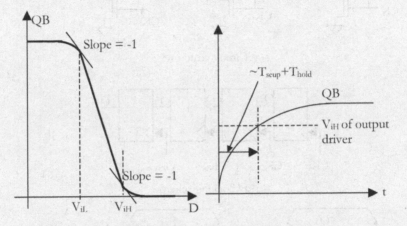

Fig. 7.13. Setup and hold times of VLSI dynamic latch

7.2.2 VLSI DYNAMIC LATCH CIRCUIT STRUCTURES

Shown in figure 7.14 are several dynamic latch circuits based on the true single-phase clock (TSPC) VLSI circuit technique.

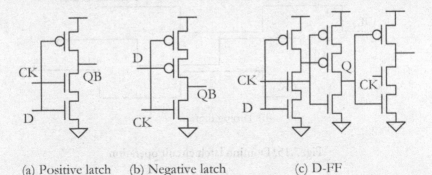

(a) Positive latch   (b) Negative latch   (c) D-FF

Fig. 7.14. VLSI TSPC latch circuit structures

195

An alternative dynamic synchronization circuit that is based on the domino circuit techniques is shown in figure 7.15.

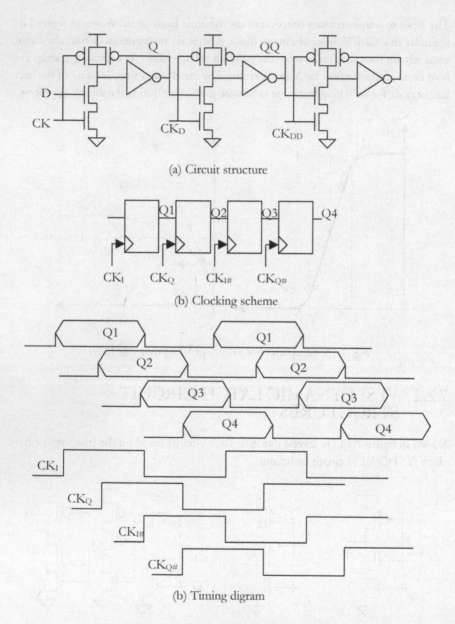

(a) Circuit structure

(b) Clocking scheme

(b) Timing digram

Fig. 7.15. Domino latch circuit operation

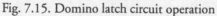

7.3 VLSI HIGH-SPEED I/O PISO AND SIPO CIRCUITS

High-speed I/O transmitter circuit is synced with the transmitter clock through the synchronization circuit. Such a synchronization operation is commonly realized using the pass gate switch controlled by the transmitter clock. Shown in figure 7.16 is a high-speed I/O sync circuit based on full-rate clock.

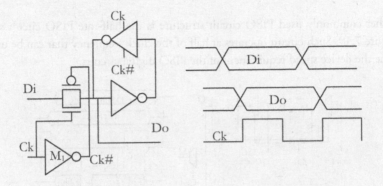

Fig. 7.16. Full-rate high-speed I/O transmitter sync circuit

In many applications, multiphase or interleave circuit architecture are commonly used to relax the speed requirement in the transmitter circuits. In such circuit architecture, only the driver is operated at full speed and other circuits are operated at lower frequency of a ratio of the data rate. A general n-phase sync circuit is shown in figure 7.17.

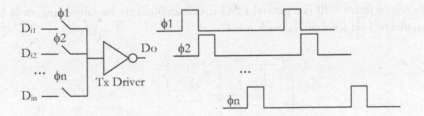

Fig. 7.17. N-phase high-speed I/O transmitter sync circuit

The parallel-in-serial-out (PISO) circuit is commonly used in the high-speed I/O circuit to convert the low-speed digital parallel bit stream to higher speed serial bit stream for data transmission. Shown in figure 7.18 is the general PISO circuit architecture for converting an N-bit parallel data to single-bit serial data of N times of clock speed.

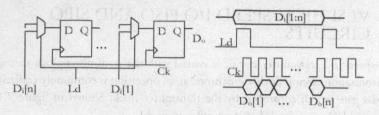

Fig. 7.18. Full-rate PISO circuit

Another commonly used PISO circuit structure is the half-rate PISO circuit shown in figure 7.19. Such circuit operates at half of the clock frequency that can be used to reduce the device speed requirement of the PISO digital circuits.

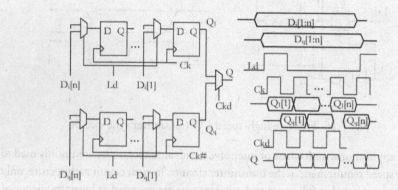

Fig. 7.19. Half-rate PISO circuit

Shown in figure 7.20 is a general PISO circuit architecture for converting an N bit parallel to 1 bit serial of Nx clock speed.

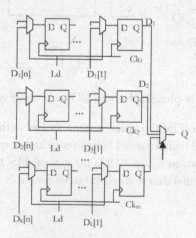

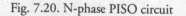

Fig. 7.20. N-phase PISO circuit

VLSI high-speed I/O PISO circuits are also widely implemented employing the multifrequency clock circuit structure as shown in figure 7.21.

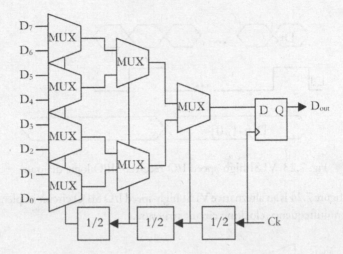

Fig. 7.21. VLSI PISO circuit implementation

This PISO function is realized based on the multiphase synchronization control clocks.

At the receiver sides of VLSI high-speed I/O circuits, high-frequency serial data stream is converted back into lower-frequency parallel data using the serial-in-parallel-out (SIPO) circuits. VLSI high-speed I/O SIPO circuits serve as part of the synchronization circuits that synchronize data stream between the VLSI high-speed I/O receiver AFE and the core circuits. SIPO circuits also realize frequency domain conversion from high-speed serial AFE domain to low-speed parallel core domain circuits. Shown in figure 7.22 is a general VLSI SIPO circuit structure, where the high-frequency serial bit stream of frequency f is converted to N-bit parallel data stream of f/N frequency.

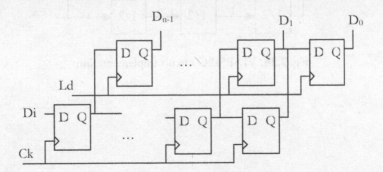

Fig. 7.22. VLSI high-speed I/O receiver SIPO circuit implementation

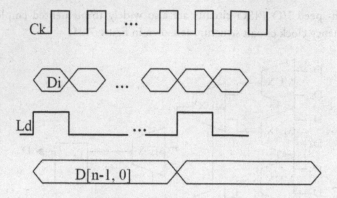

Fig. 7.23. VLSI high-speed I/O receiver SIPO logic diagram

Shown in figure 7.24 is an alternative VLSI high-speed I/O SIPO circuit implementation based on multifrequency clocking circuit structure.

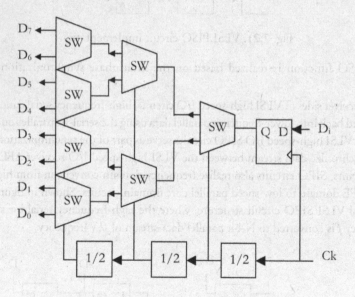

Fig. 7.24. VLSI SIPO circuit implementation

References:

[1] Payam Heydari, et al, "Design of Ultrahigh-Speed Low-Voltage CMOS CML Buffers and Latches," IEEE Trans. VLSI System, Vol. 12, No. 10, Oct. 2004.

[2] J. U. Horstmann, et. al, "Metastability Behavior of CMOS ASIC Flip-Flops in Theory and Test," IEEE JSSC, Vol. 24, No. 1, Feb. 1989.

[3] D. J. Kinniment, et. al, "Synchronization Circuit Performance," IEEE JSSC, Vol. 37, No. 2, Feb. 2002

[4] J. Lee, "High-Speed Circuit Designs for Transmitters in Broadband Data Links," IEEE JSSC, Vol. 41, No. 5, Feb. 2006.

[5] T. J. Gabara, "Metastability of CMOS Master/Slaver Flip-Flop," IEEE CICC1991.

[6] T. Yalcin, et. al, "Design of A Fully-Static Infertile Low-Power CMOS Flip-Flop," IEEE1999.

[7] C. S. Hwang, et. al, "A High-Precision Time-to-Digital Converter Using a Two-Level Conversion Scheme," IEEE Trans. On Unclear Science, Vol. 51, No. 4, August 2004.

[8] M. G. Johnson, et. al, "A Variable Delay Line PLL for CPU-Coprocessor Synchronization," IEEE JSSC, Vol. 23, No. 5, Feb. 1988.

[9] W. Yun, et. al, "New High Speed Dynamic D-Type Flip Flop For Prescaler," ISIE 2001, Korea.

[10] M. Elgamel, et. al, "Noise Tolerant Low Power Dynamic TSPCL D Flip-Flop," ISVLSI2002.

[11] R. Ginosar, "Fourteen Ways to Fool Your Synchronizer," ASYNC2003.

[12] A. Ghadin, et. al, "Comparative Energy and Delay of Energy Recovery and Square Wave Clock Flip-Flop for High-Performance and Low-Power Applications," ICM 2003.

[13] P. Zhao, et. al, "High-Performance and Low Power Conditional Charge Flip-Flop," IEEE Trans. On Very Large Scale Integration System, Vol. 12, No. 5, May 2004.

[14] S. D. Shin, et. al, "Variable Sampling Window Flip-Flop for Low Power High-Speed VLSI," IEEE Proc.—Circuits Devices Syst. Vol. 152, N0.3, June 2005.

[15] A. G.M. Strollo, et. al, "A Novel High-Speed Sense-Amplifier-Based Flip-Flop," IEEE Trans. On Very Large Scale Integration System, Vol. 13, No. 11, November 2005.

[16] B Nikolic, et al, "Sense Amplifier Based Flip-Flop," ISSCC99.

[17] C. Dike, et al, "Miller and Noise Effects in a Synchronization Flip-Flop," IEEE JSSC, vol. 34, No. 6, June 1999.

[18] J. Lee, "A 40Gb/s Clock and Data Recovery Circuit in 0.18um CMOS Technology," IEEE JSSC Vol. 38, No. 12, December, 2003.

[19] A. F. Champernowne, et al, "Latch-to-Latch Timing Rule," IEEE Tran. Computers, Vol. 39, No. 6, June 1990.

[20] M. Hansson, et al, "Comparative Analysis of Process Variation Impact on Flip-Flop Power Performance," IEEE 2007.

[21] D. J. Kinniment, et al, "Synchronization Circuit Performance," IEEE JSSC, Vol. 37, No. 2, Feb. 2002.

[22] A. Wassatsch, et al, "Scalable Counter Architecture for a Pre-Loaded 1Ghz@0.6um/5V Pre-scaler TSPC," IEEE 2001.

[23] D. Markovic, et al, "Feasibility Study of Low-Swing Clocking," IEEE Proc, 24th International Conference on Microelectronics, 2004.

[24] R. Chandrasekaran, et al, "A High-Speed Low-Power D Flip-Flop," IEEE 2005.

[25] S. Heo, et al, "Activity-Sensitive Flip-Flop and Latch Selection for Reduced Energy," IEEE Tran. VLSI Systems, Vol. 15, No. 9, Sept. 2007.

[26] J. Hu, et al, "Power-Gating Adiabatic Flip-Flop and Sequential Logic Circuits," IEEE.

[27] B. Nikolic, et al, "Sense Amplifier-Based Flip-Flop," IEEE ISSCC 1999.

[28] C. Dike, et al, "Miller and Noise Effects in a Synchronizing Flip-Flop," IEEE JSSC, Vol. 34, No. 6, June 1999.

[29] S. Hsu, et al, "A Novel High-Performance Low-Power CMOS Master0Slave Flip-Flop," IEEE 1999.

[30] J. C. Kim, et al, "A Sense Amplifier-Based CMOS Flip-Flop with an Enhanced Output Transition Time for High-Performance Microprocessors," IEEE 1999.

[31] J. C. Kim, et al, "A High-Speed 50% Power-Saving Half swing Clocking Scheme for Flip-Flop with Complementary Gate and Source Drive," IEEE 1999.

[32] B. S. Kong, et al, "Conditional-Capture Flip-Flop Technique for Statistical Power Reduction," IEEE ISSCC 2000.

[33] W. Yun et al, "New High Speed Dynamic D-Type Flip-Flop for Prescaler," ISIE 2001, Korea.

[34] M. Elgamel, et al, "Noise Tolerant Low Power Dynamic TSPCL D Flip-Flop," IEEE VLSI Symposium 2002.

[35] Y. Semiat, et al, "Timing Measurements of Synchronization Circuits," IEEE ASYNC 2003.

[36] L. Gasparini, et al, "A Digital Circuit for Jitter Reduction of GPS-Disciplined 1-pps Synchronization Signals," IEEE AMUEM 2007.

CHAPTER 8
VLSI DELAY AND PHASE COMPUTATIONAL CIRCUITS

Basic VLSI signal processing operations such as the addition, the scaling, and the integration operations can be realized in the time-domain signal processing systems. These basic signal processing circuit elements play important roles in adaptive delay time and delay variation cancellation and control loop compensation for VLSI high-speed I/O circuits.

Delay addition can be directly realized by connecting multiple VLSI delay circuit elements in series. Delay scaling operations can be implemented based on replica delay circuit elements or delay interpolation circuit elements. Time-domain integration operation, on the other hand, can either be realized employing VCO circuits or employing VLSI signal integration circuit and VLSI voltage to time conversion circuit elements.

Phase scaling operations are usually realized through various frequency division circuit techniques. Frequency division provides near-perfect scaling of the delay phase of a time-domain signal through the scaling of the reference clock period. VLSI frequency dividers are commonly used in the VLSI PLL circuits for reference clock generation operations.

8.1 BASIC VLSI DELAY PROCESSING OPERATIONS

Similar to other VLSI signal processing circuits, addition, scaling, and integration are the most fundamental operations in VLSI time-domain signal processing systems.

8.1.1 DELAY TIME ADDITION OPERATIONS

VLSI delay time addition operations can be directly realized through connecting multiple delay circuit elements in series as shown in figure 8.1.

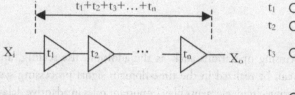
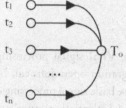

(a) VLSI delay time addition operation   (b) SFG expression

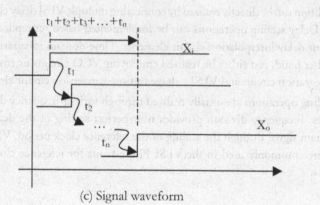

(c) Signal waveform

Fig. 8.1. VLSI delay time addition operation

Such a delay time addition operations can be mathematically expressed as

$$t_o = t_1 + t_2 + t_3 + ... + t_n \tag{8.1}$$

Delay time addition operation can also be extended to delay time subtraction operation for periodic time-domain signals (such as clocks) based on the complementary delay computation (or the early clock) concept. Under this method, subtracting a delay is equivalent to adding a complementary delay value:

$$t_1 - t_2 \rightarrow t_1 + (T - t_2)$$

(8.2)

T in the equation is the clock period of the global reference clock of the time-domain signal. Delay subtraction operation as shown in figure 8.2 has found many applications in practical clock phase signal processing circuit applications.

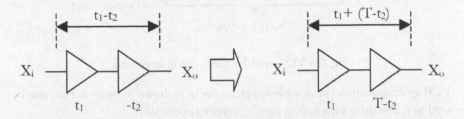

Fig. 8.2. VLSI complementary delay-based subtraction operation

8.1.2 VLSI DELAY TIME SCALING OPERATIONS

VLSI delay time scaling circuits as shown in figure 8.3 can be used to scale a given time-domain signal.

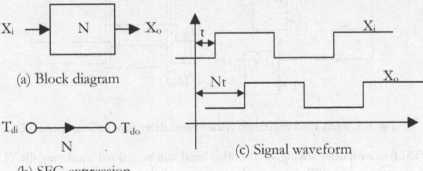

(a) Block diagram

(b) SFG expression

(c) Signal waveform

Fig. 8.3. VLSI delay time scaling operation

VLSI scaling operations can be mathematically expressed as

$$t_o = N t_i$$

(8.3)

N can typically be an integer number or fraction of integers. Delay scaling operations are commonly realized using VLSI replica delay circuit elements as shown in figure 8.4.

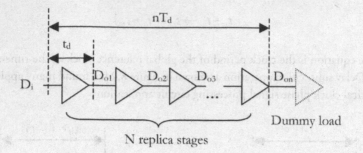

Fig. 8.4 VLSI up delay time scaling operation

VLSI synchronization circuit with polyphase clocks as shown in figure 8.5 can also be used to realize delay time scaling signal processing operations.

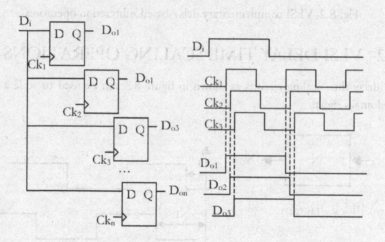

Fig. 8.5. VLSI synchronization circuit-based delay time scaling operation

VLSI fractional delay scaling, on the other hand, can be realized employing the VLSI phase interpolation (PI) circuits as shown in figure 8.6.

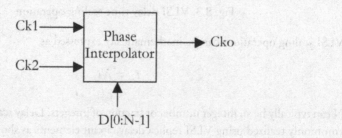

Fig. 8.6. VLSI phase interpolation circuit-based delay time scaling operation

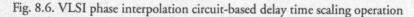

8.1.3 VLSI DELAY PHASE INTEGRATION OPERATIONS

VLSI voltage-controlled oscillator readily provides the time-domain integration operation where the output delay (or phase) is an integration of the input voltage signal as shown in figure 8.7.

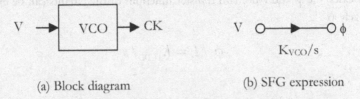

(a) Block diagram   (b) SFG expression

Fig. 8.7. VLSI delay time integration operation

The delay time integration operations can be mathematically expressed as

$$t_o = \frac{K_{VCO}}{s} V \tag{8.4}$$

Time-domain signal integration operations can also be realized in mixed-signal form in the voltage domain or digital and employing the voltage-to-time (V/T) conversion or digital-to-time (D/T) conversion circuits. Examples of VLSI time-domain signal processing circuit implementations are the loop filter circuit in the VLSI phase-locked loop (PLL) and delay-locked loop (DLL) circuits as shown in figure 8.8 and the digital loop filter realization in the phase interpolator-based VLSI high-speed I/O data recovery circuits (DRC) as shown in figure 8.9.

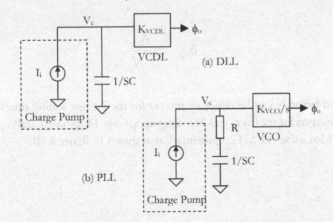

Fig. 8.8. Mixed mode time-domain signal process implementation in PLL and DLL

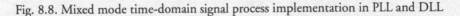

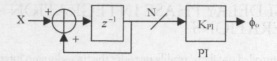

Fig. 8.9. Mixed mode time-domain signal process implementation in DRC

In these cases, the s- and z-domain transfer functions of the circuits can be expressed respectively as

$$\phi_o / I_i = K_{VCDL} / s \tag{8.5}$$

$$\phi_o / I_i = \{R + 1/(sC)\} K_{VCO} / s \tag{8.6}$$

$$\phi_o / X = K_{PI} / [z - 1] \tag{8.7}$$

8.2 VLSI DELAY PHASE SCALING CIRCUIT IMPLEMENTATIONS

One of the widely used VLSI time-domain signal computation circuit elements is the frequency divider circuit. VLSI frequency divider provides near-perfect scaling of the delay phase (not delay time!) of the periodic time-domain signal through scaling the period of the signal as

$$T \to NT \tag{8.8}$$

and

$$\delta_{CKo} = \frac{\delta_{CK}}{N} \tag{8.9}$$

The division factor N can be either an integer for the integer scaling operations or a fraction of integers for the fractional-N scaling operations. Delay phase scaling operation can be modeled using the SFG representation as shown in figure 8.10.

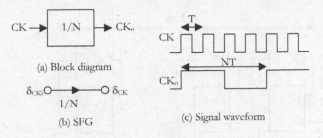

(a) Block diagram

(b) SFG

(c) Signal waveform

Fig. 8.10. VLSI frequency division operation

8.2.1 VLSI ASYNCHRONOUS FREQUENCY DIVIDER CIRCUITS

The simplest VLSI phase scaling circuit is the asynchronous binary frequency division circuit as shown in figure 8.11. The output data signal phase of the time-domain signal is divided by two as the clock period is increased to two times:

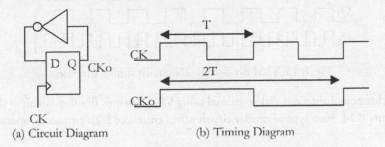

(a) Circuit Diagram   (b) Timing Diagram

Fig. 8.11. VLSI binary phase divider circuit expressions

It can be seen that such a circuit realizes a perfect delay phase divided-by-2 operation as

$$\delta_{CKo} = \frac{\delta_{CK}}{2}$$

(8.10)

High order asynchronous binary frequency dividers can be easily realized by using multiple divide-by-2 circuits as shown in figure 8.12. In general, a K stage divide-by-2 circuit stage offers the phase or frequency scaling operation with scaling factor given as the power of 2 as

$$f_o = \frac{1}{2^K} f_{ref}$$

(8.11)

209

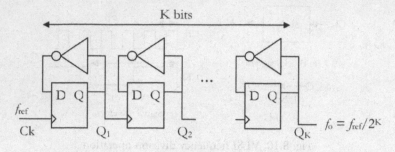

Fig. 8.12. VLSI asynchronous divided-by-2K circuit structure

The operation of the VLSI binary frequency divider circuit can be described using the logic diagram shown in figure 8.13.

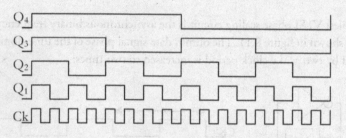

Fig. 8.13. VLSI divided by 2N circuit timing diagram

Asynchronous divider can also be realized using VLSI dynamic flip-flop circuit as shown in figure 8.14. Such type of divider circuit offers enhanced high-frequency operation.

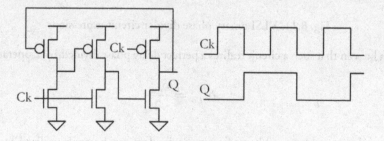

Fig. 8.14. Dynamic divide-by-2 divider

VLSI frequency divider circuit can also be realized in fully differential VLSI circuit form as shown in figure 8.15, which offers high-frequency operation with better noise rejection performance.

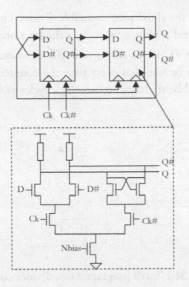

Fig. 8.15. VLSI differential CML divide-by-2 divider circuit

8.2.2 VLSI PROGRAMMABLE FREQUENCY DIVIDER CIRCUITS

VLSI programmable ratio frequency divider circuits are widely used for fraction-N frequency division operations. The programmable frequency divider circuits can be constructed based on the binary divider and a cycle-stretch circuits. A VLSI programmable ratio between $1/2^N$ and $1/(2^N+M)$ is shown in figure 8.16. Such circuit consists of an N-bit asynchronous binary divider and a feedback cycle stretch path in the first stage employing a NAND gate, where the feedback path offers the so-called cycle stretch operation since one additional clock cycle is inserted (cycle stretch) in the dividing process every time the hard-coded stretch location code is met during division counting.

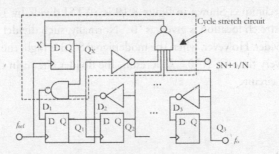

Fig. 8.16. General VLSI programmable ratio divider circuit structure

Note that in the stretch location codes, the Q_i bit must be "0" to ensure negative feedback in the circuit loop for the proper circuit operation. Therefore, the maximum number for

M is limited to 2^{N-1}. For the above programmable dividers circuit, M = 1 is commonly used in circuits in practical applications, such as the fractional-N PLL circuits.

Shown in figure 8.17 is a VLSI 1/8 (1/9) divider circuit where the stretch location code is given as "011." It can be seen that every time the counter reaches the output state "011," the next state will be stretched to two clock cycles as shown in the figure.

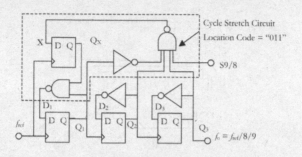

Fig. 8.17. VLSI 1/8 1/9 programmable divider circuit structure

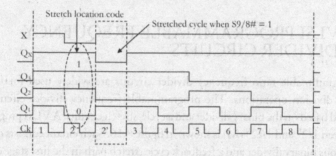

Fig. 8.18. Timing diagram for 1/8 1/9 counter based on cycle stretch

The programmable frequency dividers for other dividing ratio can be realized based on this circuit technique. Shown in figure 8.19 is a VLSI divide by 1/2 (1/3) divider where the cycle stretch location is given as "0." Normally, such divider circuit serves as a divide-by-2 divider. However, when the mode signal S2/3 is high, the divider counter inserts a cycle every time Q_1 is "0." As a result, the divider circuit in this mode serves as a divide-by-3 circuit.

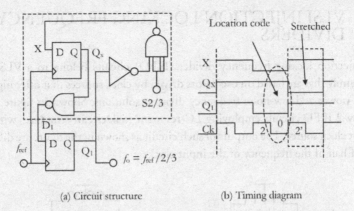

| (a) Circuit structure | (b) Timing diagram |

Fig. 8.19. VLSI 1/2 1/3 programmable divider circuit structure

Similarly, we can realize the 1/4 (1/5) divider using the circuit shown in figure 8.20, where the cycle stretch location code is set as "01."

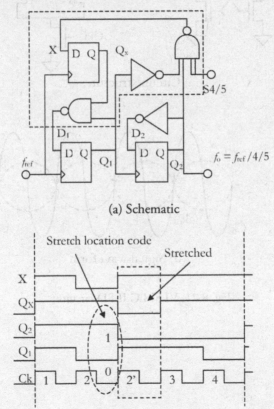

(a) Schematic

(b) Timing diagram

Fig. 8.20. VLSI 1/4 1/5 programmable divider circuit structure

8.2.3 VLSI INJECTION LOCKING FREQUENCY DIVIDERS

VLSI injection locking frequency divider (ILFD) circuits belong to a VLSI divder circuit family that are based on oscillators driven by clock sources that offer high-speed and low-power and low-noise frequency division solutions. Shown in figure 8.21 are divide-by-2 ILFD circuits employing LC-resonant tank circuit structures with Ck_i as the input clock source. The outputs of such circuit as shown in the figure are differential clocks of half of the frequency of the input clock.

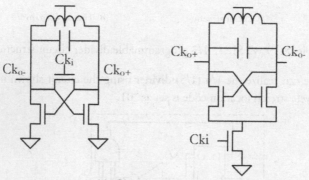

(a) Circuit structures

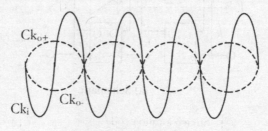

(b) Signal waveform

Fig. 8.21. VLSI LC ILFD circuits

ILFD circuit can also be used to generate polyphase clocks as shown in figure 8.22.

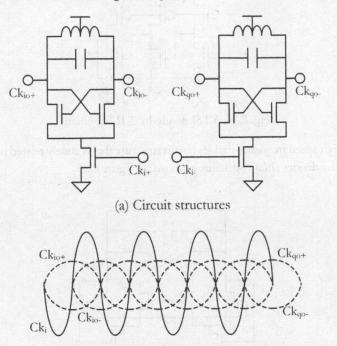

(a) Circuit structures

(b) Signal waveform

Fig. 8.22. VLSI LC I/Q ILFD circuits

Shown in figure 8.23 and figure 8.24 are alternative VLSI injection-locking frequency divider circuit implementations employing VLSI ring oscillator circuit structures.

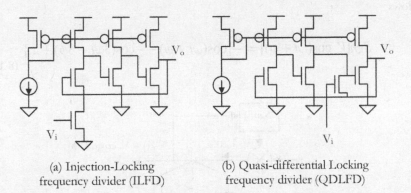

(a) Injection-Locking
frequency divider (ILFD)

(b) Quasi-differential Locking
frequency divider (QDLFD)

Fig. 8.23. VLSI injection-locking frequency divide-by-2 divider circuits

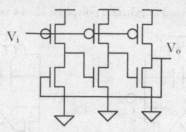

Fig. 8.24. VLSI divide-by-3 ILFD circuit

A VLSI high-speed frequency divider circuit structure that is closely related to the ILFD is the Miller divider circuit structure as shown in figure 8.25.

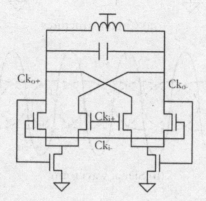

Fig. 8.25. Miller frequency divider circuit structure

Shown in figure 8.26 is the behavioral circuit model of the injection lock frequency divider, where the Sgn function in the figure can be expanded using Fourier series as follows:

$$Sgn[V_o \cos(\omega t + \phi)] = \frac{4}{\pi}[\cos(\omega t + \phi) - \frac{1}{3}\cos(3\omega t + 3\phi) + ...] \tag{8.12}$$

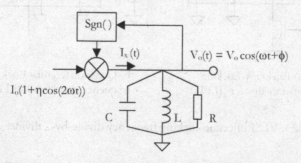

Fig. 8.26. Injection lock frequency divider circuit model

The input current to the tank circuit is given as

$$I_x(t) = I_o[1 + \eta\cos(2\omega t)] \cdot \frac{4}{\pi}[\cos(\omega t + \phi) - \frac{1}{3}\cos(3\omega t + 3\phi) + ...]$$

$$= I_o \frac{4}{\pi}[\cos(\omega t + \phi) + \frac{\eta}{2}\cos(\omega t - \phi) - \frac{\eta}{6}\cos(\omega t + 3\phi) + ...]$$

(8.13)

The above equation can also be expressed in complex form as

$$I_x(t)^* = I_o[1 + \eta\cos(2\omega t)] \cdot \frac{4}{\pi}[\cos(\omega t + \phi) - \frac{1}{3}\cos(3\omega t + 3\phi) + ...]$$

$$\approx I_o \frac{4}{\pi}e^{j(\omega t + \phi)}[1 + \frac{\eta}{2}e^{-2j\phi} - \frac{\eta}{6}e^{2j\phi}]$$

(8.14)

Note that in the above equation, higher order harmonic components are purposely ignored since it will be filtered out by the tank circuit.

The complex output voltage is given as

$$V_o e^{j(\omega t + \phi)}[sC + \frac{1}{Ls} + \frac{1}{R}]|_{s=j\omega} = I_x(t)^* = I_o \frac{4}{\pi}e^{j(\omega t + \phi)}[1 + \frac{\eta}{2}e^{-2j\phi} - \frac{\eta}{6}e^{2j\phi}]$$

(8.15)

Above equation can also be rewritten as

$$\left(\frac{\pi V_o}{4 I_o L\omega}\right)[j(\left(\frac{\omega}{\omega_o}\right)^2 - 1) + \frac{1}{Q}\left(\frac{\omega}{\omega_o}\right)] = [1 + \frac{\eta}{2}e^{-2j\phi} - \frac{\eta}{6}e^{2j\phi}]$$

$$= [1 + \frac{\eta}{3}\cos(2\phi)] - j\frac{2}{3}\sin(2\phi)$$

(8.16)

where ω_o and Q are the resonant frequency and the quality factor of the tank circuit:

$$\begin{cases} \omega_o \equiv \frac{1}{\sqrt{LC}} \\ Q \equiv \frac{1}{R}\sqrt{\frac{L}{C}} \\ \alpha \equiv [\left(\frac{\omega}{\omega_o}\right)^2 - 1] / [\frac{1}{Q}\left(\frac{\omega}{\omega_o}\right)] \end{cases}$$

(8.17)

We may further introduce two parameters as

$$
\begin{cases}
\alpha \equiv [\left(\dfrac{\omega}{\omega_o}\right)^2 - 1] / [\dfrac{2}{Q}\left(\dfrac{\omega}{\omega_o}\right)] \\[4mm]
\beta \equiv \left(\dfrac{\eta}{3}\right)\cos(2\phi)
\end{cases}
$$

$$(8.18)$$

Equation 8.16 can be expressed into real and imaginary forms as

$$
\begin{cases}
\left[\left(\dfrac{\pi V_o}{4I_o L\omega}\right)[\dfrac{1}{Q}\left(\dfrac{\omega}{\omega_o}\right)]\right] = [1+\beta] \\[4mm]
\left[\left(\dfrac{\pi V_o}{4I_o L\omega}\right)[\dfrac{1}{Q}\left(\dfrac{\omega}{\omega_o}\right)]\right]\alpha = \sqrt{\left(\dfrac{\eta}{3}\right)^2 - \beta^2}
\end{cases}
$$

$$(8.19)$$

By combining equations in 8.19, we have that

$$
\alpha[1+\beta] = \sqrt{\left(\dfrac{\eta}{3}\right)^2 - \beta^2}
$$

$$(8.20)$$

or

$$
[\alpha^2 - \left(\dfrac{\eta}{3}\right)^2] + 2\alpha^2\beta + (\alpha^2 + 1)\beta^2 = 0
$$

$$(8.21)$$

It can be seen that to ensure the real solution of β, we have that

$$
\alpha^4 > [\alpha^2 - \left(\dfrac{\eta}{3}\right)^2](\alpha^2 + 1) = \alpha^4 + (1 - \left(\dfrac{\eta}{3}\right)^2)\alpha^2 - \left(\dfrac{\eta}{3}\right)^2
$$

$$(8.22)$$

that is,

$$
\alpha \equiv [\left(\dfrac{\omega}{\omega_o}\right)^2 - 1] / [\dfrac{2}{Q}\left(\dfrac{\omega}{\omega_o}\right)] < 1 / \sqrt{\left(\dfrac{3}{\eta}\right)^2 - 1}
$$

$$(8.23)$$

It can be seen that the maximum lock-in frequency range is given as

$$\omega_{LR} \equiv 2|\omega - \omega_o| \approx 2\left(\frac{\omega_o}{Q}\right) / \sqrt{\left(\frac{3}{\eta}\right)^2 - 1} \approx 2\left(\frac{\omega_o}{Q}\right)\frac{\eta}{3} \qquad (8.24)$$

For the practical circuit implementation, both the switch MOS devices can be in the "off" state for a short period of time in the operation. If we assume the "on" time period of the device is τ, which is smaller than half of the output clock period T, the lock-in frequency range can be derived as

$$\omega_{LR} \equiv 2|\omega - \omega_o| \approx 2r^2\left(\frac{\omega_o}{Q}\right) / \sqrt{\left(\frac{3}{\eta}\right)^2 - (3 - 2r^2)^2} \qquad (8.25)$$

where

$$r \equiv \sin(\omega\tau / 2) \qquad (8.26)$$

References:

[1]   P. Heydari, et al, "A 40-Ghz Flip-Flop-Based Frequency Divider," IEEE Trans. Circuits and Systems-II: Express Briefs, Vol.53, No.12, December 2006.

[2]   M. Ali, et al, "A Multigigahertz Multimodulus Frequency Divider in 90nm CMOS," IEEE Trans. Circuits and Systems-II: Express Briefs, Vol.53, No.12, December 2006.

[3]   R. Mohanavelu, et al, "A Novel 40-Ghz Flip-Flop Frequency Divider in 0.18um CMOS," Proceedings of ESSCIRC 2005.

[4]   K. Yamaoto, et al, "A 44-uW 4.3-GHz Injection-Locked Frequency Divider with 2.3Ghz Locking Range," IEEE JSSC, Vol. 40, No.3, March 2005.

[5]   K. Yamaoto, et al, "55GHz CMOS Frequency Divider with 2.3Ghz Locking Range," IEEE March 2004.

[6]   Y. Hu, et al, "A 25-Ghz CMOS Tunable Injection Locking Quadrature LC-Divider for Multi-Channel Transceivers," Proceedings of 6th International Caribbean Conference on Devices, Circuits and Systems, Mexico, Apr. 26-28, 2006.

[7]   C. Cao, et al, "A 50-Ghz Phase-Locked Loop in 130nm CMOS," IEEE 2006 CICC.

[8]   B. Razavi, "Heterodyne Phase Locking: A Technique for High-Speed Frequency Division," IEEE JSSC, Vol. 42, No. 12, Dec. 2007.

[9]   H. Zheng, et al, "A 0.5-V 16Ghz-20Ghz Differential Injection-Locked Divider in 0.18um CMOS Process," IEEE 2007 CICC.

[10]   M. Motoyoshi, et al, "43uW 6Ghz CMOS Divider-by-3 Frequency Divider Based on Three-Phase Harmonic Injection Locking," IEEE 2006.

[11]   F. Gatta, et al, "A Direct Conversion CMOS Receiver Front-End with On-Chip LO for UMTS," ESSCIRC 2002.

[12]   A. Mazzanti, et al, "Injection Locking LC Dividers for Low Power Quadrature Generation," IEEE CICC2003.

[13]   Q. Du, et al, "A Low Phase Noise DLL Clock Generator with A Programmable Dynamic Frequency Divider," IEEE CCECE/CCGEI2006.

[14]   Y. Moon, et, al, "A Divide-by-16.5 Circuit for 10-Gb Ethernet Transceiver in 0.13um CMOS," IEEE JSSC Vol. 40, No. 5, May 2005.

[15]   R. Mohanavehu, "A Novel Ultra High-Speed Flip-Flop-Based Frequency Divider," IEEE ISCAS 2004.

[16]   A. Mazzanti, et al, "Injection Locking LC Dividers for Low Power Quadrature Generation," IEEE CICC 2003.

[17]   K. H. Tsai, et al, "A 39.2-45.5Ghz Frequency Divider using Switched Cross-Coupled Pair," IEEE 2007.

[18]   K. Sengupta, et al, "Maximum Frequency of Operation of CMOS Static Frequency Dividers: Theory and Design Techniques," IEEE 2006.

[19]   K. H. Chen, et al, "A Sub-1V Low Power High-Speed Static Frequency Divider," IEEE 2007.

[20] R. Nonis, et al, "A Design Methodology for MOS Current-Mode Logic Frequency Dividers," IEEE Tran. Circuits and Systems—I: Regular Papers, vol. 54, no. 2, Feb. 2007.

[21] X. J. Mao, et al, "New Frequency Divider with 8 Output Phase for Phase Switching Pre-scaler," IEEE 2006.

[22] J. C. Chen, et al, "Ultra-Low-Voltage CMOS Static Frequency Divider," IEEE 2005.

[23] W. F. Chung, et al, "CMOS RF Analog Frequency Divider using Switched Capacitors," IEEE 2005.

[24] P. Heydari, et al, "A 40-Ghz Flip-Flop-Based Frequency Divider in 0.18um CMOS," IEEE ESSCIRC 2005.

[25] J. M. C. Wong, et al, "A 1-V 2.5-mW 5.2-Ghz Frequency Divider in a 0.35um CMOS Process," IEEE JSSC, vol. 38, no. 10, Oct. 2003.

[26] E. Tournier, et al, "High-Speed Dual-Modulus Prescaler Architecture for Programmable Digital Frequency Dividers," Electronics Letters, Nov. 22, 2001.

[27] B. Chang, et al, "A 1.2 Ghz CMOS Dual-Modulus Prescaler using New Dynamic D-Type Flip-Flop," IEEE JSSC, vol. 31, no. 5, May, 1996.

[28] M. Ali, et al, "A Multigigahertz Multimodulus Frequency Divider in 90-nm CMOS," IEEE tran. Circuits and Systems—II: Express Briefs, vol. 53, no. 12, Dec. 2006.

[29] B. A. Floyed, et al, "Sub-Integer Frequency Synthesis Using Phase-Rotating Frequency Dividers," IEEE TCAS.

[30] Ken Yamamoto, et al, "55Ghz CMOS Frequency Divider with 3.2Ghz Locking Range," IEEE 2004.

[31] B. Razavi, et al, "CMOS Transceivers at 60 Ghz and Beyond," IEEE 2007.

[32] J. Midtgaard, et al, "Fully Integrated 1.7Ghz, 188 dBc/Hz FoM, 0.8V, 320uW LC-Tank VCO and Frequency Divider," 2005 VLSI Circuits Digest of Technical Papers.

[33] M. Fujishima, et al, "High-Performance Frequency Dividers Utilizing Differential Locking," IEEE 2004.

[34] C. van der Bos, et al, "Frequency Division Using an Injection-Locking Relaxation Oscillator," IEEE 2002.

[35] F. Gatta, et al, "A Direct Conversion CMOS Receiver Front-End with On-Chip LO for UMTS," IEEE ESSCIRC 2002.

[36] C. Cao, et al, "A 50 Ghz Phase-Locked Loop in 130nm CMOS process," IEEE CICC 2007.

[37] L. Zhang, et al, "A Double-Balanced Injection-Locked Frequency Divider for Tunable Dual-Phase Signal Generation," IEEE.

[38] S. Cheng, et al, "A Fully Differential Low-Power Divide-by-8 Injection-Locked Frequency Divider Up to 18Ghz," IEEE JSSC, vol.42, no. 3, March 2007.

[39] S. L. Jang, et al, "Circuit Techniques for CMOS Divide-by-Four Frequency Divider," IEEE Microwave and Wireless Components Letters, vol.17, no.3, March 2007.

[40] M. Motoyoshi, et al, "43uW 6Ghz CMOS Divide-by-3 Frequency Divider Based on Three-Phase Harmonic Injection Locking," IEEE 2006.

[41] S. H. Lee, et al, "A Low Voltage Divide-by-4 Injection Locked Frequency Divider with Quadrature Output," IEEE Microwave and Wireless Components Letters, vol.17, no.5, May 2007.

[42] A. Q. Safarian, et al, "A Study of High-Frequency Regenerative Frequency Dividers," IEEE 2005.

[41] S. H. Lee, et al, "A Low Voltage Divide-by-4 Injection Locked Frequency Divider with Quadrature Output," IEEE Microwave and Wireless Components Letters, vol.17, no.5, May 2007.

CHAPTER 9
VLSI PHASE DETECTOR CIRCUITS

Phase detector (PD) circuits are commonly used in VLSI high-speed I/O circuits for the delay time or phase detection operations. There are two major VLSI phase detector circuit families, including clock phase detectors (CPD) and the data phase detectors (DPD). VLSI CPD circuits are commonly used to compare the phases of the two periodic (clock) signals. VLSI CPD circuits are key circuit elements of VLSI PLL and DLL circuits. VLSI DPD circuits, on the other hand, are commonly used to compare the delay phase of a data signal with respect to a clock reference. VLSI DPD circuits are commonly used in the high-speed I/O data recovery circuits. In practical applications, the outputs of the VLSI phase detector circuits are in discrete-time signal forms that are usually lowpass filtered to create voltage domain loop control signals, where the lowpass filtered phase detector output is proportional to the delay phase differences of the two input signals.

VLSI time measurement circuits (TMCs) belong to a VLSI time-domain signal processing circuit family that is closely related to the phase detector circuits. Time measurement circuits are widely used in the VLSI design for testing (DFT) circuit implementations, and that can be used for the built-in self-test (BIST) of the high-speed I/O circuits.

9.1   VLSI PHASE DETECTOR CIRCUIT MODEL

A PD circuit can be mathematically modeled as the phase subtraction operation. The outputs of the PD circuits in general are synchronous or asynchronous discrete-time signals. The average (or lowpass filtered) PD circuit output signals are proportional to the input clock phase difference that can usually be mathematically expressed as

$$\overline{Y} = K_{PD} \cdot (\phi_1 - \phi_2) + Y_o \equiv K_{PD} \cdot [\Delta\phi + \phi_o]$$

(9.1)

where K_{PD} is defined as the phase detection gain. Y_o and ϕ_o are the DC output and phase offset values respectively.

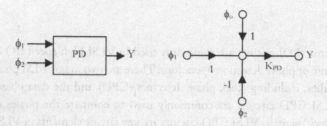

Fig. 9.1. Phase detector circuit model

In practical VLSI high-speed I/O circuit applications, the following are desired:

- Equation 9.1 is required to be linear and monotonic near the PD circuit operation point. Such requirement is used to ensure the loop stability of the PLL, DLL, or data recovery circuits. Such requirements can also be used to avoid the dead zone issue in the phase locking operation.

- The phase offset parameter ϕ_o is usually selected to be either zero or $\pi/2$ for phase synchronization or for phase offset phase synchronization operations respectively.

- K_{PD} and Y_o are desired to be independent of the input signal nonidealities, such as the duty cycle, slew rate and the amplitude variations, and the power supply. Such requirements can be used to ensure low noise operation of the PD circuits.

9.2   VLSI CLOCK PHASE DETECTOR CIRCUITS

Clock phase detector (CPD) circuits are widely used in VLSI PLL and DLL circuits for sensing the phase difference between the input reference clock and the feedback clock signals. The outputs of the PD circuits are either analog or digital signals based on the analog or digital phase detection implementations respectively.

There are several widely used VLSI CPD circuit implementations for detecting the phase difference of two clock signals.

9.2.1 ANALOG MULTIPLIER PHASE DETECTOR

CPD for sinusoidal clock inputs signals can be realized based on the analog multiplier circuit as shown in figure 9.2.

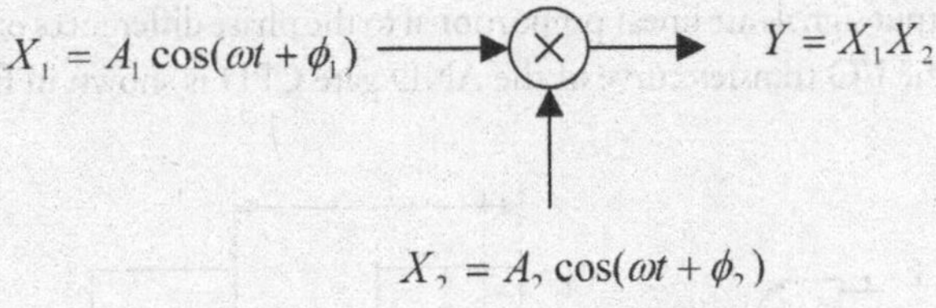

$$X_1 = A_1 \cos(\omega t + \phi_1) \longrightarrow \bigotimes \longrightarrow Y = X_1 X_2$$

$$X_2 = A_2 \cos(\omega t + \phi_2)$$

Fig. 9.2. Analog multiplier phase detector

$$Y = X_1 \cdot X_2 = \frac{A_1 A_2}{2}[\cos(\phi_1 - \phi_2)] + \frac{A_1 A_2}{2}[\cos(2\omega t + \phi_1 + \phi_2)]$$

(9.2)

Since the high-frequency components in the output of the analog multiplier CPD circuits can be eliminated through lowpass filtering operation, the low-frequency component of phase detector output can be expressed as function of the phase difference of the two input clock signals as

$$\overline{Y} = \frac{A_1 A_2}{2}[\cos(\phi_1 - \phi_2)]$$

(9.3)

The input to output transfer function of the analog multiplier phase detector circuit can be plotted using the I/O transfer curve as shown in figure 9.3.

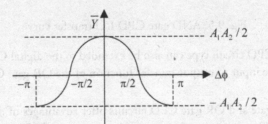

Fig. 9.3. Analog multiplier phase detector I/O transfer curve

Such analog multiplier CPD can typically offer higher operation speed compared with other implementation. However, the analog CPD circuit implementation suffers from high static power consumption, poor noise immunity since the gain of the PD depends on the amplitude of the signal. In addition, it cannot detect the frequency error of the inputs.

9.2.2 AND GATE PHASE DETECTOR

An AND logic gate shown in figure 9.4 can be used to realize digital multiplication function of two digital signals. Such circuit structure can therefore be used as CPD for phase detection of two digital clock signals. For the AND gate CPD circuits, the average values of the output signals are linear proportional to the phase differences of the two digital clock inputs. The I/O transfer curve of the AND gate CPD is shown in figure 9.5.

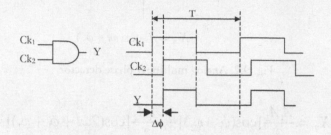

Fig. 9.4. AND gate CPD circuit

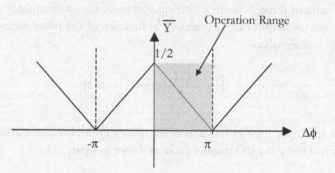

Fig. 9.5. AND gate CPD I/O transfer curve

The AND gate CPD circuit type can also be extended to the digital OR gate as shown in figure 9.6. The input to output transfer function of the OR gate CPD is shown in figure 9.7.

Both the AND gate and OR gate CPD circuits offer advantages of design simplicity, low power, and high operation frequency. However, such CPD circuits suffer from major limitations for high sensitivity to the input signal nonideality, such as the clock duty-cycle distortion, and offset errors.

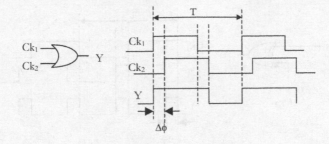

Fig. 9.6. OR gate CPD circuit

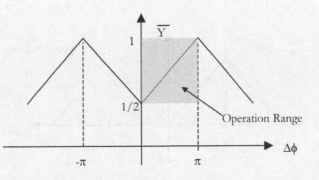

Fig. 9.7. OR gate CPD I/O transfer curve

9.2.3 XOR PHASE DETECTOR

The XOR CPD circuit shown in figure 9.8 provides a circuit improvement to the AND CPD circuit. For the XOR CPD circuits, the dc values of the output signals are linear proportional to the phase differences of the two digital input signals. The I/O transfer curve of the XOR CPD is shown in figure 9.9.

The dc offset of the above XOR CPD circuit can be eliminated employing the fully differential circuit structure, such as the Gilbert circuit structure shown in figure 9.10. The Gilbert cell CPD circuit offers transfer curve similar to the XOR CPD with removed dc offset as shown in figure 9.11 as a result of the fully differential input and output configuration.

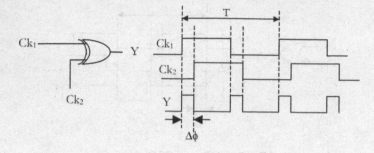

Fig. 9.8. XOR phase detector circuit

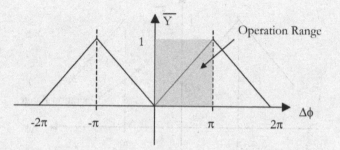

Fig. 9.9. XOR phase detector I/O transfer curve

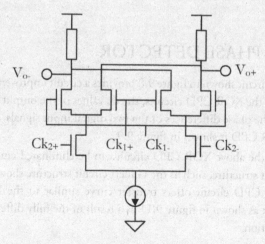

Fig. 9.10. Gilbert cell CPD circuit

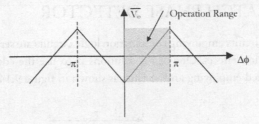

Fig. 9.11. Gilbert cell CPD I/O transfer curve

It is important to note that XOR CPD circuit, like other logic gate CPDs, is sensitive to the clock duty-cycle distortion.

9.2.4 DOUBLE-XOR PHASE DETECTOR

Shown in figure 9.12 is a VLSI double-XOR CPD circuit structure. In the double-XOR CPD, the input Ck_1 is compared with I/Q clocks in phase detection.

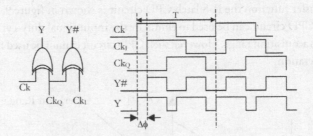

Fig. 9.12. Double XOR phase detector circuit

The I/O transfer function of such double-OXR CPD circuit is shown in figure 9.13.

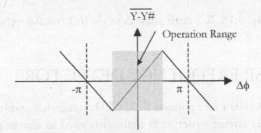

Fig. 9.13. Double-XOR phase detector I/O transfer curve

It can be seen that double-XOR CPD circuit has a zero phase offset. Note that the double-XOR CPD circuit can also be extended to double-AND or Double-OR CPD circuit structures.

9.2.5 R-S LATCH PHASE DETECTOR

The above CPD circuits employing combination logic circuits are sensitive to the clock duty-cycle distortion effects. One of the solutions is to use the edge-triggered CPD circuit implemented employing an R-S latch as shown in figure 9.14.

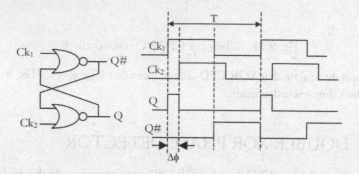

Fig. 9.14. R-S latch phase detector

The I/O transfer function the R-S latch CPD circuit is shown in figure 9.15. The R-S latch circuit CPD circuit can be used to eliminate the input signal duty-cycle sensitivity and improves acquisition range. However, such CPD circuit cannot be used for harmonic detection operation.

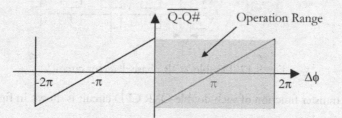

Fig. 9.15. R-S latch phase detector I/O transfer curve

9.2.6 PHASE FREQUENCY DETECTOR

Shown in figure 9.16 is a very popular VLSI CPD circuit that employs two resettable D-FFs. Such CPD circuit structure is commonly used in charge pump VLSI PLL circuits, where the phase detector circuit is used to generate PWM pulses, with the pulse widths equal to the phase difference between the clock inputs. It is important to note that such CPD circuits also offer phase and frequency detection (PFD) function.

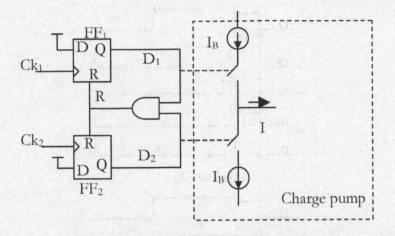

Fig. 9.16. VLSI phase frequency detector circuit

The operation of the PFD circuit using the time diagram shown in figure 9.17 can be described as follows:

- When the clock Ck_1 leads the clock Ck_2, the output of the FF_1 is first asserted to "1" at the rising edge of clock Ck_1. FF_1 output is reseted to "0" at the rising edge of clock Ck_2. As a result, the time duration (or pulse width) for FF_1 to stay high ("1") equals the delay difference between clock Ck_1 and clock Ck_2.

- When the clock Ck_2 leads the clock Ck_1, the output of the FF_2 is first asserted to "1" at the rising edge of clock Ck_1. FF_2 output is reseted to "0" at the rising edge of clock Ck_1. As a result, the time duration (or pulse width) for FF_2 to stay high ("1") equals the delay difference between clock Ck_2 and clock Ck_1.

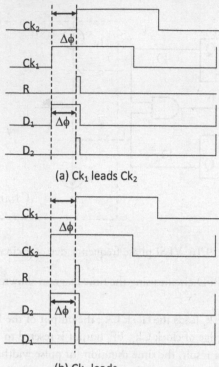

(a) Ck₁ leads Ck₂

(b) Ck₂ leads

Fig. 9.17. Timing diagrams of the VLSI phase frequency detector circuits

For the PFD circuit, the two outputs are used as pseudo-differential outputs of the phase detection.

The state diagram of the phase frequency detector circuit is shown in figure 9.18.

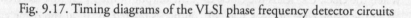
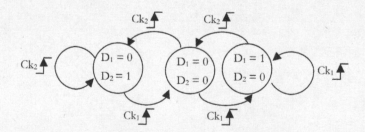

Fig. 9.18. State diagram of the VLSI phase detector circuit

As can be seen that when there is frequency difference between the two input clock signals, the output signals from the PFD detector circuit will have nonzero signal outputs. Therefore, the PFD circuit can be seen offering the frequency difference "memory" capability. Such a phase and frequency detection operation can be explained using the PFD I/O transfer function as shown in figure 9.19.

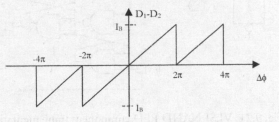

Fig. 9.19. I/O transfer function of VLSI phase frequency detector circuit

The I/O transfer function of the PFD circuit can be expressed as

$$\bar{I} = \frac{I_B}{2\pi}\phi_i \quad (-2\pi \leq \phi_i < 2\pi)$$

(9.4)

The phase detection gain of the CPD circuit is given as

$$K_{PD} = \frac{I_B}{2\pi} \quad (-2\pi \leq \phi_i < 2\pi)$$

(9.5)

In the VLSI circuit implementation, the D-FF in the PFD circuit can be directly realized using the CMOS NOR logic gates as shown in figure 9.20.

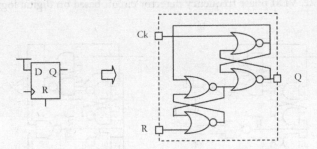

Fig. 9.20. VLSI NOR gate PFD circuit component implementation

The alternative VLSI D-FF implementation based on the NAND gates is shown in figure 9.21.

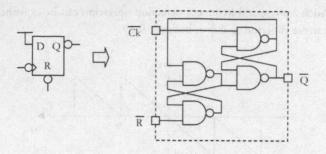

Fig. 9.21. VLSI NAND PFD component implementation

Based on the NAND gate D-FF circuits, the PFD circuit can be realized in VLSI circuit as shown in figure 9.22. It is important to note that the NAND PFD circuit is operated at the falling edges of the clock inputs. The detailed VLSI NAND gate PFD circuit implementation is shown in figure 9.23.

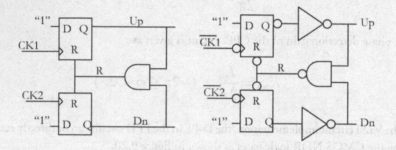

Fig. 9.22. VLSI phase frequency detector circuit based on digital logic gates

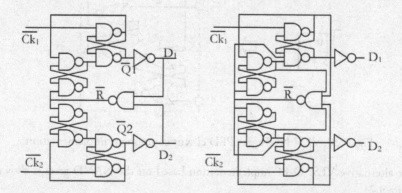

Fig. 9.23. VLSI phase frequency detector circuit implementations

VLSI D-FF in the PFD circuits can also be realized based on VLSI dynamic circuit structures as shown in figure 9.24 for high-frequency circuit applications.

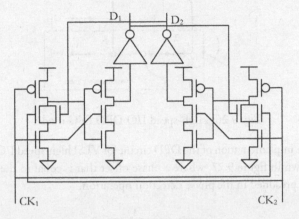

Fig. 9.24. VLSI Dynamic PFD circuit implementation

9.3   VLSI DATA PHASE DETECTOR CIRCUITS

Data phase detector (DPD) circuits are commonly used in VLSI high-speed I/O data recovery circuits (DRCs) as shown in figure 9.25, where the PDP circuit is used to extract the phase information of the random input data stream.

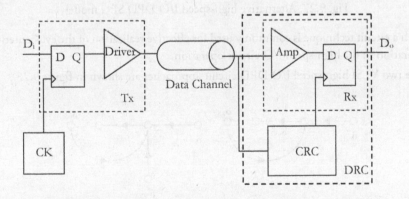

Fig. 9.25. DPD within the DRC-based high-speed I/O circuits

The operation of such DPD circuit can be described using the SFG model as shown in figure 9.26, where the input data phase is extracted and compared with receiver reference clock phase.

235

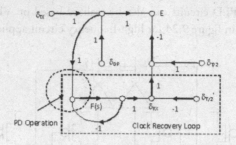

Fig. 9.26. High-speed I/O DPD SFG model

An alternative implementation of the DPD circuit for VLSI high-speed I/O data recovery circuit is shown in figure 9.27, where a phase offset that is equal to the eye-centering phase shift is preadded in the phase detection operation.

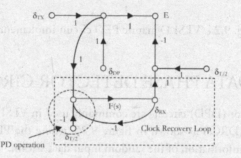

Fig. 9.27. Alternative high-speed I/O DPD SFG model

Such a circuit technique is commonly used for effective realization of the eye-centering operation in the high-speed I/O circuit operation.

The two VLSI high-speed I/O DPD circuit approaches are shown in figure 9.28.

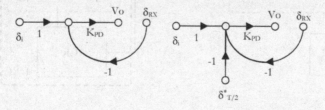

(a) Direct phase detection   (b) Phase-shifted phase detection

Fig. 9.28. Basic approaches for data phase detection

In the direct phase detection architecture, the input data edge is directly compared with the rise (or fall) edge of the receiver sampling clock. While in the shifted phase detection, the edge of the input data signal is compared with a half-clock period eye-centering phase shifted receiver sampling clock. The mathematical expression for the above phase detection architectures is given as

$$\overline{V}_o = \begin{cases} K_{PD}\overline{(\delta_i - \delta_{RX})} & \textit{Direct DPD} \\ K_{PD}\overline{(\delta_i - \delta_{RX} - \delta_{T/2}^*)} & \textit{Shifted DPD} \end{cases} \tag{9.6}$$

Similar to the CPD circuit, the DPD circuit can also be modeled using the phase detector gain K_{PD}.

9.3.1 BASIC XOR DPD CIRCUIT

VLSI DPD circuit can be realized based on the XOR gate and D-FF as shown in figure 9.29, where the data stream has a maximum frequency that is half of the clock frequency. Such a DPD circuit has a PWM signal output with the duty cycle equal to the delay time (or phase) difference of the data stream with respect to the clock reference.

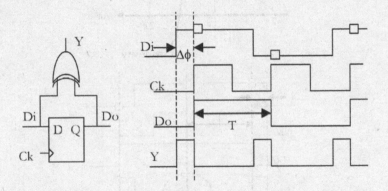

Fig. 9.29. XOR DPD circuit structure

The I/O transfer function of the basic XOR DPD can be expressed as

$$\overline{Y} = \frac{V_m}{2\pi}\overline{\Delta\phi} + Y_o \qquad 0 < \Delta\phi < 2\pi \tag{9.7}$$

The transfer function of the basic XOR DPD is shown in figure 9.30.

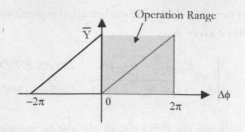

Fig. 9.30. Basic XOR DPD I/O transfer function curve

It is important to note that when there is no transition in the data signal, the XOR DPD output for that clock period will be zero as shown in figure 9.31. The basic XOR DPD will fail to properly detect the input data signal phase and results in data pattern dependent phase offset.

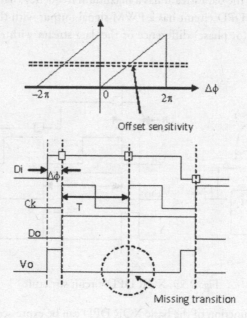

Fig. 9.31. Limitation on basic DRC phase detector circuit

9.3.2 PSEUDO-DIFFERENTIAL PHASE DETECTOR

The data pattern sensitivity problem of basic XOR DPD can be effectively resolved using the pseudo differential DPD circuit employing two basic XOR DPD circuits as shown in figure 9.32.

In the pseudo-differential DPD circuit structure, the second XOR stage is used to generate a reference to compensate for the data pattern dependent dc offset of the first

XOR DPD stage. As a result, the data pattern sensitivity of the basic XOR DPD circuit can be resolved.

The transfer function and the phase detection gain of the pseudo-differential DPD can be expressed as

$$\overline{Y_+ - Y_-} = \frac{V_m}{2\pi}\overline{(\Delta\phi - \pi)} \quad 0 < \Delta\phi < 2\pi$$

(9.8)

$$K_{PD} = \frac{V_m}{2\pi}$$

(9.9)

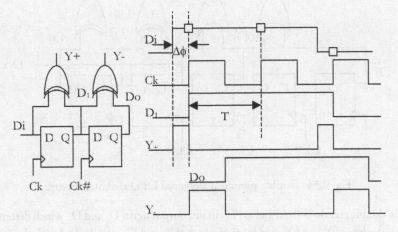

Fig. 9.32. Pseudo-differential DPD circuit structure

The transfer function plot of the pseudo-differential DPD is shown in figure 9.33.

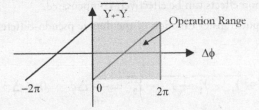

Fig. 9.33. Transfer function of pseudo-differential DPD circuit

9.3.3 DOUBLE PSEUDO-DIFFERENTIAL PHASE DETECTOR

The pseudo-differential DPD circuit structure is sensitive to the duty-cycle distortion of the clock signal since both the ck and ck# signals are used in the circuit. The duty-cycle sensitivity effect significantly limits the use of the pseudo-differential DPD circuit in practical high-speed I/O application.

A double pseudo-differential DPD circuit shown in figure 9.34 can be used to resolve the clock duty-cycle sensitivity issue of the pseudo-differential DPD circuit.

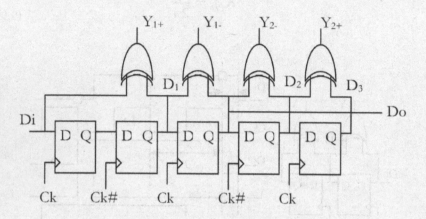

Fig. 9.34. Double pseudo-differential DPD circuit structure

In the double pseudo-differential DPD circuit, output signal D_1 and D_2, which determine the trial edges of Y_{1+} and Y_2 and leading edges of Y_{1-} and Y_{2-}, are clocked with the falling edge of the clock. On the other hand, the leading edges of Y_{1+} and Y_2 and trail edges of Y_{1-} and Y_{2+} are clocked with the rising edge of the clock. Since both the rising and falling clock edges are used for both the output polarities as shown in figure 9.35, the duty-cycle distortion effects can be effectively compensated.

The input to output transfer function of the double pseudo-differential DPD circuit is given as

$$\begin{cases} \overline{Y} \equiv \overline{Y_{1+} - Y_{1-} + Y_{2+} - Y_{2-}} = \dfrac{V_m}{2\pi}\Delta\phi & 0 < \Delta\phi < 2\pi \\[2em] K_{PD} = \dfrac{V_m}{2\pi} \end{cases}$$

$$(9.10)$$

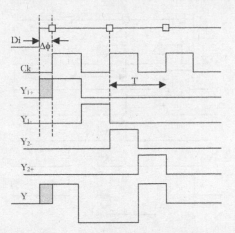

Fig. 9.35. Double pseudo-differential DPD timing diagram

The major performance limitation to the double pseudo-differential phase detector circuit is the increased circuit complexity and latency.

The transfer function of the double pseudo-differential DPD circuit is shown in figure 9.36.

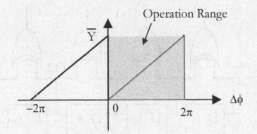

Fig. 9.36. Transfer function of double pseudo-differential DPD circuit

Shown in figure 9.37 is the detailed analysis of the duty-cycle distortion compensation effect of the double pseudo-differential DPD circuit.

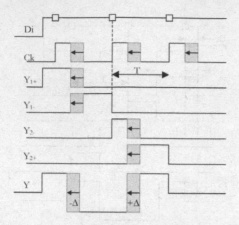

Fig. 9.37. Double pseudo-differential DPD timing diagram

An alternative double pseudo-differential DPD circuit is shown in figure 9.38. This DPD circuit provides reduced circuit component in circuit implementation compared with double pseudo-differential DPD circuit. Such circuit also offers reduced phase detection latency by half-a-clock period.

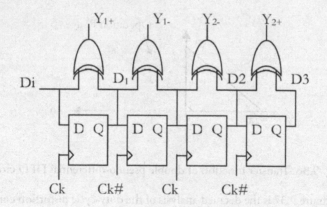

Fig. 9.38. Alternative double pseudo-differential DPD circuit structure

The timing diagram of this phase detect circuit is shown in figure 9.39. It can be seen that this circuit offers similar duty-cycle distortion compensation as the double pseudo-differential DPD circuit.

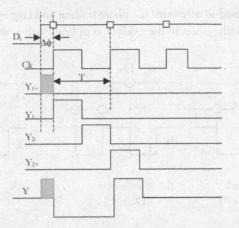

Fig. 9.39. Timing diagram of alternative double pseudo-differential DPD.

9.3.4 DIGITAL PHASE DETECTOR

For analog DPD circuits, the transition of the phase detector output can occur in continuous-time-domain based on the relationship of the input data and the receiver sampling clock phases. Such signal provides a pulse width modulation (PWM) representation of the input signal phase difference. On the other hand, the transition times of the digital DPD circuits are quantized, which can only occur at the rise (or falling) edges of the receiver sampling clock.

Shown in figure 9.40 is a digital high-speed I/O DPD circuit (also called Alexander phase detector). The output of the digital DPD circuit provides the pulse code modulation (PCM) representation of the input signal phase difference.

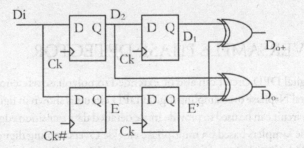

Fig. 9.40. The Alexander DPD circuit structure

The Alexander phase detector belongs to the 2x oversampling digital phase detector circuits where the input data stream is sampled twice at per clock period (one at the rise edge of the clock and one at the falling edge of the clock). Three data adjacent points (called triplet) are used to determine the phase leading and lagging information of the input data stream with respect to the receiver sampling clock as shown in figure 9.41.

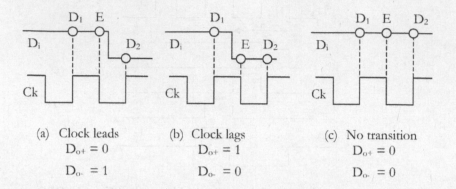

(a) Clock leads
$D_{o+} = 0$
$D_{o-} = 1$

(b) Clock lags
$D_{o+} = 1$
$D_{o-} = 0$

(c) No transition
$D_{o+} = 0$
$D_{o-} = 0$

Fig. 9.41. Logic decision of Alexander DPD circuit

The transfer function of the Alexander DPD circuit with data transition is shown in figure 9.42.

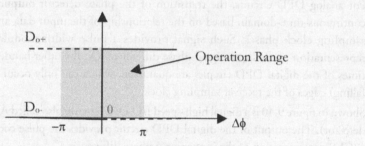

Fig. 9.42. Transfer function of double Alexander DPD circuit

9.3.5 OVERSAMPLE PHASE DETECTOR

Single-bit digital DPD circuit can also be extended to polyphase reference clock design cases. A general N-phase oversampling digital DPD circuit is shown in figure 9.43. Such digital DPD circuit can be used to provide more detailed data transition edge information using multiple samplers based on multiphase clocks. Oversampling digital DPD offers improved S/N in signal phase detection compared with the Alexander DPD circuit.

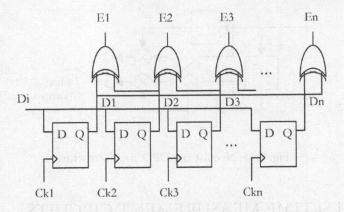

Fig. 9.43. The oversampling phase detector circuit structure

For this oversampling phase detector, the position of "1" at the output field as shown in figure 9.44 represents the phase value of the phase detector circuit.

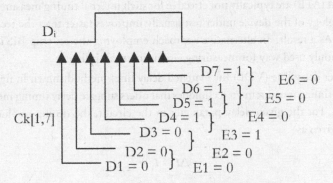

Fig. 9.44. The oversampling phase detector circuit structure

9.3.6 NYQUIST RATE PHASE DETECTOR

Most commonly used VLSI DPD for high-speed I/O circuits are based on certain types of oversampling and signal crossover detection. There has been a reported family of data phase detection based on the Nyquist spaced signal sampling without using zero-crossing. In these DPD circuit techniques, phase information are extracted based on the estimates, which are products of sampled signal or error pattern and a weighting pattern whose components are functions of the data symbols. The expected value of the estimates is used to extract the phase information for using in the high-speed I/O data recovery loop.

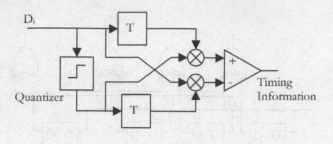

Fig. 9.45. Nyquist rate DPD circuit structure

9.4 VLSI TIME MEASUREMENT CIRCUITS

Accurate measurement of high-speed I/O internal timing parameters, such as setup time, hold time, jitter margin, etc., are extremely difficult due mainly to their small magnitudes (usually in the order of picoseconds). Most existing techniques using automatic test equipment (ATE) are typically not effective for such internal timing measurement since the technology of the device under test usually improves faster than the technology of the ATE. As a result, an alternative approach employing the on-chip BIST circuits is the commonly used way for measuring circuit internal timing.

The vernier delay line (VDL) (or matched delay line) method shown in figure 9.46 is one of the timing measurement techniques that offers subgate delay timing measurement resolution. For the given delay parameter in the circuit, the timing resolution of the circuit is given as

$$\Delta t = t_1 - t_2 \tag{9.11}$$

That can be significantly smaller than the gate delay of the circuit.

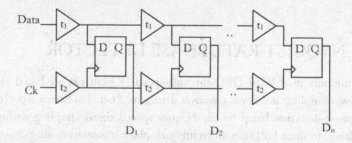

Fig. 9.46. Subgate timing resolution sampling circuit structure

An alternative approach for accurate timing measurement based on the time amplifier circuit technique is shown in figure 9.47. Such circuit technique can be used effectively to improve the timing measurement of the on-chip VLSI circuits.

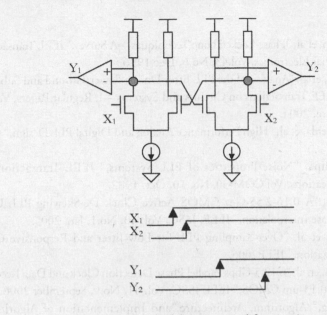

Fig. 9.47. VLSI delay time amplifier circuit

References:

[1] G.C. Hsieh, et al, "Phase-Locked Loop Techniques—A Survey," IEEE Transactions on Industrial Electronics, Vol.43, No.6, Dec.1996.

[2] P. Heydari, et al, "Analysis of the PLL Jitter Due to Power/Ground and Substrate Noise," IEEE Transactions on Circuits and Systems—I: Regular Papers, Vol 51, No. 12, Dec. 2004.

[3] T. M. Almeida, et al, "High Performance Analog and Digital PLL Design," IEEE 1999.

[4] V. F. Kroupa, "Noise Properties of PLL Systems," IEEE Transactions on Communications, Vol COM-30, No. 10, Oct. 1982.

[5] A. Maxim, "A 0.16-2.55-Ghz CMOS Active Clock De-Skewing PLL Using Analog Phase Interpolation," IEEE, JSSC, Vol. 40, No.1, Jan. 2005.

[6] M. Inoue, et al, "Over-Sampling PLL for Low-Jitter and Responsive Clock Synchronization," IEEE2006.

[7] Y. Ohtomo, et, al, "A 12.5-Gbps Parallel Phase Detection Clock and Data Recovery Circuit in 0.13-um CMOS," IEEE JSSC, Vol. 41, No.9, September 2006.

[8] H. J. Song," Algorithm, Architecture, and Implementation of Algorithmic delay-Locked Loop Based Data Recovery Circuit for High-Speed Serial Data Communication," 2001 IEEE SOCC.

[9] M. Papamichail, et, al, "Linear Range Extension of a Phase-Frequency-Detector with Saturated Output," ISCAS 2006.

[10] J. K. Kwon, "A 5-Gb/s 1/8-Rate CMOS Clock and Data Recovery Circuit," ISCAS 2004.

[11] F. Centurelli, et, al, "Robust Three-State PFD Architecture with Enhanced Frequency Acquisition Capabilities," ISCAS 2004.

[12] Mi. Chan, et, al, "An Improved Band-Bang PLL Employing a Quaternary Phase Detector," IEEE 2006.

[13] K. Ware, et, al, "A 200 Mhz CMOS Phase-Locked Loop with Dual Phase Detector," IEEE JSSC, Vol. 24, No. 6, December, 1989.

[14] B. G. Kim, et al, "A 250 Mhz-2Ghz Wide Range Delay-Locked Loop," IEEE JSSC Vol. 40, No.6, June 2005.

[15] K. Minami, et, al, "A 1Ghz Portable Digital Delay-Locked Loop with Infinite Phase Capture Ranges," IEEE ISSCC 2000.

[16] S. Butala, et, al, "A CMOS Clock Recovery Circuit for 2.5-Gb/s NRZ Data," IEEE JSSC Vol. 36, No. 3, March 2003.

[17] C. Kim, et, al, "Low-Power Small-Area +/-7.28ps Jitter 1 Ghz DLL-Based Clock Generator," IEEE ISSCC 2002.

[18] C. K. Ong, et, al, "A Scalable On-Chip Jitter Extraction Technique," IEEE VTS 2004.

[19] M. C. Tsai, et, al, "An All Digital High-Precision Built-In Delay Time Measurement Circuit," IEEE VTS 2008.

[20] M. Oulmane, et, al, "A CMOS Time Amplifier For Femto-Second Resolution Timing Measurement," IEEE ISCAS 2004.

[21] N. Abaskharoun, et, al, "Strategies for On-Chip Sub-Nanosecond Signal Capture and Timing Measurements," IEEE 2001.

[22] A. H. Chan, et, al, "A Deep Sub-Micron Timing Measurement Circuit Using A Single-Stage Vernier Delay Line," IEEE CICC 2002.

[23] R. C. H. van de Beek, et, al, "A 2.5-10-Ghz Clock Multiplier Unit with 0.22ps RMS Jitter in Standard 0.18um CMOS," IEEE JSSC, vol 39, no.11 2004.

[24] J. Cheng, et, al, "High-Speed Phase Detector Design," IEEE 2004.

[25] Y. Sumi, et, al, "Dead-Zone-Less PLL Frequency Synthesizer by Hybrid Phase Detectors," IEEE 1999.

[26] C. S. Huang, et, al, "A High-Precision Time-to-Digital Converter Using a Two-Level Conversion Scheme," IEEE Tran. Nuclear Science, vol. 51, no. 4, August 2004.

[27] B. Razavi, "Challenges in the Design of High-Speed Clock and Data Recovery Circuits," IEEE Communication Magazine, August 2002.

[28] S. Desgrez, et, al, "A New MMIC Sampling Phase Detector Design for Space Applications," IEEE JSSC, vol 38, no. 9, Sept. 2003.

[29] W. Rosenkrane, "Phase-Locking Loops with Limiter Phase Detectors in Presence of Noise," IEEE 1982.

[30] S. Soliman, et, al, "An Overview of Design Techniques for CMOS Phase Detectors," IEEE 2002.

[31] J. Savoj, et, al, "A 10-Gh/s CMOS Clock and Data Recovery Circuit with a Half-Rate Linear Phase Detector," IEEE JSSC, vol 36, no. 5, May. 2001.

[32] C. Y. Yang, et, al, "Fast-Switching Frequency Synthesizer with a Discriminator-Aided Phase Detector," IEEE JSSC, vol 35, no. 10, Oct. 2000.

[33] M. Ramezani, et, al, "An Improved Bang-Bang Phase Detector for Clock and Data Recovery Applications," IEEE 2001.

[34] A. R. Shahani, et, al, "Low-Power Dividerless Frequency Synthesis Using Aperture Phase Detector," IEEE JSSC, vol 33, no. 12, Dec. 1998.

[35] T. H. Tofil et, al, "Analysis of Parameter-Independent PLLs with Bang-Bang Phase-Detectors," IEEE 1998.

[36] R. Karlquist, et, al, "A Frequency Agile 40 Gb/s Half Rate Linear Phase Detector for Data Jitter Measurement," IEEE CSIC 2005 Digest.

[37] R. Ahola, et, al, "A Novel Phase Detector with No Dead Zone and a Chargepump with Very Wide Output Voltage Range," IEEE.

[38] A. Reisenzahn, et, al, "Phase-Synchronization in UWB Receiver with Sampling Phase Detectors," IEEE 2005.

[39] S. Mehrmanesh, et, al, "A Comprehensive Bang-Bang Phase Detector Model for High Speed Clock and Data Recovery Systems," IEEE 2005.

[40] A. Reisenzahn, et, al, "Ultra-Wideband Sampling Down Converter with Sampling Phase Detector," IEEE 2005.

[41] A. Rezayee, et, al, "A 9-16 Gb/s Clock and Data Recovery Circuit with Three-State Phase Detector and Dual-Path Loop Architecture," IEEE 2003.

[42] R. C. Halgren, et, al, "Improved Acquisition in Phase-Locked Loops with Sawtooth Phase Detectors," IEEE Tran. Communications, vol. 30, no. 10, Oct. 1982.

[43] M. Renaud, et, al, "A CMOS Three-State Frequency Detector Complementary to an Enhanced Linear Phase Detector for PLL, DLL, or High Frequency Clock Skew Measurement," IEEE 2003.

[44] M. Ramezeni, et, al, "Jitter Analysis of a PLL-Based CDR with a Bang-Bang Phase Detector," IEEE 2002.

[45] A. Reisenzahn, et, al, "A 10 Gb/s CDR with a Half-Rate Bang-Bang Phase Detector," IEEE 2003.

[46] M. A. Abas, et, al, "Built-in Time Measurement Circuits—A Comparative Design Study," IEEE IET Comput. Digit. Tech. 2007.

[47] M. Oulmane, et, al, "A CMOS Time Amplifier for Femto-Second Resolution Timing Measurement," IEEE 2004.

[48] M. C. Tsai, et, al, "An All Digital High-Precision Built-in Delay Time Measurement Circuit," IEEE 2008.

[49] E. R. Ruotsalainen, et, al, "Time Interval Digitization with an Integrated 32-Phase Oscillator," IEEE 2005.

[50] A. H. Chan, et, al, "A Deep Sun-Micro Timing Measurement Circuit Using a Single-State Vernizer Delay Line," IEEE CICC 2002.

[51] N. Abaskharoun, et, al, "Strategies for On-Chip Sub-Nanosecond Signal Capture and Timing Measurements," IEEE 2001.

[52] K. H. Mueller, et, al, "Timing Recovery in Digital Synchronous Data Receivers," IEEE Tran. On Communication, vol. Com-24, No.4, May 1976.

CHAPTER 10
VLSI PHASE-LOCKED LOOP CIRCUITS

Phase-locked loops (PLL) circuits are widely used in VLSI high-speed I/O circuits. Key applications of VLSI PLL circuits include the frequency (clock) synthesis, the phase (delay) synthesis, and the delay variation (jitter) filtering. In frequency synthesis applications, PLL circuits are used to translate the off-chip reference clock (usually at lower frequency) to on-chip clocks with frequency for specific I/O circuit data rates. In the phase synthesis applications, the PLL circuits are usually used associated with the data and clock recovery operations in high-speed I/O circuits, where the receiver reference clocks are regenerated based on the phase of the received data stream such that sampling clock are aligned to the center of the received data eye patterns for optimal data capture. In the jitter filtering applications, PLL circuits are used as clock signal conditioning circuits to eliminate the high-frequency jitter contents within the reference clocks.

VLSI PLL circuits suffer from various voltage and timing variation effects that can significantly impact the VLSI high-speed I/O circuit performance. Timing uncertainties of PLL circuits are mainly contributed from the jitter and clock feedthrough effect of input reference clocks, the added noises from the core circuits of the PLLs, such as the VCO circuits, the phase detector circuits, and the clock distribution circuits. These jitter effects consist of both the random jitter components from the thermal noise effects of the devices and the deterministic jitter components from the cross talk noises and power supply noises.

10.1 VLSI PLL PROTOTYPE CIRCUIT AND SFG MODELS

PLL circuits are commonly used in the VLSI high-speed I/O transmitter and the receiver circuits as shown in figure 10.1 for reference clock generation.

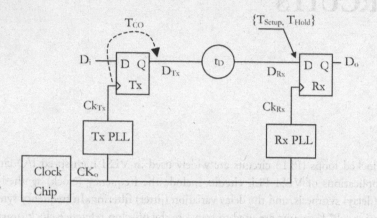

Fig. 10.1. Typical VLSI PLL circuit applications

The functionality of VLSI high-speed I/O PLL circuit can be modeled in both the time-domain and the s-domain based on the delay phase transfer function and the SFG representation as shown in figure 10.2 and figure 10.3 as

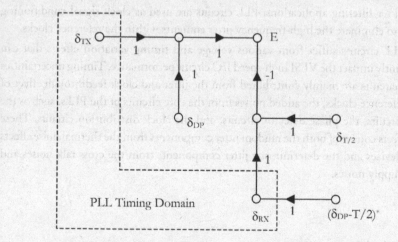

Fig. 10.2. Timing domain model of PLL within the high-speed I/O circuit

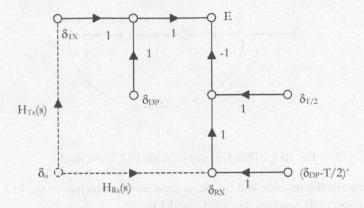

Fig. 10.3. PLL-based s-domain high-speed I/O circuit model

Based on the SFG model, the timing jitter from the input reference clock is multiplied by the jitter transfer function of the PLL circuit, which can be directly included in the s-domain I/O SFG model. It is expected that the input reference timing jitter will experience certain amplification or attenuation effects based on the PLL jitter transfer characteristics.

10.1.1 PLL S-DOMAIN AND SFG MODELS

A typical PLL circuit consists of a phase (or phase-frequency) detector (PD/PFD), a loop filter (LPF), a voltage-controlled oscillator (VCO), and one or multiple frequency dividers as shown in figure 10.4.

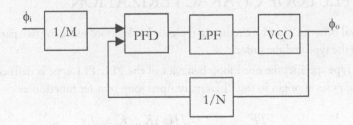

Fig. 10.4. PLL circuit block diagram

The PLL can be modeled using the linear s-domain SFG model as shown in figure 10.5, where ϕ_i and ϕ_o are the input and output clock phases respectively. ϕ_e is the phase error of the phase detector, and 1/M and 1/N are the input and feedback frequency dividing ratio respectively. K_{PD} and K_{VCO} are the phase detection gain and the VCO gain respectively. H(s) is the loop filter transfer function. Such SFG model provides a mixed-signal presentation of the PLL circuit using both the phase and voltage signal representations.

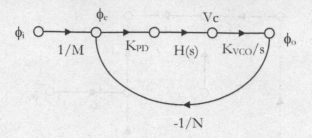

Fig. 10.5. SFG of linear s-domain PLL linear model

The phase transfer function and the phase error transfer function of the PLL can be derived respectively based on above SFG model as

$$\frac{\phi_o}{\phi_i} = \frac{N}{M} \frac{\dfrac{H(s)K_{PD}K_{VCO}/N}{s}}{1 + \dfrac{H(s)K_{PD}K_{VCO}/N}{s}} \tag{10.1}$$

$$\frac{\phi_e}{\phi_i} \equiv \frac{\phi_i/M - \phi_o/N}{\phi_i/M} = \frac{1}{1 + \dfrac{H(s)K_{PD}K_{VCO}/N}{s}} \tag{10.2}$$

10.1.2 PLL LOOP CHARACTERIZATION

The general behavior of a PLL circuit can typically be described using the two parameters known as the type and the order.

The PLL type specifies the open loop behavior of the PLL. PLL type is defined as the number of poles at origin in the PLL circuit open loop transfer function as

$$TF_{Open\_loop} = H(s)K_{PD}K_{VCO}/s \tag{10.3}$$

The PLL order, on the other hand, is defined by the number of poles in its closed-loop transfer function as

$$TF_{Close\_Loop} \equiv \frac{\phi_o}{\phi_i} = \frac{N}{M} \frac{\dfrac{H(s)K_{PD}K_{VCO}/N}{s}}{1 + \dfrac{H(s)K_{PD}K_{VCO}/N}{s}} \tag{10.4}$$

The steady state tracking error of the PLL can be determined by the PLL type employing the final value theorem as

$$\phi(t)|_{t\to\infty} \equiv (s \cdot \phi_e(s))|_{s\to 0}$$

$$= (\frac{1}{1+\dfrac{H(s)K_{PD}K_{VCO}/N}{s}} \frac{\phi_i}{M} \cdot s)|_{s\to 0} \propto (s^{(Type+1)} \cdot \phi_i(s))|_{s\to 0}$$

$$(10.5)$$

Note that PLL type directly impacts the tracking performance of the PLL circuits with respect to the input phase changes:

• For step phase input:

$$\phi_i(t) = \begin{cases} 0 & t < 0 \\ A & t \geq 0 \end{cases}$$

$$(10.6)$$

$$\phi_i(s) = \frac{A}{s}$$

$$(10.7)$$

$$\phi_e(t)|_{t\to\infty} \propto s^{Type}|_{s\to 0}$$

$$(10.8)$$

It can be seen that for PLL to track the step phase input signal, the type of the PLL is required to be equal or higher than 1.

• For step velocity phase input:

$$\phi_i(t) = \begin{cases} 0 & t < 0 \\ At & t \geq 0 \end{cases}$$

$$(10.9)$$

$$\phi_i(s) = \frac{A}{s^2}$$

$$(10.10)$$

$$\phi_e(t)|_{t\to\infty} \propto s^{Type-1}|_{s\to 0}$$

$$(10.11)$$

It can be seen that for PLL to track the velocity phase input signal, the type of the PLL is required to be equal or higher than 2.

• For step acceleration phase input, we have that

$$\phi_i(t) = \begin{cases} 0 & t < 0 \\ At^2 & t \geq 0 \end{cases} \qquad (10.12)$$

$$\phi_i(s) = \frac{A}{s^3} \qquad (10.13)$$

$$\phi_e(t)\big|_{t \to \infty} \propto s^{Type-2}\big|_{s \to 0} \qquad (10.14)$$

It can be seen that for PLL to track the acceleration phase input signal, the type of the PLL is required to be equal or higher than 3.

Generally speaking, for PLL to track a high order phase input signal, the type of the PLL is required to be one order higher than the phase input signals.

The locations of the PLL closed-loop poles as shown in figure 10.6 directly impact the stability and phase signal transfer function of the PLL circuit. It can be derived that a PLL is stable if, and only if, all poles of the closed-loop PLL transfer function are within the left side of the s-plane.

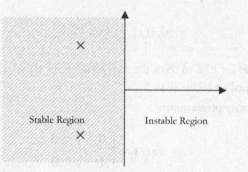

Fig. 10.6. Stability criteria of the PLL circuit

It can be seen that for second-order PLL, the relative locations of the complex conjugate poles directly impact the quality factor Q (or damping factor ζ) of the PLL circuit. And therefore, it directly impacts the jitter transfer function of the PLL circuits.

10.1.3 FIRST-ORDER PLL

The SFG of a first-order PLL is shown in figure 10.7. Such PLL is characterized by a zeroth-order loop filter using a scaling (gain) element.

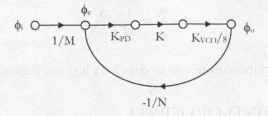

Fig. 10.7. SFG of first-order PLL

First-order PLL circuit can be normalized by introducing a PLL loop bandwidth ω_o and the normalization parameters as

$$\begin{cases} \phi_e \equiv \phi_i / M - \phi_o / N \\ S \equiv \dfrac{s}{\omega_o} \\ \omega_o \equiv K_{PD} \cdot K \cdot K_{VCO} / N \end{cases}$$

(10.15)

The prototype s-domain transfer function based on the normalized first-order PLL circuit can be expressed as

$$H(S) = \frac{\phi_o}{\phi_i} = \frac{N}{M} \frac{1}{1+S}$$

(10.16)

The prototype SFG and the Bode plot of the normalized PLL are shown as figure 10.8.

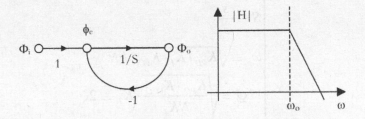

Fig. 10.8. Normalized SFG model of the first-order PLL

It can be seen that a first-order PLL serves as a phase domain first-order lowpass filter where the bandwidth of the PLL is determined by the loop gain of the PLL circuit as given in equation 10.15.

The phase transfer function of the first-order PLL is given as

$$\frac{\phi_e}{\phi_i} = \frac{1}{M}\frac{S}{1+S}$$

$$(10.17)$$

It can be seen that the phase transfer function has a highpass frequency response.

10.1.4 SECOND-ORDER PLL

A typical second-order VLSI PLL circuit typically employs a first-order loop filter consisting of a proportional gain term and the integral gain term as shown in figure 10.9.

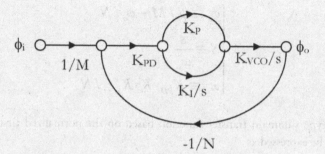

Fig. 10.9. SFG of typical second-order PLL circuit

The second-order PLL can be normalized based on the SFG manipulation as shown in figure 10.10 by introducing a set of the normalized parameters as

$$
\begin{cases}
\phi_e \equiv \phi_i / M - \phi_o / N \\[2mm]
S \equiv \dfrac{s}{\omega_o} \\[2mm]
\omega_o \equiv \sqrt{K_{PD} \cdot K_I \cdot K_{VCO} / N} \\[2mm]
1/Q \equiv \sqrt{\dfrac{K_{PD} \cdot K_{VCO}}{NK_I}} \cdot K_P \equiv 2\zeta
\end{cases}
$$

$$(10.18)$$

258

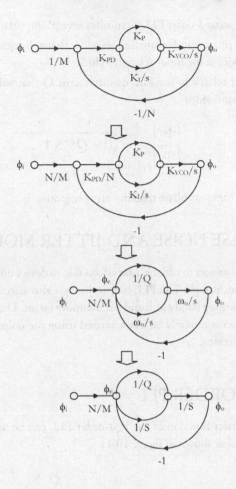

Fig. 10.10. Normalization of second-order PLL SFG

The prototype input to output and the tracking phase transfer functions of second-order PLL can be expressed respectively as

$$\frac{\phi_o}{\phi_i} = \frac{N}{M} \frac{\frac{1}{Q}S + 1}{S^2 + \frac{1}{Q}S + 1}$$

$$(10.19)$$

$$\frac{\phi_e}{\phi_i} \equiv \frac{\phi_o - \phi_i}{\phi_i} = \frac{1}{M} \frac{S^2}{S^2 + \frac{1}{Q}S + 1}$$

$$(10.20)$$

It can be seen that a second-order PLL circuit has several important properties:

- A lowpass input phase transfer function response that will reject the high frequency jitter contents within the input reference clock

- A peaking effect related to the PLL quality factor Q that will amplify the phase noise near the bandwidth

$$\left| \frac{\Phi_o}{\Phi_i} \right|_{S=1} = \sqrt{1 + Q^2} > 1$$

(10.21)

- A second-order highpass phase tracking error response

10.2 PLL PHASE NOISE AND JITTER MODELS

VLSI PLL circuits are subject to various jitter effects due to device noise, supply/substrate noises, and clock feedthrough. The PLL output jitter is also affected by the PLL loop parameters such as the loop bandwidth and the damping factor. The jitter performances of the PLL circuits can commonly be characterized using the noise source models and the jitter transfer functions.

10.2.1 FIRST-ORDER PLL

The phase noise transfer function of the first-order PLL can be derived based on the prototype SFG model as shown in figure 10.11

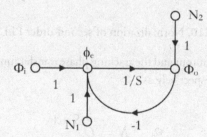

Fig. 10.11. Normalized noise SFG of first-order PLL

where the N_1 noise term is mainly contributed from noise from the input signal path, noise of the input (1/M) and the feedback (1/N) dividers, and noise of loop filter.

The N_2 term is mainly contributed from the phase noise of the VCO circuit. The noise transfer functions with respect to the noise source are given as

$$\Phi_o \big|_{\Phi_i=0} = \frac{1}{1+S} N_1 + \frac{S}{1+S} N_2$$

(10.22)

It can be seen from the plot shown in figure 10.12 that the noise transfer function for N_1 has a lowpass frequency response and the noise transfer function for N_2 has a highpass frequency response. As a result, only low-frequency noise in the N_1 term is critical to the PLL operation. On the other hand, for N_2 noise term, only high-frequency noise will impact the PLL performance.

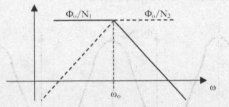

Fig. 10.12. Responses of first-order PLL normalized noise transfer functions

The phase tracking error ϕ_e due to the two noise sources can be expressed as

$$f_e \big|_{\Phi_i=0} = \frac{S}{1+S} N_1 + \frac{S}{1+S} N_2$$

(10.23)

10.2.2 SECOND-ORDER PLL

The noise transfer functions of second-order PLL prototype circuit can be expressed based on three major noise sources as shown in figure 10.13.

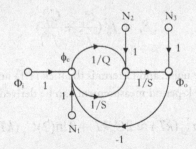

Fig. 10.13. Normalized noise SFG of second-order PLL

where N_1 represents the noise in the input signal path, the noise of the input (1/M), and the feedback (1/N) divider circuits. Phase noise source N_2 represents noise in the PLL loop filter circuit. Phase noise source N_3 represents the phase noise of VCO circuit.

The normalized noise transfer functions of the second-order prototype PLL circuit can be expressed as

$$\Phi_o = \frac{1}{S^2 + \frac{1}{Q}S + 1}N_1 + \frac{S}{S^2 + \frac{1}{Q}S + 1}N_2 + \frac{S^2}{S^2 + \frac{1}{Q}S + 1}N_3$$

(10.24)

It can be seen that the noise transfer functions have the lowpass, bandpass, and highpass responses respectively as shown in figure 10.14.

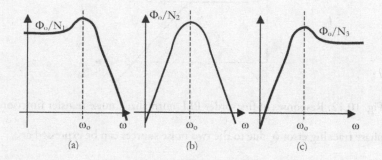

Fig. 10.14. Noise transfer functions of second-order PLL circuit

10.2.3 PLL PHASE NOISE AND JITTER MODELS

The internal phase noises of a free-running VCO circuit are mainly contributed from the white frequency noise and the flicker noise with the phase noise spectrum density given in the form as

$$S_f \approx f_o^2 \left(\frac{C_w}{f^2} + \frac{C_f}{f^3}\right)$$

(10.25)

The C_w term is the white noise, and C_f term is the flick (1/f) noise. The variance of the adjacent period jitter for k-period measurement can be derived as

$$\sigma_{\Delta J}^2(kT) \approx 2C_w kT + 8\ln(2)C_f(kT)^2$$

(10.26)

Specifically, the period jitter (PJ) due to the white noise can be expressed as

$$s_J \approx \sqrt{C_w T} \equiv k\sqrt{T}$$

(10.27)

10.2.3 PLL Phase Noise and Jitter Models

For a first-order PLL with loop bandwidth ω_o and transfer function given as

$$H(s) = \frac{1}{s/\omega_o + 1},$$

(10.28)

AJ, PJ, and CCJ due to VCO white noise can be respectively expressed as

$$\sigma_A^2 \approx \frac{C_w}{2\omega_o}$$

(10.29)

$$\sigma_J^2(kT) \approx 2\sigma_A^2\{1 - e^{-\omega_o kT}\}$$

(10.30)

$$\sigma_{\Delta J}^2(kT) \approx 6\sigma_A^2\{1 - \frac{4}{3}e^{-\omega_o kT} + \frac{1}{3}e^{-2\omega_o kT}\}$$

(10.31)

For the second-order PLL with natural frequency ω_o, damping factor ζ, and the closed loop transfer function as

$$H(s) = \frac{2\varsigma(s/\omega_o) + 1}{(s/\omega_o)^2 + 2\varsigma(s/\omega_o) + 1}$$

(10.32)

The white-noise- and flicker-noise-induced jitters are given as

$$\sigma_A^2 \approx \frac{C_w}{4\varsigma\omega_o} + \frac{C_f}{\omega_o^2}f(\varsigma)$$

(10.33)

where $f(\zeta)$ is a monotonic, decreasing function of ζ.

For PLL with maximally flat closed loop transfer function, $\zeta = 0.707$, the white-noise-induced jitters can be expressed as

$$\sigma_A^2 \approx \frac{\sqrt{2}C_w}{4\omega_o}$$

(10.34)

$$\sigma_J^2(kT) \approx 2\sigma_A^2\{1 - \sqrt{2}e^{-\frac{\omega_o kT}{\sqrt{2}}}\cos(\frac{\pi}{4} + \frac{\omega_o kT}{\sqrt{2}})\}$$

(10.35)

The flicker-noise-induced jitter of second-order PLL are given as

$$\sigma_A^2 \approx \frac{\pi C_F}{2\omega_o^2}$$

(10.36)

$$\sigma_J^2(kT) \approx 2\sigma_A^2 \begin{cases} 0.6(\omega_o kT) & \omega_o kT < 2 \\ 1 & \omega_o kT > 2 \end{cases}$$

(10.37)

10.3 VLSI PLL CIRCUIT IMPLEMENTATION

Second-order PLL circuits are the most commonly used VLSI PLL circuit structures in VLSI high-speed I/O circuits. A typical VLSI second-order PLL circuit can be constructed using a first-order loop filter employing a proportional path and an integration path as shown in figure 10.15.

Key design parameters of the loop filter circuit for second-order PLL are the proportional loop filter gain factor K_P and the integral gain factor K_I of the loop filter, the phase detection gain K_{PD}, the VCO gain K_{VCO}, and the divider ratios. Major design tasks of the VLSI PLL circuit include

- selection of proper VCO circuit structure and device sizing for proper frequency tuning range, VCO gain, and noise performance;
- selection of charge pump current IB for desired phase detection gain;
- selection of proper dividers for desired input to output frequency ratio;
- implementation of PFD circuit;
- selection of proper loop filter circuit for desired loop bandwidth and quality (damping) factor; and
- optimization of PLL circuit for desired jitter performance, power dissipation, and layout area.

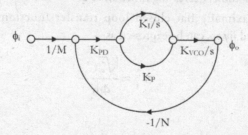

Fig. 10.15. SFG model of VLSI second-order PLL circuits

10.3.1 PHASE FREQUENCY DETECTOR

Clock phase (or phase frequency) detector circuits are widely used in the VLSI PLL circuit to extract the phase difference between the input reference and the feedback clock as shown in figure 10.16.

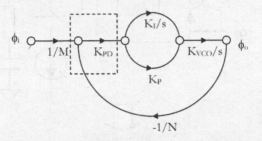

Fig. 10.16. SFG model of phase detector circuit

The phase-to-voltage (current) conversion and scaling operation of the VLSI phase detector circuits are modeled using the detection gain K_{PD}:

$$\overline{V}_o = K_{PD} \cdot \Delta\phi \qquad (10.38)$$

Shown in figure 10.17 is a phase detection circuit implementation for VLSI charge pump-based PLL circuits.

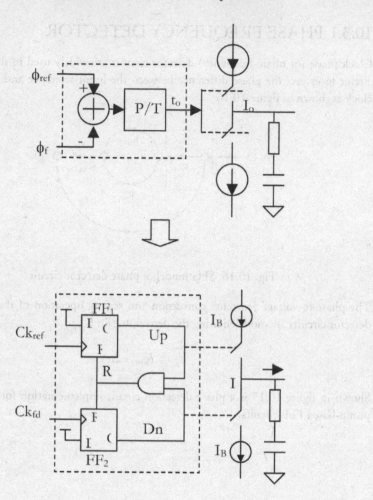

Fig. 10.17. VLSI phase detector circuit implementation for charge pump-based PLL

The PFD gain parameters of the circuit can be calculated based on the bias current as

$$\bar{I} = \frac{I_B}{2\pi} \phi_i \quad (-2\pi \leq \phi_i < 2\pi)$$

(10.39)

$$K_{PD} = \frac{I_B}{2\pi} \quad (-2\pi \leq \phi_i < 2\pi)$$

(10.40)

Shown in figure 10.18 is a typical VLSI PFD implementation employing NAND logic gates.

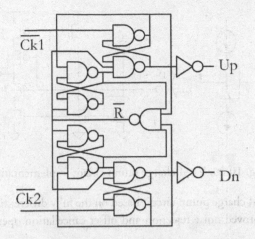

Fig. 10.18. VLSI phase frequency detector circuit implementation

10.3.2 CHARGE PUMP AND LOOP FILTER

Charge pump and loop filter circuit in a second-order VLSI PLL are usually implemented together with the phase frequency detector circuit. Such circuit structures provide seamless integration of VLSI signal processing circuit operations for the signal format conversion, the switching noise filtering, and the loop frequency response compensation. Shown in figure 10.19 is the SFG model of a typical charge pump/loop filter circuit for the second-order charge pump-based VLSI PLL circuits.

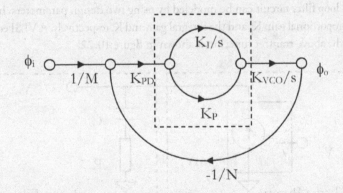

Fig. 10.19. SFG model of charge pump and first-order loop filter circuits

A VLSI circuit implementation of charge-pump circuit is shown in figure 10.20, such circuit realizes the time-to-current (T/I) conversion for the phase detector circuit.

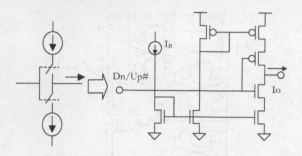

Fig. 10.20. VLSI charge pump circuit implementation

An enhanced VLSI charge pump circuit based on the fully differential circuit structure that provides improved noise rejection and offset cancellation operation is shown in figure 10.21.

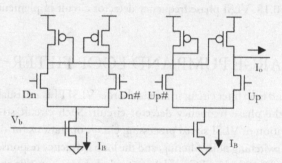

Fig. 10.21. Fully differential VLSI charge pump circuit implementation

The VLSI loop filter circuit can be modeled by using two design parameters, including the filter proportional gain K_p and the integral gain and K_I respectively. A VLSI equivalent circuit of the above transfer function is shown in figure 10.22.

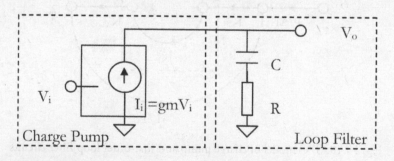

Fig. 10.22. VLSI passive RC loop filter circuit structure

The frequency domain transfer function of such circuit can be expressed as

$$H(s) \equiv \frac{V_o}{I_i} \equiv (K_P + \frac{K_I}{s}) = (R + \frac{1}{Cs})$$

(10.41)

$$\Rightarrow \begin{cases} K_P = R \\ K_I = 1/C \end{cases}$$

(10.42)

Two alternative VLSI loop filter circuit implementations that can be used to minimize the charge injection effects are shown in figure 10.23.

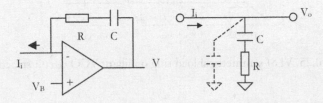

(a) Active-RC loop filter circuit   (b) Charge injection glitch filtering

Fig. 10.23. Circuit enhancements of VLSI loop filter circuits

10.3.3 VCO CIRCUIT

VLSI voltage-controlled oscillator (VCO) circuit is used for clock regeneration operation. The s-domain SFG model of the VCO circuit is shown in figure 10.24.

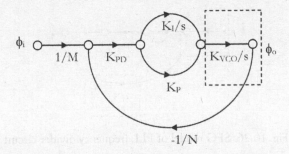

Fig. 10.24. S-domain SFG model of the VLSI VCO circuits

The s-domain model of VLSI VCO circuit is given as

$$\frac{\phi_o(s)}{V_i(s)} = \frac{K_{VCO}}{s}$$

(10.43)

The VCO gain K_{vco} can be derived based on the VCO frequency tuning curve from the implemented VCO circuit. One commonly used VLSI VCO circuit structure is the ring oscillator-based VLSI VCO circuit employing the MOS symmetrical active load and replica biasing technique as shown in figure 10.25.

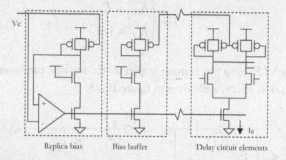

Fig. 10.25. VLSI symmetrical load ring oscillator VCO circuit structure

10.3.4 FREQUENCY DIVIDER

Frequency dividers are used in the VLSI PLL circuits for the linear phase and frequency scaling signal processing operations:

$$f_o = \frac{N}{M} f_{ref}$$

(10.44)

The SFG model of PLL frequency divider circuit is shown in figure 10.26.

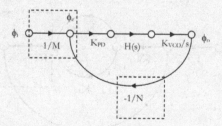

Fig. 10.26. SFG model of PLL frequency divider circuit

Binary frequency dividers are the most commonly used VLSI frequency divider circuit structures in PLL circuit applications. Such types of VLSI circuits offer the phase or frequency scaling operation with scaling factor given as the power of 2.

$$f_o = \frac{1}{2^K} f_{ref}$$

(10.45)

The binary frequency divider can easily be constructed using D-FF circuits that have the inverted feedback as shown in figure 10.27. Multiple stages of same divide-by-2 VLSI circuit structures can be used to realize high-scaling ratios.

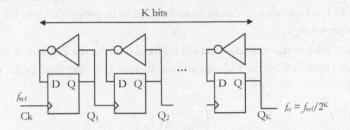

Fig. 10.27. VLSI divided by 2K circuit structure

10.4 VLSI ADAPTIVE BANDWIDTH PLL CIRCUITS

For conventional PLL architectures as shown in figure 10.28, K_{PD}, K_I, K_P and K_{VCO} are selected to be constants independent of the input clock frequency. Such PLL circuit has fixed ω_n and ζ are independent of ω_{ref}.

$$\omega_o \equiv \sqrt{K_{PD}K_I K_{VCO}/N}$$

$$1/Q \equiv 2\zeta = \frac{K_P}{K_I}\sqrt{K_{PD}K_I K_{VCO}/N}$$

$$(10.46)$$

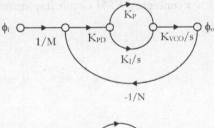

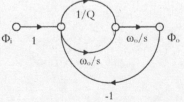

Fig. 10.28. The SFG and normalized SFG of second-order PLL circuits

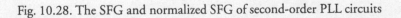

Fixed bandwidth PLL circuit usually has narrower frequency operation range that may suffer from performance degradation when the operation frequency is deviated from its optimal operation frequency. Such issue can be resolved by using the adaptive bandwidth PLL circuit. In the adaptive bandwidth PLL circuit approach, the Q (or ζ) factor of PLL is kept constant, while ω_o is designed to be proportional to the reference clock frequency ω_{ref}.

One way to achieve the adaptive bandwidth PLL is to make the PLL circuit parameters adaptively controlled by the PLL circuit bias current I_B as specified in the following relations:

$$\begin{cases} \omega_{ref} \propto \left(I_B\right)^{0.5} \\ K_{PD} \propto \left(I_B\right)^{1} \\ K_{VCO} \propto \left(I_B\right)^{0} \\ K_P \propto \left(I_B\right)^{-0.5} \\ K_I \propto \left(I_B\right)^{0} \end{cases}$$

(10.47)

Based on the above constraints, the nature frequency and the quality factor Q of the PLL are given respectively as

$$\begin{cases} \omega_o \equiv \sqrt{K_{PD}K_I K_{VCO} / N} \propto \left(I_B\right)^{0.5} \propto \omega_{ref} \\ 1/Q \equiv 2\zeta = \dfrac{K_P}{K_I}\sqrt{K_{PD}K_I K_{VCO} / N} \propto \left(I_B\right)^{0} \end{cases}$$

(10.48)

Shown in figure 10.29 is a conceptual VLSI circuit implementation of the adaptive bandwidth PLL circuit.

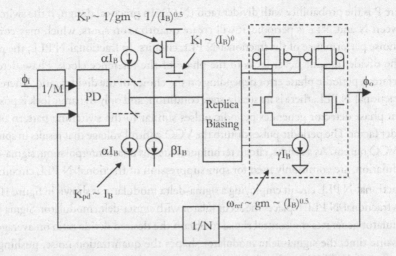

$$K_P \sim 1/gm \sim 1/(I_B)^{0.5}$$
$$K_I \sim (I_B)^0$$

Fig. 10.29. Adaptive bandwidth PLL circuit architecture

10.5 VLSI FRACTIONAL-N PLL CIRCUITS

Integer-N PLL circuits have output clocks with frequencies the integer multiple of the reference clock frequency. The resolutions of the integer-N PLL circuit for frequency synthesis applications are typically limited to the reference frequency. In practical design, the loop bandwidth of the PLL is typically set to a factor of 10, lower than the reference clock frequency to ensure loop stability. Lower reference frequency and PLL bandwidth are required if better resolutions are required. This has been the fundamental disadvantage of the integer-N PLL circuits. Low PLL bandwidth may cause significant increase in the lock time. Furthermore, since PLL has a highpass response for the VCO noise, reduction in the bandwidth causes inadequate suppression of the VCO phase noise. Therefore, a wide loop bandwidth for the PLL is desired to improve lock time and the noise suppression. For an integer-N PLL circuit, one of the design trade-offs is between the frequency resolution and the performance in noise and lock time.

The fractional-N PLLs that have output frequencies the fractions of the reference frequency can be used to overcome the shortcomings of the integer-N PLL circuits by providing finer frequency resolution than reference frequency without a decrease in loop bandwidth. In addition, in fractional-N PLL circuit, a dual modulus is usually used and the divider ratio of the PLL alternates between N and N+1 by selecting the appropriate modulus divider. The divider ratio is dynamically controlled by a programmable accumulator, which creates a periodic switching pattern. Under the condition of the overflow of the accumulator, the divider ratio switches from N to N+1. The average divide ratio is given as

$$N_{effective} = N + P(N+1) \tag{10.49}$$

where P is the probability with divider ratio (N+1). In practical design, if the switching between N and N+1 is periodic, it will create quantization spurs, which may corrupt the noise performance of the fractional-N PLL circuits. In fractional-N PLL, the phase of the divider output never locks to the phase of the reference clock. Phase detector generates a periodic phase error depending on the choice of the divider ratio. Therefore, in fractional-N PLL, there is no steady-state condition, and only dynamic lock is possible when phase detector generates periodic pulses similar to the switching pattern of the divider factor. The periodic pulses disturb the VCO control voltage that results in spurs in the VCO output. As a results, circuit techniques, such as phase interpolation, sigma-delta modulation, are commonly used for spur suppression of fractional-N PLL circuits.

A fractional-N PLL circuit employing a sigma-delta modulator is shown in figure 10.30. This fractional-N PLL replaces the accumulator with sigma-delta modulator. Sigma delta modulator generates the control signal to obtain the desired divide ratio on average. At the same time, the sigma-delta modulator shapes the quantization noise, pushing the spurs out of the PLL loop filter bandwidth. Note that the performance improvement of the sigma-delta fractional-N PLL comes with two main drawbacks: First, the PLL loop bandwidth has to be restricted in order to have suppression of increased high-frequency noise due to sigma-delta noise shaping. Second, a higher linearity is required for some of the critical blocks in PLL to avoid any spur generation.

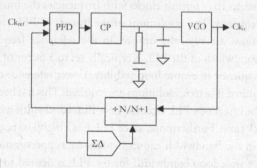

Fig. 10.30. Fractional-N PLL based on sigma-delta modulator

10.6 SPREAD SPECTRUM CLOCKING (SSC)

Spread spectrum clocking (SSC) techniques are widely used in VLSI high-speed I/O circuits to reduce electromagnetic interference (EMI) effects that are usually associated with high current switching of high-speed data, especially multiple transmitter data transmissions. In high-speed I/O circuits, the transmitted data stream itself contains energy of its synchronous clock. When the clock is pure and stable, a sharp and strong peak of spectrum can cause strong EMI radiation around the electronic devices. EMI effects can be minimized by introducing slow wander within a certain amount of frequency region known as spread spectrum clocking (SSC) as shown in figure 10.31.

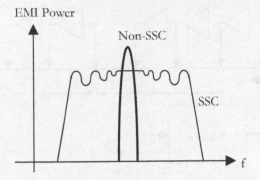

Fig.10.31. SSC technique

There are three typically used SSC circuit techniques, including the frequency divider modulation, the VCO modulation and the multiphase clock combination and digital processing. Shown in figures 10.32, 10.33, and 10.34 are VLSI SSC generator circuit examples of these circuit techniques.

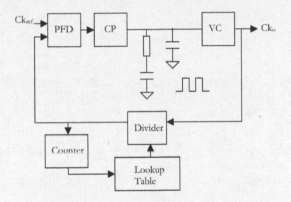

Fig. 10.32. Divider modulation SSC circuit

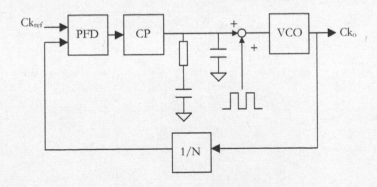

Fig. 10.33. VCO modulation SSC circuit

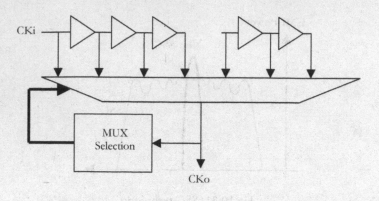

Fig. 10.34. multiphase clock combination SCC circuit

References:

[1]   G. C. Hsieh, et al, "Phase-Locked Loop Techniques—A Survey," IEEE Transactions on Industrial Electronics, Vol.43, No.6, Dec.1996.

[2]   P. Heydari, et al, "Analysis of the PLL Jitter Due to Power/Ground and Substrate Noise," IEEE Transactions on Circuits and Systems—I: Regular Papers, Vol 51, No. 12, Dec. 2004.

[3]   T. M. Almeida, et al, "High Performance Analog and Digital PLL Design," IEEE 1999.

[4]   J. Kim, et, al, "Design of CMOS Adaptive-Bandwidth PLL/DLLs: A Generation Approach," IEEE Trans. Circuits and System—II: Analog And Digital Signal Processing, Vol.50, 11, November, 2003.

[5]   M. Kozak, et, al, "Rigorous Analysis of Delta-Sigma Modulator for Fractional-N PLL Frequency Synthesis," IEEE Trans. Circuits and Systems, regular papers, Vol.50, No. 6, June, 2004.

[6]   K. Woo, et, al, "Fast-Lock Hybrid PLL Combining Fractional-N and Integer-N Modes of Differing Bandwidth," IEEE JSSC, Vol.43, No.2, February, 2008.

[7]   G. Nash, "Phase-Locked Loop Design Fundamentals," Application Note AN535, Motorola, 1994.

[8]   "Phase Noise" Application Note, CX72300/1/2, Conextant, 2001.

[9]   J. G. Maneatis, "Low-Jitter Process-Independent DLL and PLL Based on Self-Biased Techniques," IEEE, JSSC, Vol. 31, No.11, Jan. 1996.

[10]   V. F. Kroupa, "Noise Properties of PLL Systems," IEEE Transactions on Communications, Vol COM-30, No. 10, Oct. 1982.

[11]   A. Maxim, "A 0.16-2.55-Ghz CMOS Active Clock De-Skewing PLL Using Analog Phase Interpolation," IEEE, JSSC, Vol. 40, No.1, Jan. 2005.

[12]   M. Inoue, et al, "Over-Sampling PLL for Low-Jitter and Responsive Clock Synchronization," IEEE2006.

[13]   M. Kokubo, et al, "Spread-Spectrum Clock Generator for Serial ATA using Fractional PLL Controlled by DS Modulator with Level Shifter," ISSS2005.

[14]   K. B. Hardin, et al, "Investigation into the Interference Potential of Spread Spectrum Clock Generation to Broadband Digital Communication," IEEE Trans. Electromagnetic Compatibility, Vol. 45, No.1, 2003.

[15]   Y. Lee, at el, "Electromagnetic Interference Mitigation by Using a Spread-Spectrum Approach," IEEE Trans. Electromagnetic Compatibility, Vol. 44, No.2, 2002.

[16]   H. H. Chang, et, al, "A Spread Spectrum Clock Generator with Triangular Modulation," IEEE JSSC, Vol.38, No.4, April, 2003.

[17]   M. Badaroglu, et, al, "Digital Ground Bounce Reduction by Supply Current Shaping and Clock Frequency Modulation," IEEE Trans. Computer-Aided Design of Integrated Circuits and Systems, Vol.24, No. 1, January 2005.

[18]   H. R. Lee, et, al, "A Low Jitter 5000ppm Spread Spectrum Clock Generator for Multi-Channel SATA Transceiver in 0.18um CMOS," IEEE ISSCC 2005

[19] J. Kim, et, al, "Spread Spectrum Clock Generator with Delay Cell Array to Reduce Electromagnetic Interference," IEEE Trans. Electromagnetic Compatibility, Vol. 47, No.4, 2005.

[20] S. Damphousse, et, al, "All Digital Spread Spectrum Clock Generator for EMI Reduction," IEEE JSSC, Vol.42, No.1, January 2007.

[21] C. Barrett, "Fractional/Integer-N PLL Basics," Technical Brief SWRA029, Texas Instrument, 1999.

[22] M. H. Perrott, et, al, "A Modeling Approach for sigma-Delta Fractional-N Frequency Synthesizers Allowing Straightforward Noise Analysis," IEEE JSSC Vol. 37, No.8, August 2002.

[23] G. Kolumban, et, al, "Frequency Domain Analysis of Double Sampling Phase-Locked Loop," ISCAS2000.

[24] A. Mehrotra, "Noise Analysis of Phase-Locked Loops," IEEE2000.

[25] K. Ware, et, al, "A 200 Mhz CMOS Phase-Locked Loop with Dual Phase Detector," IEEE JSSC, Vol. 24, No. 6, December, 1989.

[26] J. G. Maneatis, et, al, "Self-Biased High-Bandwidth, Low-Jitter 1-to-4096 Multiplier Clock-Generator PLL," ISSCC 2003.

[27] C. Zhang, et, al, "An Experimental Study of Phase Noise in CMOS Phase-Locked Loops Considering Different Noise Sources," IEEE2006.

[28] J. Kim, et, al, "A 20-GHz Phase-Locked Loop for 40-Gb/s Serializing Transmitter in 0.13um CMOS," IEEE JSSC, Vol. 41, No. 4, April 2006.

[29] D. C. Lee, "Analysis of Jitter in Phase Locked Loops," IEEE Trans. Circuits and System-II: Analog and Digital Signal Processing, Vol. 49, No.11, Nov. 2002.

CHAPTER 11
VLSI DELAY-LOCKED LOOP CIRCUITS

VLSI delay-locked loop (DLL) circuits can be viewed as a special family of VLSI phase-locked loop (PLL) circuits that are constructed based on the VLSI VCDL circuits. The feedback loop in the DLL circuit is used to ensure the absolute delay time of the VCDL is adaptively matched to a reference independent of the PVT conditions.

Typical DLL circuit applications in VLSI high-speed I/O circuits include clock distribution delay compensation, datapath delay compensation, clock phase manipulation, multiphase clock generation, and clock multiplication.

VLSI DLL circuits, in principle, can be used almost where the VLSI PLL circuits are used. However, DLL circuits are more popular and beneficial for these applications that only require the phase or delay adjustment and not for frequency synthesis. There has been a great deal of interests in various DLL circuits in the recent years for data and clock signal delay compensation in VLSI high-speed I/O circuit applications. In these applications, DLL circuits are used as alternative to PLL circuits to provide precise spaced timing edge even in the presence of PVT variations. DLL circuits are preferred circuits in those applications because of their unconditional stability, lower phase-error accumulation effects, and fast locking times.

11.1 DLL CIRCUIT ARCHITECTURES

A basic VLSI DLL circuit structure is shown in figure 11.1. It consists of a voltage-controlled delay line (VCDL), a phase detector (PD), and a loop filter (LPF).

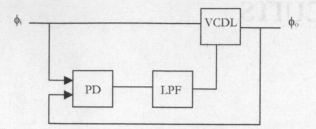

Fig. 11.1. DLL circuit architecture

In the basic VLSI DLL circuit, the DLL loop ensures the output signal phase ϕ_o of the VCLD is phase aligned with the input signal phase ϕ_i. In many high-speed I/O applications, the master-slaver DLL (MS-DLL) circuit configurations that employ two or more VCDL circuits based on replica delay circuit elements as shown in figure 11.2 are commonly used.

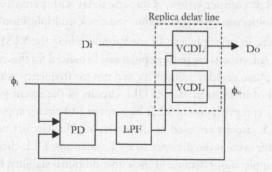

Fig. 11.2. Master-slaver DLL circuit architecture

In the MS-DLL circuit, both periodic and nonperiodic signals can be used as the reference signal of the DLL circuit. Since the master and slaver VCDL circuits in the DLL circuits share the same control signal and the slaver VCDL circuit is a replica of the master VCDL circuit, the slaver VCDL delay is compensated independent of the PVT conditions when the master VLDL circuit loop is locked to the reference signal.

In the MS-DLL circuit structure shown in figure 11.3, the master and slaver VCDL can be selected to have different delay line lengths. As a result, the delay of the slaver VCDL is a scaled version of the master VCDL delay as

$$\begin{cases} \phi_o - \phi_i = T \\ Y - X = \dfrac{M}{N}T \end{cases}$$

$$(11.1),$$

where T is the total delay of the master VCDL delay time of the DLL circuit.

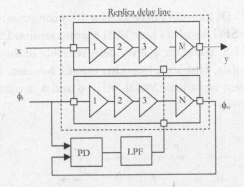

Fig. 11.3. Delay synthesis DLL circuit architecture

In the practical VLSI circuit applications, the MS-DLL circuits are commonly used for delay synthesis operation, where the control loop of DLL circuit can be realized in either digital or analog circuit forms. Shown in figure 11.4 is a dual loop DLL circuit architecture that is based on an analog DLL core circuit and a fully digital second-level DLL circuit.

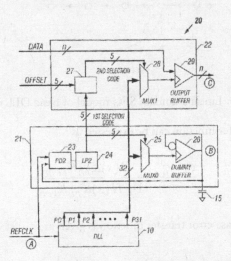

Fig. 11.4. Dual loop master-slaver DLL architecture

In this DLL circuit structure, the analog DLL core circuit is used to generate accurate delay reference, and the digital DLL circuit is used to control the delay times for all slaver delay lines. The digital DLL circuit includes a master DLL structure for delay reference, and multiple-slaver DLL consists of multiple replica slaver delay lines for signal- or clock-delay control. Programmable offset can be introduced into the slaver delay line in such DLL circuit structure.

11.2 DLL CIRCUIT S-DOMAIN AND SFG MODELS

VLSI high-speed I/O DLL circuits can be used for delay compensation of both the data or clock signals. The SFG model of a basic VLSI continuous-time DLL circuit structure is shown in figure 11.5. Where the ϕ_r is the delay reference of the DLL circuit, K_{PD} is the phase detection gain, H(s) is the loop filter transfer function, K_{VCDL} is the VCDL gain, and ϕ_s is the zero input static VCDL delay. ϕ_i and ϕ_o are the input and output delay phases respectively.

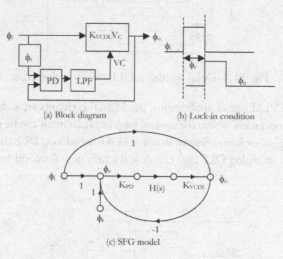

(a) Block diagram   (b) Lock-in condition

(c) SFG model

Fig. 11.5. Linear s-domain SFG model of basic DLL circuit

The DLL phase transfer function can be expressed as

$$\frac{\phi_o - \phi_i}{\phi_r} = \frac{H(s)K_{PD}K_{VCDL}}{1 + H(s)K_{PD}K_{VCDL}} \qquad (11.2)$$

The DLL tracking phase error transfer function is given as

$$\frac{\phi_e}{\phi_r} \equiv \frac{\phi_o - \phi_i - \phi_r}{\phi_r} = \frac{1}{1 + H(s)K_{PD}K_{VCDL}} \qquad (11.3)$$

The following can be seen based on equations 11.2 and 11.3:

- The input to the output phase transfer function represents an all-pass response, implying the input to output passing-through behavioral of the basic DLL circuit. This effect is resulted from the fact that signal (or clock) is not regenerated within PLL circuits.

- The delay reference to output phase transfer function is lowpass transfer function, implying the bandwidth limitation effect of DLL circuits. It can be seen that DLL will not be able to lock to the delay reference change with frequency beyond the DLL bandwidth.

- The static delay phase error transfer function has a highpass transfer function, implying that at high frequency beyond the DLL loop bandwidth signal will pass through the DLL circuit. In practical applications, this term is usually contributed from the PVT corners of the circuit as well as the jitter injection from the supply and other cross talk noises.

- The track phase error transfer function of the DLL circuit is a highpass transfer function.

The SFG model of the master-slaver DLL circuit structures is shown in figure 11.6. In this model, the VCDL gain of the slaver delay line is scaled proportionally from the master delay line considering identical delay elements of different stages are used for the master and slaver delay line. In addition, the static delay phase is also scaled to reflect the delay stage number differences.

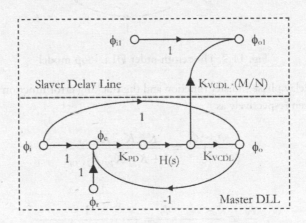

Fig. 11.6. SFG model of linear s-domain master-slaver DLL circuit

Based on the MS-DLL circuit model, the input to output delay transfer function of the slaver delay line can be expressed as

$$\frac{(\phi_{o1} - \phi_{i1})}{\phi_r} = \frac{M}{N}\frac{(\phi_o - \phi_i)}{\phi_r} = \frac{M}{N}\left(\frac{H(s)K_{PD}K_{VCDL}}{1 + H(s)K_{PD}K_{VCDL}}\right) \tag{11.4}$$

Such a transfer function represents a lowpass transfer response for the slaver line delay with respect to delay time reference, implying the slaver delay line will be locked to a ratio of the reference delay value within the DLL loop bandwidth and signal will pass through with static delay at high frequency:

$$\frac{(\phi_{o1} - \phi_{i1})}{\phi_r}\Big|_{\omega \to 0} = \frac{M}{N} \tag{11.5}$$

11.2.1 ZEROTH-ORDER DLL

The s-domain model of the zeroth-order DLL loop is shown in figure 11.7

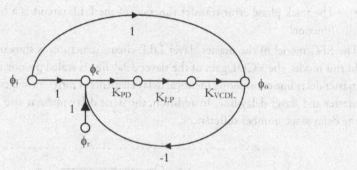

Fig. 11.7. The zeroth-order DLL loop model

The output delay phase transfer function and the tracking error function of such DLL circuit are given respectively as

$$\frac{\phi_o - \phi_i}{\phi_r} = \frac{K_{LP}K_{PD}K_{VCDL}}{1 + K_{LP}K_{PD}K_{VCDL}} \tag{11.6}$$

$$\frac{\phi_e}{\phi_r} \equiv \frac{\phi_o - \phi_i - \phi_r}{\phi_r} = \frac{1}{1 + K_{LP}K_{PD}K_{VCDL}} \tag{11.7}$$

It can be seen that the tracking error is basically limited by the loop gain of the DLL loop as it is inversely proportional to the open loop gain of the DLL.

11.2.2 FIRST-ORDER DLL

An s-domain first-order DLL loop can be realized using a lossless integrator in the DLL loop filter as shown in figure 11.8.

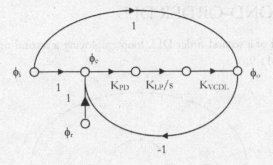

Fig. 11.8. The first-order DLL loop model

The output delay phase transfer function and the tracking error function for the first-order DLL loop are given respectively as

$$\frac{\phi_o - \phi_i}{\phi_r} = \frac{\dfrac{K_{LP}K_{PD}K_{VCDL}}{s}}{1 + \dfrac{K_{LP}K_{PD}K_{VCDL}}{s}}\phi_r \tag{11.8}$$

$$\frac{\phi_e}{\phi_r} \equiv \frac{\phi_o - \phi_i - \phi_r}{\phi_r} = \frac{1}{1 + \dfrac{K_{LP}K_{PD}K_{VCDL}}{s}} \tag{11.9}$$

It can be seen that such DLL circuit type can offer zero static tracking phase error as

$$\frac{\phi_e}{\phi_r}\Big|_{s\to 0} \equiv \frac{(\phi_o - \phi_i - \phi_r)}{\phi_r}\Big|_{s\to 0} = \frac{1}{1 + \dfrac{K_{LP}K_{PD}K_{VCDL}}{s}}\Big|_{s\to 0} = 0 \tag{11.10}$$

By introducing the bandwidth of the first-order DLL loop as ω_o, the tracking phase error frequency response can be expressed as

$$\begin{cases} \omega_o = K_{PD}K_{LP}K_{VCDL} \\ \phi_e = \dfrac{s/\omega_o}{1 + s/\omega_o}\phi_r \end{cases} \tag{11.11}$$

It can be seen that the tracking error has a highpass frequency response with the cutoff frequency equal to the open loop gain.

11.2.3 SECOND-ORDER DLL

The SFG model of a second-order DLL loop employing a second-order loop filter is shown in figure 11.9.

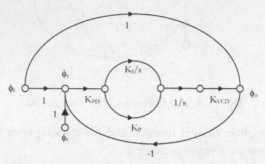

Fig. 11.9. SFG model of second-order DLL loop

The output delay phase transfer function and the tracking error function for the second-order DLL loop are given respectively as

$$\frac{\phi_o - \phi_i}{\phi_r} = \frac{1}{\left(s/\omega_o\right)^2 + \dfrac{1}{Q}\left(s/\omega_o\right) + 1} \tag{11.12}$$

$$\frac{\phi_e}{\phi_r} \equiv \frac{\phi_o - \phi_i + \phi_r}{\phi_r} = \frac{\left(s/\omega_o\right)^2}{\left(s/\omega_o\right)^2 + \dfrac{1}{Q}\left(s/\omega_o\right) + 1} \tag{11.13}$$

where

$$\begin{cases} \omega_o \equiv \sqrt{K_{PD}K_I K_{VCDL}} \\ \dfrac{1}{Q} \equiv K_P \sqrt{K_{PD}K_{VCDL}/K_I} \end{cases} \tag{11.14}$$

are the DLL loop bandwidth and the damping factor. It can be seen that the tracking error has a highpass frequency response. Therefore, the DLL circuit can provide phase tracking for delay signal up to the bandwidth of the DLL circuit.

11.3 DLL PHASE ERRORS AND JITTER MODEL

VLSI DLL circuit output jitter can be modeled based on the input signal noise (N_1), the time reference noise (N_2), the VCDL control voltage noise (N_3), and the VCDL circuit noise (N_4) as shown in figure 11.10.

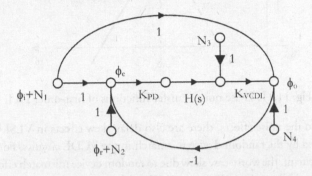

Fig. 11.10. DLL phase noise and jitter SFG model

The phase noise transfer function of the DLL for the above-stated noise sources can be expressed as

$$\phi_o = N_1 + \frac{N_4 + K_{VCDL}N_3}{1 + H(s)K_{PD}K_{VCDL}} + \frac{H(s)K_{PD}K_{VCDL}N_2}{1 + H(s)K_{PD}K_{VCDL}} \qquad (11.15)$$

For the first-order DLL circuit, we have that

$$\phi_o = N_1 + \frac{K_{VCDL}N_3 + N_4}{1 + \dfrac{K_{LP}K_{PD}K_{VCDL}}{s}} + \frac{\dfrac{K_{LP}K_{PD}K_{VCDL}}{s}}{1 + \dfrac{K_{LP}K_{PD}K_{VCDL}}{s}}N_2 \qquad (11.16)$$

It can be seen that if phase noise sources are totally uncorrelated, phase noise N_1 from the input signal source has an all-pass frequency response. It can also be seen that phase noise transfer functions for N_3 and N_4 noise terms have the first-order highpass responses, and phase noise transfer function for N_2 has a first-order lowpass transfer function. Their noise transfer functions are shown in figure 11.11.

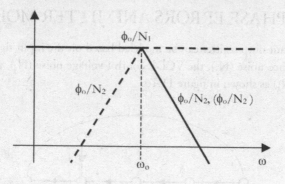

Fig. 11.11. Phase noise transfer functions of first-order DLL

In addition to the noise effects, there are also delay skew effects in VLSI DLL circuit that are caused by the random device mismatch in the VCDL circuits. For an N-stage master DLL circuit, the worst-case skew due to random device mismatch effects typically occurs in the middle of the VCDL circuit. The delay uncertainty of the VCDL circuit within a DLL can be approximately expressed as

$$\left(\Delta t_{RMS}\right)_{N/2} \approx \frac{\sqrt{N}}{2}\left(\Delta t_{RMS}\right)_{cell}$$

(11.17)

It can also be derived that the general phase error at the m-th stage output due to random device mismatch for an N-stage VCDL within a closed loop DLL is approximately given as

$$\left(\Delta t_{RMS}\right)_{m} \approx \sqrt{\frac{N(N-m)}{N}}\left(\Delta t_{RMS}\right)_{cell}$$

(11.18)

On the other hand, for an N-stage slaver VCDL in DLL circuit, the delay uncertainty at the output of the m-th stage of a slaver VCDL is given as

$$\left(\Delta t_{RMS}\right)_{m} \approx \sqrt{m}\left(\Delta t_{RMS}\right)_{cell}$$

(11.19)

The worst delay uncertainty typically occurs at the last stage of the VCDL, and it can be approximately expressed as

$$\left(\Delta t_{RMS}\right)_{M} \approx \sqrt{M}\left(\Delta t_{RMS}\right)_{cell}$$

(11.20)

11.4 VLSI DLL CIRCUIT IMPLEMENTATION

For the first-order VLSI DLL circuit employing a first-order loop filter using an integration as shown figure 11.12, the key design parameters include the phase detector gain K_{PD}, the loop filter integral gain factor K_I, and the VCDL gain K_{VCDL}. Key design tasks of the VLSI DLL circuit are then given as follows:

- Selection of VCDL circuit structure and device sizing for proper delay tuning range, gain, and noise performance
- Selection of proper charge pump current I_B for desired phase detection gain
- Selection of reference delay time
- Implementation of PD circuit
- Selection of proper loop filter circuit for desired loop bandwidth, and quality (damping) factor
- Optimization of DLL circuit for desired jitter performance, power dissipation, and layout area

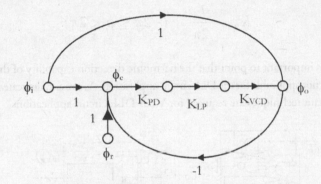

Fig. 11.12. SFG model of VLSI first-order DLL circuits

11.4.1 PHASE DETECTOR

Clock phase detector circuits as shown in figure 11.13 are typically used in the VLSI DLL circuit to extract the phase difference between the input reference and the feedback clock or data signals.

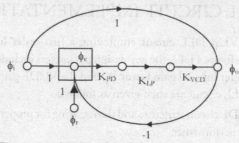

Fig. 11.13. SFG model of phase detector circuit

VLSI PFD circuits shown in figure 11.14 that were used for charge pump PLL can also be used for charge pump DLL circuit. The detection gain K_{PD} of the phase detection circuit can be expressed as

$$\bar{I} = \frac{I_B}{2\pi}\Delta\phi \quad (-2\pi \leq \phi_i < 2\pi)$$

(11.21)

$$K_{PD} = \frac{I_B}{2\pi} \quad (-2\pi \leq \phi_i < 2\pi)$$

(11.22)

However, it is important to point that the harmonic detection capability of this PFD for PLL circuit is not applicable for the DLL circuit application. As a result, dedicated harmonic detection circuit techniques are required for VLSI DLL circuit applications.

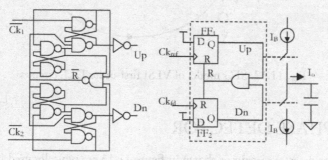

Fig. 11.14. VLSI PD circuit for charge pump DLL

11.4.2 CHARGE PUMP AND LOOP FILTER

Charge pump and loop filter circuit in VLSI DLL circuit is usually implemented together with the phase frequency detector circuit to provide seamless integration of VLSI signal processing circuit operations for the signal format conversion, the switching noise filtering, and the loop frequency response compensation.

Shown in figure 11.15 is the SFG model of a typical charge pump/loop filter circuit for the charge pump-based VLSI DLL circuits.

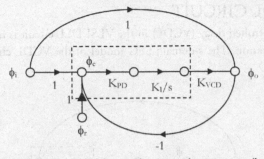

Fig. 11.15. SFG model of first-order DLL charge pump/loop filter

A VLSI circuit implementation of VLSI DLL charge pump circuit is shown in figure 11.16.

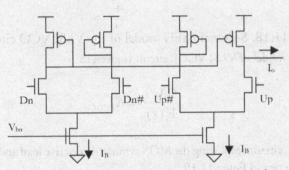

Fig. 11.16. Fully differential VLSI charge pump circuit implementation

The VLSI DLL loop filter circuit is behaviorally determined by the integral gain parameter K_I. A VLSI circuit implementation of the DLL loop filter transfer function is shown in figure 11.17. Such circuit realizes the filter with the frequency transfer function as

$$H(s) \equiv \frac{V_o}{I_i} \equiv \frac{K_I}{s} = \frac{1}{Cs}$$

(11.23)

$$\Rightarrow K_I = 1/C$$

(11.24)

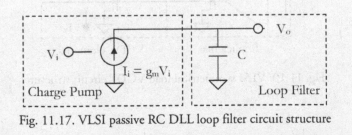

Fig. 11.17. VLSI passive RC DLL loop filter circuit structure

11.4.3 VCDL CIRCUIT

VLSI voltage-controlled delay (VCDL) in the VLSI DLL circuit is used for the delay regeneration operation. The s-domain SFG model of the VCDL circuit is shown in figure 11.18.

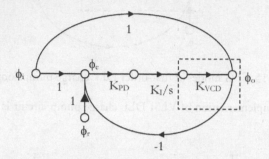

Fig. 11.18. S-domain SFG model of the VLSI VCO circuits

The s-domain model of VLSI VCDL circuit is given as

$$\frac{\phi_o(s)}{V_i(s)} = K_{VCDL}$$

$$(11.25)$$

A VLSI VCDL circuit employing the MOS symmetrical active load and replica biasing technique is shown in figure 11.19.

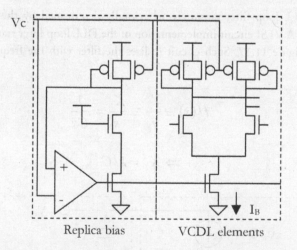

Fig. 11.19. VLSI symmetrical load VCDL circuit structure

References:

[1]   G. Jovannovic, et al," Delay Locked Loop with Linear Delay Element," Serbia and Montenegro, Nis, September 28-30, 2005.

[2]   J. G. Maneatis, "Low-Jitter Process-Independent DLL and PLL Based on Self-Biased Techniques," IEEE JSSC, Vol. 31, No. 11, Nov. 19911.

[3]   B. W. Garlepp, et al, "A Portable Digital DLL for High-speed CMOS Interface Circuits," IEEE JSSC, Vol. 34, No. 5, May 1999.

[4]   S-S Hwang, et al, "A DLL Based 10-320 Mhz Clock Synchronizer," ISCAS 2000, May 28-31, Geneva, Switzerland.

[5]   C. Kim, et al, "A Low-Power Small-Area +/- 7.28ps Jitter 1-Ghz DLL-Base Clock Generator", IEEE JSSC, Vol. 37, No. 11, Nov. 2002.

[6]   Y Arai, "A High-Resolution Time Digitizer Utilizing Dual PLL Circuits," IEEE2004.

[7]   H. J. Song, "Digital Delay Locked Loop for Adaptive De-skew Clock Generation," US Patent, No. 6275555, 2001.

[8]   H. J. Song, "Programmable De-Skew Clock Generation based on Dual Digital Delay Locked Loop Structure," SOCC2001.

[9]   J. Kim, et, al, "Design of CMOS Adaptive-Bandwidth PLL/DLLs: A Generation Approach," IEEE Trans. Circuits and System—II: Analog And Digital Signal Processing, Vol.50, N0.11, November, 2003.

[10] J. G. Maneatis, "Low-Jitter Process-Independent DLL and PLL Based on Self-Biased Techniques," IEEE, JSSC, Vol. 31, No.11, Jan. 1996.

[11] B. G. Kim, et al, "A 250Mhz-2-Ghz Wide-Range Delay-Locked Loop", IEEE JSSC, Vol. 40, No.6, June 2005.

[12] Y. Moon, et, al, "An All-Analog Multiphase Delay-Locked Loop Using a Replica Delay Line for Wide-Range Operation and Low-Jitter Performance," IEEE JSSC, Vol.35, No.3, March 2000.

[13] K. Minami, et, al, "A 1Ghz Portable Digital Delay-Locked Loop with Infinite Phase Capture Ranges," IEEE ISSCC 2000.

[14] T. Matano, et, al, "A 1-Gb/s/pin 512-Mb DDRII SDRAM Using a Digital DLL and a Slew-Rate-Controlled Output Buffer," IEEE JSSC Vol.38, No. 5, May 2003.

CHAPTER 12
VLSI ON-CHIP IMPEDANCE TERMINATION CIRCUITS

VLSI high-speed I/O transmitter and receiver circuits are typically physically located in separate chips within the electronic systems, connected through an electrical channel, usually consisting of the chip package, binding wires, connectors, and motherboard traces. Proper termination of high-speed I/O channel is very critical since the channel reflection and ringing-induced jitters are among the major sources of VLSI high-speed I/O circuit delay uncertainty effects. Most VLSI high-speed I/O circuit standards have specific channel characteristic impedances with 50Ω as a widely used impedance value.

Ringing effects usually occur in improperly terminated short channels, where the total length of all stubs on a line is not short enough or there is a discontinuity in impedance, such as a connector, in the channel. Ringing effects may cause false signal edge due to signal amplitude variation. Reflections, on the other hand, usually occur in improperly terminated long channels. Reflection effects cause timing problems due to the inherent property of voltage doubling on an open transmission line. Both the ringing and reflection effects can be effectively minimized using the proper impedance termination and slew rate control circuit techniques.

12.1 HIGH-SPEED I/O TERMINATION CONFIGURATIONS

Old-generation I/O circuits usually employed off-chip impedance termination to minimize the channel ringing and reflection effects. Off-chip termination method usually offers the flexibility in the PCB design with externally tunable termination resistance on the PCB of the I/O circuits. However, off-chip impedance termination schemes typically suffer from major drawbacks of higher cost, larger form factor, and limited impedance termination accuracy that are due to uncorrelated PVT variations of on-chip circuits. As a result, on-chip impedance termination circuits or sometimes called on-die termination (ODT) circuits are widely adopted in most high-speed I/O circuits to provide adaptive impedance termination to compensate for the PVT variations.

Most high-speed I/O circuits are typically based on two basic impedance termination configurations, including receiver (or far end) termination and transmitter (or near end) terminations.

12.1.1 RX TERMINATION

For the receiver impedance termination circuit configurations, the termination resistors are placed at the high-speed I/O receiver side within or next to the receiver AFE circuits. Receiver impedance termination can usually be realized in four basic circuit configurations, including the power line terminations, the Thevenin termination, the V_{TT} termination, and the AC termination as shown in figure 12.1. The power line terminations are usually the simplest and the most commonly used receiver termination configurations, where the high-speed I/O channel is directly terminated to the power supply or ground lines with a termination resistor of the value equal to the transmission line impedance of the channel. Since the power supply lines typically have low source impedance, the termination resistor can be directly connected to the supply line without introducing significant errors. In practical applications, either the ground or Vcc can be selected based on the receiver circuit structure for the best noise performance. For example, if the NMOS input pair is used in the receiver amplifier as shown in figure 12.2, the Vcc termination is the typically preferred circuit configuration for best noise immunity. On the other hand, the ground termination as shown in figure 12.3 is usually preferred if the receiver amplifier employs the PMOS input pair.

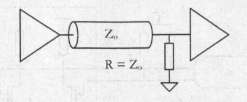

(a) Power line termination

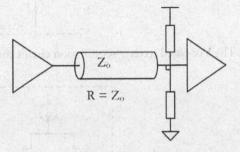

(b) Thevenin termination

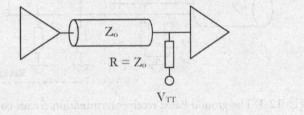

(c) Vtt termination

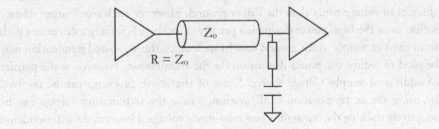

(d) Ac termination

Fig. 12.1. The receiver termination circuit configurations

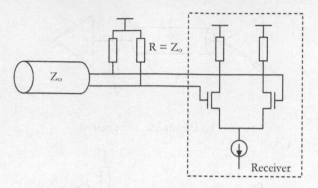

Fig. 12.2. The Vcc-based receiver termination circuit configuration

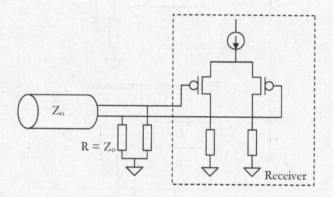

Fig. 12.3. The ground-based receiver termination circuit configuration

The Thevenin and V_{TT} termination configurations are typically used to terminate the channel to voltage other than the Vcc or ground. However, such termination scheme suffers from the high power dissipation penalty since there is usually a dc current path from the Vcc supply to the ground line. In such cases, the V_{TT}-based termination may be used to reduce the power dissipation for the termination, however, at the penalty of additional supply voltage source. Some of the above problems can be resolved by using the ac termination configuration, where the termination voltage can be adaptively tracking the transmitter common-mode voltage. However, the ac termination configuration usually requires a large ac coupling capacitor, which cannot be easily integrated on-chip.

12.1.2 TX TERMINATION

Shown in figure 12.4 are two basic VLSI transmitter termination configurations. In figure 12.4a, a serial termination resistor is used with a voltage mode transmitter circuit. For the circuit shown in figure 12.4b, a parallel termination resistor is used with a current mode transmitter circuit.

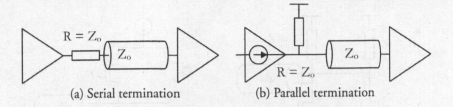

(a) Serial termination   (b) Parallel termination

Fig. 12.4. The transmitter termination configurations

Note that for both serial and parallel termination configurations, the effective output impedance of the transmitter driver must be included in the effective transmitter output impedance.

Shown in figure 12.5 is a VLSI current mode transmitter circuit structure where the resistor R serves as both the load of the transmitter driver stage and the channel termination resistor.

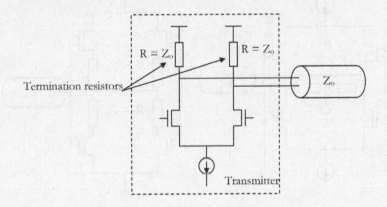

Fig. 12.5. The transmitter parallel termination configuration

Shown in figure 12.6 are four commonly used VLSI high-speed I/O differential transmitter impedance termination circuit structures.

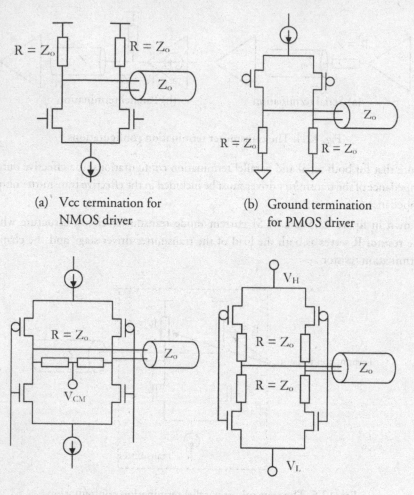

(a) Vcc termination for
NMOS driver

(b) Ground termination
for PMOS driver

(c) Termination for current mode
push-pull driver

(d) Termination for voltage
mode push-pull driver

Fig. 12.6. VLSI on-chip transmitter termination circuit structures

12.2 VLSI TUNABLE RESISTOR CIRCUITS

In most VLSI high-speed I/O circuits, the termination resistors are also typically implemented within the I/O transceiver circuit. In these design, the termination resistors are usually based on either MOS VCR or linear resistor bank.

Shown in figure 12.7 is a VLSI continuously tunable resistor implementation based on the MOS VCR circuit.

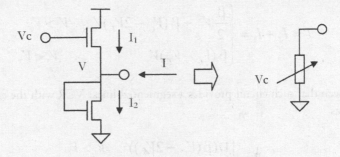

Fig. 12.7. Grounded MOS VCR based on MOS in saturation

The behavior of such voltage-controlled resistor (VCR) can be analyzed based on the MOS device I-V characteristics as

$$I = I_2 - I_1 = \frac{\beta}{2}(V_{gs2} - V_T)^2 - \frac{\beta}{2}(V_{gs1} - V_T)^2$$

$$= \beta(V - \frac{V_C}{2})(V_C - 2V_T)$$

(12.1)

It can be seen that such circuit serves as a linear voltage-controlled resistor (VCR) with equivalent resistance given as

$$R \equiv (\frac{dI}{dV})^{-1} = \frac{1}{\beta(V_C - 2V_T)}$$

(12.2)

In an alternative VLSI VCR circuit implementation shown in figure 12.8, a saturation mode MOS device is used in parallel with an identical linear mode MOS device.

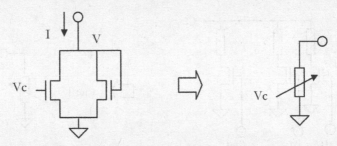

Fig. 12.8. Grounded MOS VCR based on saturation and linear MOS pair

The I-V characteristics of such MOS VCR can be derived using the two MOS I-V characteristics as

$$I = I_2 + I_1 = \begin{cases} \dfrac{\beta}{2}V_T^2 + \beta(V_C - 2V_T)V & V > V_T \\ \beta(V_C - V_T)V & V < V_T \end{cases}$$

(12.3)

It can be seen that such circuit provides a segmented linear VCR with the equivalent resistance as

$$R = \begin{cases} 1/(\beta(V_C - 2V_T)) & V > V_T \\ 1/(\beta(V_C - V_T)) & V < V_T \end{cases}$$

(12.4)

It can be seen that both above VLSI VCR implementations offer fair linearity VCR solutions. In some applications, the MOS devices as shown in figure 12.9 can be directly used as MOS VCR for high-speed I/O circuit application if lower resistor linearity is allowed.

Fig. 12.9. Grounded MOS VCR based on simple MOS device

For VLSI high-speed I/O circuits that require high termination resistor linearity, the programmable linear resistor bank is the most commonly used circuit structure. Such circuit structure provides the digitally tunable linear resistor, which can be easily implemented in standard digital CMOS technologies.

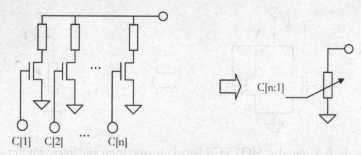

Fig. 12.10. Digitally controllable resistor based on resistor bank

12.3 VLSI ADAPTIVE IMPEDANCE TUNING CIRCUITS

Most recent high-speed I/O circuit typically employs adaptively controlled termination resistance to compensate for the PVT variations. Shown in figure 12.11 is a VLSI high-speed I/O transmitter impedance tuning circuit. In this circuit, an external high-precision resistor is used as a reference. The negative feedback loop is used to adaptively adjust the effective impedance of the VCR to match the external resistor reference. The replica circuit is then used to force the effective output impedance of the transmitter to match the external reference resistance.

$$R = R_{ext} \tag{12.5}$$

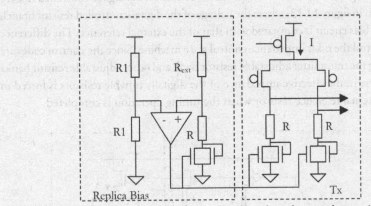

Fig. 12.11. VLSI adaptive impedance tuning loop based on analog tuning structure

Shown in figure 12.12 is an alternative VLSI transmitter impedance tuning circuit implementation.

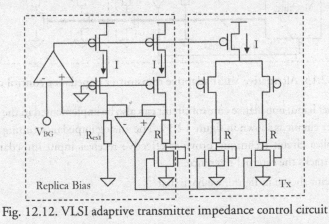

Fig. 12.12. VLSI adaptive transmitter impedance control circuit

In such circuit, the bias current of transmitter is adaptively controlled by the current control loop employing a bandgap voltage reference and an external accurate reference resistor as

$$I = \frac{V_{BG}}{R_{ext}}$$

(12.6)

The output impedance of the transmitter is adaptively controlled by the impedance tuning loop as

$$R = R_{ext}$$

(12.7)

The above tuning circuit can also be implemented based on the digital-controlled resistor bank as shown in figure 12.13, where the voltage of the digital-controlled resistor branch in the replica bias circuit is compared with that of the external reference. The difference is used to control the n-bit impedance control state machine. Since the control codes are shared among the transmitter adjustable resistor bank and other adjustable resistor banks in the I/O circuit, the effective impedance of the digitally tunable resistors is forced to match the external reference resistor when the tuning operation is completed.

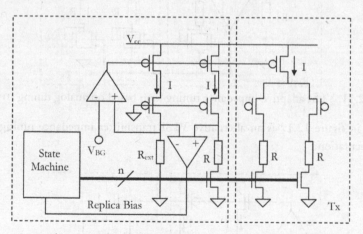

Fig. 12.13. Alternative VLSI adaptive transmitter impedance control circuit

On the other hand, impedance control circuit can also be implemented in the high-speed I/O receiver circuit as shown in figure 12.14. The analog impedance tuning loop based on the replica circuit technique provides effective receiver input impedance, which adaptively tracks the external reference.

For the circuit shown in figure, we have that

$$I = \frac{R_1 + R_2}{R_1 + R_2 + R_3 + R_4} \frac{V_{cc}}{R_{ext}}$$

(12.8)

$$R_{M1} = R_{ext}$$

(12.9)

$$R_{M4} = \frac{R_3 + R_4}{R_1 + R_2} R_{ext}$$

(12.10)

$$R_{Rx} = R_{M1} // R_{M4} = \frac{R_3 + R_4}{R_1 + R_2 + R_3 + R_4} R_{ext}$$

(12.11)

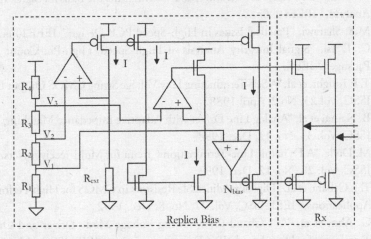

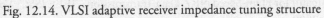

Fig. 12.14. VLSI adaptive receiver impedance tuning structure

References:

[1] H. J. Song, "Dual Mode Transmitter," US Patent No. 6531896B1, 2003.

[2] H. J. Song, "Dual Mode Transmitter with Adaptively Controlled Slew Rate and Impedance Supporting Wide Data Rates," SOCC2001.

[3] Y S. Kim, et al, "A 4-Gb/s/Pin Current Mode 4-Level Simultaneous Bidirectional I/O with Current Mismatch Calibration," ISCAS2006.

[4] E. Takahashi, et al, "8 Gbps Parallel Data Transmission with Adaptive I/O Circuit," IEEE2006.

[5] F. Zarkeshvari, et al, "High-Speed Serial I/O Trends, Standards, and Techniques," 2004 1st International Conference on Electrical and Electronics Engineering.

[6] S. Kim, et al, "A 6-Gbps/pin 4.2mW/Pin Half-Duplex Pseudo-LVDS Transceiver," IEEE2006.

[7] J. H. Kim, et al, "A 4-Gb/s/Pin Low Power Memory I/O Interface Using 4-Level Simultaneous Bi-Direction Signaling," IEEEE JSSC, Vol. 40, N0.1, July, 2005.

[8] R. H. G. Cuny, et al, "SPICE and IBIS Modeling Kits the Basis for Signal Integrity Analysis," IEEE1996.

[9] M.S. Sharawi, "Practical Issues in High-Speed PCB Design," IEEE2004.

[10] C. T. Tsi, "Signal Integrity Analysis of High-Speed, High-Pin-Count Digital Packages," IEEE.

[11] T. F. Knight, et al, "A Self-Terminating Low Voltage Swing CMOS Output Driver," JSSC, Vol.23, No.2, April 1988.

[12] B. Nauta, et al, "Analog Line Driver with Adaptive Impedance Matching," IEEE JSSC, Vol. 33, No. 12, Dec. 1998.

[13] M. Dolle, "A Dynamic Line-Termination Circuit for Multi-receiver Nets," IEEE JSSC, Vol. 28, No. 12, Dec. 1993.

[14] T. J. Gabara, et al, "Digitally Adjustable Resistors in CMOS for High-Performance Applications," IEEE JSSC, Vol. 27, No. 8, Dec. 1992.

[15] G. Ahn, et al, "A 2-Gbaud 0.7-V Swing Voltage-Mode Driver and On-Chip Terminator for High-Speed NRZ Data Transmission," IEEE JSSC, Vol. 35, No.6, June. 2000.

[16] C. S. Choy, et al, "Design Procedure of Low-Noise High-Speed Adaptive Output Drivers," 1997 IEEE International Symposium on Circuits and Systems, July 9-12, 1997, Hong Kong.

[17] M. I. Montrose, et al, "Analysis on the Effectiveness of Clock Trace Termination Methods and Trace Lengths on a Printed Circuit Board," IEEE1996.

[18] S. Vishwanthalah, et al, "Dynamic Termination Output Driver for a 600MHZ Microprocessor," IEEE JSSC, Vol. 35, No. 11, Nov. 2000.

[19] W. G. Rivard, et al, "A 1.2um CMOS Differential I/O Systems Capable of 400Mbps Transmission Rates," IEEE.

[20] T. P. Thomas, et al, "Four-Way Processor 800MT/s Front Side Bus with Ground Reference Voltage Source I/O," 2002 Symposium on VLSI Circuits Digest of Technical Papers.

[21] T. F. Knight, et al, "A Self-Terminating Low-Voltage Swing CMOS Output Driver," IEEE JSSC Vol.23, No.2, April 1988.

[22] A. Boni, et al, "LVDS I/O Interface for Gb/s-Per-Pin Operation in 0.35-um CMOS," IEEE JSSC Vol. 36, No.4, April 2001.

[23] A. Boni, "1.2-Gb/s True PECL 100k Compatible I/O Interface in 0.35-um CMOS," IEEE JSSC Vol.36, No.6, June 2001.

[24] J. Kudoh, et al, "A CMOS Gate Array with Dynamic-Termination GTL I/O Circuits," IEEE 1995.

[25] J. Griffin, et al, "Large-Signal Active Resistor Output Driver," IEEE 1999.

[26] J. J. Nam, et al, "A UTMI-Compatible Physical Layer USB2.0 Transceiver Chip," IEEE 2003.

CHAPTER 13
VLSI HIGH-SPEED I/O EQUALIZATION CIRCUITS

Equalization circuits are typically used in VLSI high-speed I/O to compensate for the intersymbol interference (ISI) effects. VLSI equalization circuits can be implemented in various forms, either in the digital or in the analog domains, within the transmitter or the receiver circuits, with the feed-forward or feedback configurations, or in the linear or nonlinear circuits. Practical VLSI high-speed I/O circuits usually combine multiple equalization techniques to achieve optimal circuit performances.

High-speed I/O equalization circuits can be implemented based on either time-domain or frequency-domain circuit approaches. Time-domain approaches rely on the solution to minimize the spread of the data symbol propagated through the high-speed I/O channel. Frequency-domain approaches, on the other hand, rely on the solution to improve the channel bandwidth.

Transmitter equalization based on the de-emphasis techniques typically offers a simple VLSI high-speed I/O circuit equalization solution. However, transmitter equalizations usually suffer from maximum signal swing limitation and therefore are not suitable for highly loss channels. Both linear and nonlinear receiver equalization techniques, such as the feed-forward equalizers (FFEs) and decision feedback equalizers (DFEs), are typically used in such design cases.

In some applications, the adaptive equalization techniques are used to compensate for unknown or time-varying channel responses. Such equalization techniques are usually based on certain adaptive control algorithms, such as the least-mean-square (LMS), the sign-sign LMS (SS-LMS), etc.

13.1 INTERSYMBOL INTERFERENCE EFFECTS

In digital data communication, intersymbol interference (ISI) effects are typically caused by the settling time of the data signal beyond its bit time (usually specified by the UI). Common root causes of the ISI effects are the channel bandwidth limitation, VLSI circuit bandwidth limitation, and cross talk effects.

Shown in figure 13.1 is a lowpass RC ISI model where the bandwidth of the circuit is comparable to the data signal frequency.

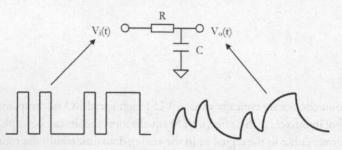

Fig. 13.1. ISI effect of lowpass RC circuit

In such bandwidth-limited circuit, the slew rate of the output signal is subjected to the RC time constant limitation such that signal will not be able to settle within the bit time. As a result, each data bit will ramp up (or down) at different initial values depending on the previous data pattern. Consequently, the delay time of the signal for each data bit as measured by the 50% swing crossover will be different, depending on the previous data pattern. Such delay variation effect has been discussed in previous chapter as the data-dependent jitter (DDJ) effect.

13.1.1 RC CIRCUIT MODEL OF ISI EFFECTS

For the lowpass RC network shown in figure 13.1 with ideal digital input, the output waveform can be derived analytically by solving the first-order differential equation. Shown in figure 13.2 is the output waveform for an input bit stream ... 111 ... 100 ... 00 ... that contains N_{2k} "1" and N_{2k+1} "0," respectively, where the initial voltage of the output waveform is assumed to be Y_{2k} and the clock period (or UI) to be T.

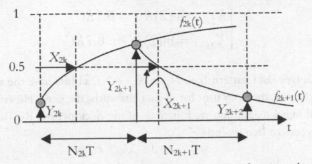

Fig. 13.2. Output waveform of the RC lowpass circuit for given data stream

The normalized output signal with respect to the "1" and "0" data stream can be expressed as

$$\begin{cases} f_{2k}(t) = 1 - (1 - Y_{2k})e^{\frac{-t}{RC}} \\ f_{2k+1}(t) = Y_{2k+1}e^{\frac{-(t-N_{2k}T)}{RC}} \end{cases} \tag{13.1}$$

The final voltages with respect to the rising and falling signal waveforms can be derived as function of the initial voltage, the RC time constant, the data clock period T, and the data pattern parameter N_{2k} and N_{2k+1} as

$$\begin{cases} Y_{2k+1} = 1 - (1 - Y_{2k})e^{-N_{2k}(\frac{T}{RC})} \\ Y_{2k+2} = Y_{2k+1}e^{-N_{2k+1}(\frac{T}{RC})} \end{cases} \tag{13.2}$$

The general delay time with respect to the rising and falling signal can be derived as

$$\begin{cases} X_{2k} = \ln(2)RC + RC\ln(1 - Y_{2k}) \\ X_{2k+1} = \ln(2)RC + RC\ln(Y_{2k+1}) \end{cases} \tag{13.3}$$

It is important to note the following:

- For very high RC circuit bandwidth (T/RC>>1), Y_{2k} = 0 and Y_{2k+1} = 1, implying each bit will be able to settle within the data UI. As a result, the delay time of the circuit is data pattern independent:

$$\begin{cases} X_{2k} = \ln(2)RC = 0.7RC \\ X_{2k+1} = \ln(2)RC = 0.7RC \end{cases}$$

$$(13.4)$$

- For clock-type data pattern (i.e., $N_{2k} = N_{2k+1} = N$), all bits have the same settling time. As a result, the delay time for all data bits will be the same. However, the delay time and signal magnitude are functions of the clock period, RC time constant, and data pattern frequency as

$$\begin{cases} Y_{2k} = 1 - Y_{2k+1} = e^{-N\frac{T}{RC}} / (1 + e^{-N\frac{T}{RC}}) \\ X_{2k+1} = X_{2k} = 0.7RC - RC\ln(1 + e^{-N\frac{T}{RC}}) \end{cases}$$

$$(13.5)$$

- For single-pulse pattern (also known as lone-pulse or impulse-response pattern) (i.e., $Y_{2k} = 0$ and $N_{2k} = 1$), both the signal magnitude and actual bit time will be significantly reduced. In addition, the center of the pulse will experience a pushed-out effect with respect to the nominal data bits. Therefore, such data pattern represents one of the worst-case conditions in data transmission through bandwidth-limited circuit. The magnitude and bit time for lone-pulse signal propagating through the lowpass RC circuit can be expressed as

$$\begin{cases} Y_{2k+1} = 1 - e^{-(\frac{T}{RC})} \\ X_{2k+1} - X_{2k} = T + RC\ln(Y_{2k+1}) \end{cases}$$

$$(13.6)$$

Shown in figure 13.3 are the simulation results of output signal magnitude, bit time, and pulse-out effect of the lone pulse input. It can be seen that there are significant magnitude attention and push-out effects associated to the bandwidth-limited channels.

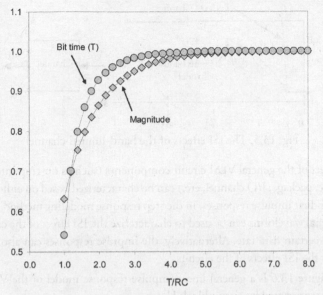

Fig. 13.3. Simulated single-pulse magnitude and bit time

13.1.2 GENERAL MODEL OF ISI EFFECTS

Bandwidth limitation effects are commonly observed in VLSI high-speed I/O circuits, mainly due to frequency-dependent channel attenuation. Such effect introduces the ISI and DDJ during data transmission as shown in figure 13.4.

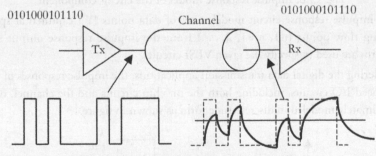

Fig. 13.4. High-speed I/O channel ISI effect

Similar to the lowpass RC circuit, the ISI effect in high-speed I/O channel can be described as the settling of data transition beyond the bit time as shown in figure 13.5.

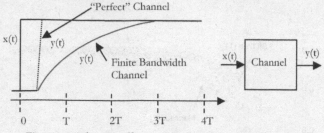

Fig. 13.5. The ISI effects of the band-limited channel

The ISI effect of the general VLSI circuit components (such as on-chip interconnects, active device, package, I/O channel, etc.) can be characterized based on either their step responses or their impulse responses. In the step response modeling method, the output ramping signal waveforms can be used to characterize the ISI effect of the circuits with respect to a certain data rate. Alternatively, the impulse responses can also be used to characterize the ISI effects of the circuits.

Shown in figure 13.6 is a general linear impulse response model of the VLSI circuit component represented by circuit block H.

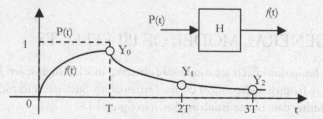

Fig. 13.6. Impulse response model of the circuit component

Under impulse response circuit model, a set of data points $\{Y_n\}$ captured at specific sampling time points (nT, n =1, 2, . . .) from the impulse response output signal waveform are used to specify the given VLSI circuits.

Considering the digital data transmission applications, the impulse responses of VLSI high-speed I/O circuits, including both the on-chip circuits and the channel, can be approximately modeled in discrete-time form as shown in figure 13.7.

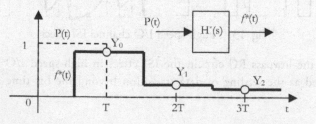
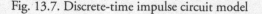

Fig. 13.7. Discrete-time impulse circuit model

Mathematically, the above modeling method can be viewed as an s-to-z conversion between the s-domain and z-domain impulse response models of the VLSI high-speed I/O circuits as shown in figure 13.8.

Fig. 13.8. S-to-z conversion between the s-domain and z-domain models

It is interesting to see that the z-domain model of the general VLSI circuit component represents a finite-impulse-response (FIR) filter as shown in figure 13.9 as

$$H(z) \approx Y_0 + Y_1 z^{-1} + Y_2 z^{-2} + \ldots \tag{13.7}$$

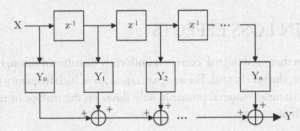

Fig. 13.9. The FIR model of linear VLSI component

13.2 HIGH-SPEED I/O CHANNEL BANDWIDTH LIMITATION EFFECTS

Skin loss and dielectric losses are two major high-speed I/O channel band-limitation effects for metallic media, such as PCB traces and cables. Such effects typically result in higher channel attenuation at higher signal frequency. Shown in figure 13.10 is a typical I/O frequency responses curve.

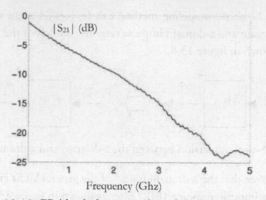

Fig. 13.10. FR4 backplane trace loss characteristics example

It can be seen that higher than 20dB attenuation (i.e., 10x reduction) can be observed for signal rate beyond 5Gb/s.

13.2.1 SKIN LOSS EFFECTS

It has been seen that for dc signal, current is uniformly distributed throughout the cross section of the conductor channel. For ac signal, especially at high frequency (>100 MHz), current in the channel material primarily only flows on the surface of the conductor with skin depth of

$$\delta = 1/\sqrt{\pi f \mu \sigma}$$

(13.8)

As a result, the effective resistances per unit length for various signal frequencies for the round conductor of radius r and thin strip of width W are given respectively as

$$\begin{cases} R(f) = (1/2r)\sqrt{f\mu/\pi\sigma} \\ R(f) = (1/2W)\sqrt{\pi f \mu/\sigma} \end{cases}$$

(13.9)

Generally speaking, the skin effect resulted resistance can be expressed using general physical dimension d and constant K_s as

$$R(f) = K_s \sqrt{f}/d$$

(13.10)

In addition, skin loss (attenuation) is proportional to the square root of frequency and the length of the trace as

$$A(f,l) = e^{-\frac{R(f)}{Z_o}l} \qquad (13.11)$$

13.2.2 DIELECTRICAL LOSS EFFECTS

Dielectric loss is due to the energy loss in the dielectric of the channel. Dielectric loss has a frequency response that is proportional to the frequency of the signal as

$$R(f) = K_d f \qquad (13.12)$$

Consequently, the attenuation for a transmission line of length l caused by the dielectric loss can in general be expressed as

$$A(f,l) = e^{-\frac{R(f)}{Z_o}l} \qquad (13.13)$$

13.2.3 ON-CHIP CIRCUIT BANDWIDTH LIMITATION EFFECTS

In addition to the VLSI high-speed I/O channel, VLSI on-chip circuits, such as interconnects, and the buffer circuit may also suffer from the bandwidth limitation effects at very high data rate. The effective bandwidth of such circuit can usually be characterized by the effective RC time constant of the circuit.

13.3 VLSI EQUALIZATION CIRCUIT MODELS

VLSI equalization circuits are commonly used to compensate for the ISI effects in VLSI high-speed I/O circuits. Various equalization implementation methods differ from each other based on the ISI compensation algorithms and circuit architectures.

13.3.1 NRZ SIGNAL IMPULSE RESPONSE MODEL

Non-return-to-zero (NRZ) signal is the most commonly used data format for VLSI high-speed I/O circuits. For VLSI differential NRZ data representation, where 1 represents the data bit "1" and -1 represents the data bit "0," a general NRZ bit stream can be decomposed mathematically into the positive and negative impulse signal as shown in figure 13.11 as

- For bit = "1" with the time duration as (kT< t < kT+T), the bit signal can be expressed based on the impulse signal as

$$P_k^{(1)} = P(k) \equiv h(t - kT) - h(t - kT - T)$$

(13.14)

where P(k) represents an impulse signal with normalized magnitude and pulse duration of T, starting at kT. h(t) is the step signal giving as

$$h(t) = \begin{cases} 0 & t < 0 \\ 1 & t > 0 \end{cases}$$

(13.15)

- For bit = "0" with time duration (kT < t < kT+T), we have that

$$P_k^{(0)} = (-1)^{(1+0)} P(k) = -P(k)$$

(13.16)

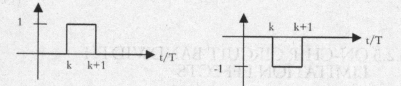

Fig. 13.11. Basic pulse functions of NRZ data stream

Based on this approach, a given data stream can be decomposed based on the linear combination of the defined impulse functions as

$$V(t) \equiv \{d_k\} = \sum_{k=\infty}^{\infty} P_k^{(d_k)}$$

(13.17)

Shown in figure 13.12 is the impulse decomposition of data stream X="01000101 . . .":

$$X = "01000101..." = P_0^{(0)} + P_1^{(1)} + P_2^{(0)} + P_3^{(0)}$$
$$+ P_4^{(0)} + P_5^{(1)} + P_6^{(0)} + P_7^{(1)} + ...$$

(13.18)

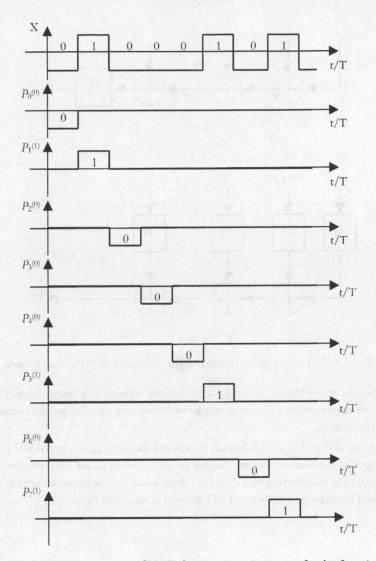

Fig. 13.12. Decomposition of NRZ data stream using sum of pulse functions

13.3.2 CHANNEL IMPULSE RESPONSE MODEL

For a linear channel with a transfer function H, the following two system diagrams shown in figure 13.13a and figure 13.13b are equivalent. That is, that the response of the channel to the NRZ data stream can be expressed by the linear combination of the channel's response to the individual pulse function.

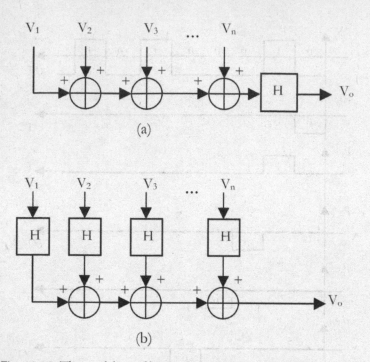

(a)

(b)

Fig. 13.13. The modeling of linear channel impact to NRZ data stream

Based on the above model, it can be seen that the impact of a linear channel to the NRZ data stream can be fully characterized by its impulse signal responses as discussed in the last section.

It can be seen that due to the channel bandwidth-limiting effect, individual pulse at the output of the channel will not be able to settle within its bit time window and it will impact the neighboring bits. This ISI effects based on the impulse decomposition mode of a band-limited high-speed I/O channel is shown in figure 13.14.

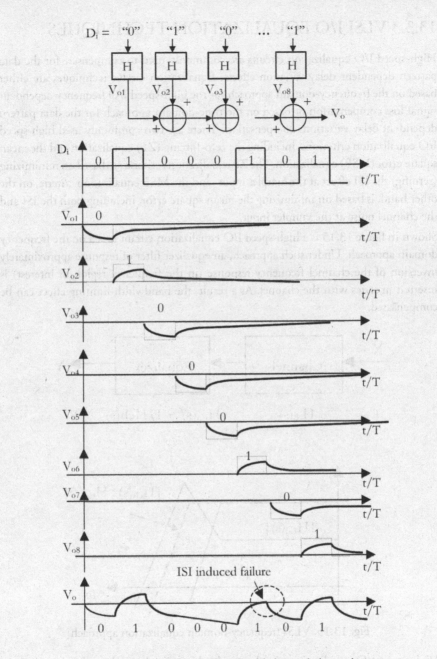

Fig. 13.14. ISI effects of bandwidth-limited channel

13.3.3 VLSI I/O EQUALIZATION TECHNIQUES

High-speed I/O equalization circuits are commonly used to compensate for the data pattern dependent delay variation effects. Equalization circuit techniques are either based on the frequency-domain approach for the high-speed I/O frequency dependent signal loss compensation or based on the time-domain approach for the data pattern dependent delay variations compensation. There are two commonly used high-speed I/O equalization criterions, including the zero-forcing (ZF) equalization and the mean square error (MSE) equalization. The ZF equalizations criteria are based on minimizing (zeroing) the ISI effect at the sampler input, and the MSE equalization criteria, on the other hand, is based on minimizing the mean square error, including both the ISI and the channel noise at the sampler input.

Shown in figure 13.15 is a high-speed I/O equalization circuit based on the frequency domain approach. Under such approach, an equalizer filter of response approximately inversion of the channel frequency response (in the frequency region of interest) is inserted in series with the channel. As a result, the bandwidth-limiting effect can be compensated.

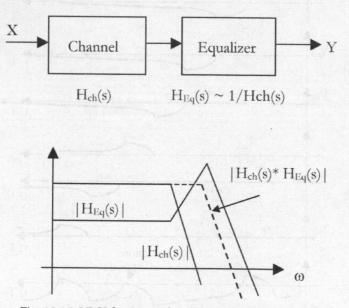

Fig. 13.15. VLSI frequency-domain equalization approach

High-speed I/O equalization circuit can also be realized based on the time-domain approach shown in figure 13.16, where the signal is predistorted through the equalization circuit to compensate for the ISI effect in the channel.

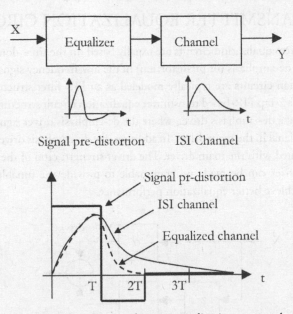

Fig. 13.16. VLSI time-domain equalization approach

An alternative time-domain equalization approach that is based on the data pattern dependent delay adjustment to compensate for the ISI resulted DDJ effects is shown in figure 13.17.

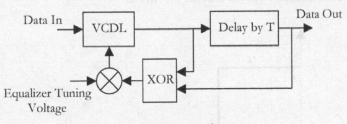

Fig. 13.17. DDJ equalization

13.4 VLSI EQUALIZER CIRCUIT IMPLEMENTATIONS

For a given equalization algorithm, VLSI equalization circuits can be implemented based on different architectures, such as the linear equalization, the decision feedback equalization (DFE), etc. In addition, VLSI high-speed I/O equalization circuits can be implemented either within the transmitter or within the receiver circuits. For unknown or time-varying channel, the equalization circuits can also be made adaptive.

13.4.1 TRANSMITTER EQUALIZATION CIRCUITS

VLSI transmitter equalization circuits are usually based on the time-domain approach employing the de-emphasis (or predistortion) of the low-frequency signal component. Such equalization circuits are typically modeled as an FIR filter structure. Shown in figure 13.18 is a 2-tap FIR-based transmitter equalization circuit structure employing a main driver and a de-emphasis driver, where the de-emphasis driver signal is a delayed version of the signal in the main driver. In addition, the de-emphasis driver has opposite polarity compared with the main driver. The driver strength ratio of the main and the de-emphasis driver can be made programmable to provide the tunable equalization capability to achieve better equalization performance.

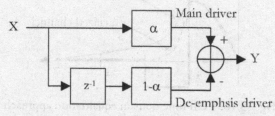

Fig. 13.18. Transmitter de-emphasis equalizer structure

The impulse response of such transmitter equalization circuit is shown in figure 13.19. It can be seen that such signal predistortion provides a way to compensate for the data symbol spread resulted from the channel ISI effects.

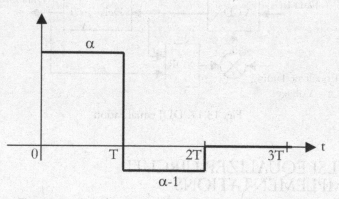

Fig. 13.19. Impulse response of 2-tap transmitter equalizer

Shown in figure 13.20 is a VLSI circuit implementation of transmitter equalization circuit.

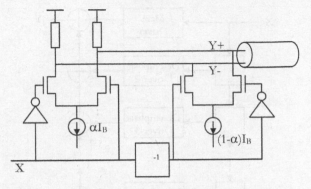

Fig. 13.20. VLSI Transmitter de-emphasis equalizer circuit

For the 2-tap equalization circuit shown above, the transfer function of the equalization circuit can be expressed as

$$H_{Eq}(z) = C[\alpha - (1-\alpha)z^{-1}]$$

(13.19)

The frequency response of such equalization circuit is shown in figure 13.21.

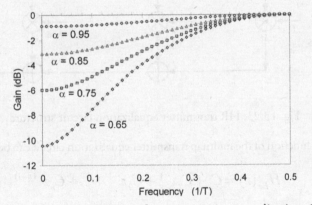

Fig. 13.21. Frequency response of 2-tap transmitter equalization circuit

It can be seen that such equalization circuit has a highpass frequency response. Therefore, such time-domain approach, in fact, is equivalent to the frequency-domain approach for using the highpass filter for ISI effect compensation.

The 2-tap transmitter equalization circuit can also be extended into multitap transmitter equalization circuit structure as shown in figure 13.22.

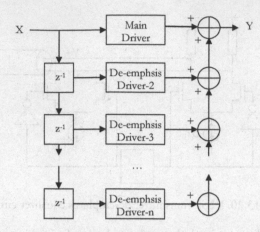

Fig. 13.22. Transmitter de-emphasis equalizer structure

The general z-domain model of such transmitter equalization circuit is shown in figure 13.23.

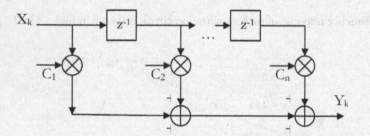

Fig. 13.23. FIR transmitter equalization circuit structure

The transfer function of the multitap transmitter equalization circuit can be expressed as

$$H_{Eq}(z) = C_1 + C_2 z^{-1} + C_3 z^{-2} + ... + C_n z^{-(n-1)}$$

(13.20)

13.4.2 RECEIVER EQUALIZATION CIRCUITS

There are several commonly used receiver equalization circuits for VLSI high-speed I/O circuit applications, such as the passive, the active continuous-time, the split-path amplifier, the discrete-time, and the continuous-time active receiver equalization circuit structures.

Shown in figure 13.24 are two VLSI passive receiver equalization circuit structures that are based on the frequency-domain approach employing the highpass RC or RL filter circuit structures.

The passive RC equalization circuit shown in figure 13.24a provides the frequency transfer function as

$$H(j\omega) = \frac{jRC\omega}{1 + jRC\omega}$$

(13.21)

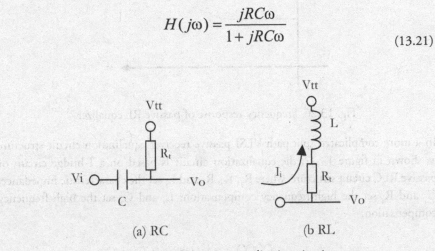

(a) RC   (b RL

Fig. 13.24. VLSI passive equalization circuits

Such equalizer circuit has a highpass frequency response with zero dc gain and unity high frequency gain as shown in figure 13.25.

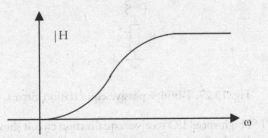

Fig. 13.25. Frequency response of passive RC equalizer

Similarly, for the passive RL equalization circuit shown in figure 13.24b, the transfer function of the equalization circuit is given as

$$H(j\omega) = R + jL\omega$$

(13.22)

Such circuit provides a resistive impedance termination at dc and the increased impedance at higher frequency. As a result, the equalization circuit also has a highpass frequency response as shown in figure 13.26:

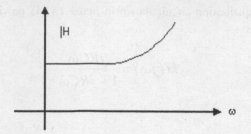

Fig. 13.26. Frequency response of passive RL equalizer

In a more complicated split path VLSI passive receiver equalization circuit structure as shown in figure 13.27, the equalization circuit is based on a T-bridge circuit of passive RLC circuit structure, where R_1, R_2, R_3 and L_1 set the characteristic impedance; C_1 and R_4 set the high-frequency compensation; L_2 and C_2 set the high-frequency compensation.

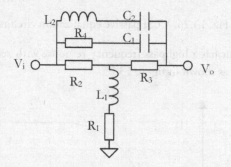

Fig. 13.27. T-bridge passive equalization circuit

The split path VLSI high-speed I/O receiver equalization circuit shown in figure 13.28 provides two signal paths for the signals with a high-frequency path to compensate for high-frequency signal loss in bandwidth-limited channel and a low-frequency or wideband path circuit to match the delay of the circuit to the high-frequency circuit path.

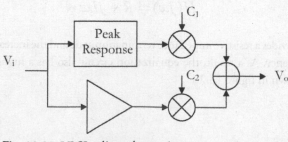

Fig. 13.28. VLSI split path equalization circuit structure

The output of the split path equalization circuit is the weighted sum of the two signal paths. The frequency response of the split path equalization circuit is then a highpass filter that provides enhanced high-frequency signal components.

Shown in figure 13.29 is an alternative VLSI split path equalization circuit implementation. In such circuit, the highpass circuit is constructed using a highpass RC network. The use of the programmable gain amplifiers provides the programming capability for the equalization response tuning.

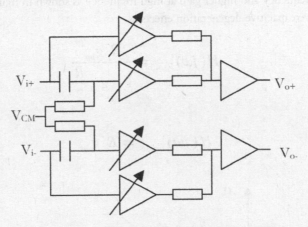

Fig. 13.29. VLSI split path equalization circuit

VLSI split path equalization circuit can also be implemented with VLSI circuit shown in figure 13.30, where the source degeneration impedance of the transconductor structure consists of a resistor and a capacitor. In such circuit, the resistor provides the low-frequency signal path and the capacitor provides the high-frequency path. Such circuit provides a compact split path VLSI amplifier that can be used to compensate for the ISI effect.

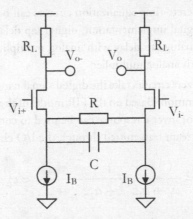

Fig. 13.30. VLSI source degeneration equalizer circuit structures

The transfer function of such source degeneration equalization circuit is given as

$$H(j\omega) = \frac{R_L g_{mi}}{1 + g_{mi}\frac{R}{2}/(1 + jRC\omega)}$$

(13.23)

where g_{mi} is the transconductance of the input MOS device. Such circuit has a lower gain at low frequency and higher gain at high frequency as shown in figure 13.31 due to the resistive/capacitive degeneration effects as

$$H(j\omega)|_{\omega \to 0} = \frac{R_L g_{mi}}{1 + g_{mi}\frac{R}{2}}$$

(13.24)

$$H(j\omega)|_{\omega \to \infty} = R_L g_{mi}$$

(13.25)

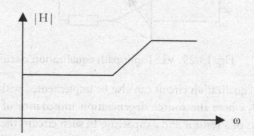

Fig. 13.31. Frequency response of source degeneration equalizer

Discrete-time receiver equalization circuits are also widely used in VLSI high-speed I/O circuit applications. Discrete-time equalization circuits can be implemented in several ways, including fully digital implementation, digital tap delay with multiplication to DAC, serial sampling analog tap delay with analog multiplier, and parallel sampling analog tap delay line with analog multiplier.

The discrete-time equalizers circuits take the digital signal nature of the high-speed I/O data stream into consideration. Based on the FIR model of the channel, a discrete-time filter (analog or digital) of inverse response can be used to compensate for the channel loss to the digital data stream transmitted through the I/O circuit.

$$H_{EQ}(z) = \frac{1}{H_{CH}(z)} = \frac{1}{a_0 + a_1 z^{-1} + a_2 z^{-2} + ...} \approx c_0 + c_1 z^{-1} + c_2 z^{-2} + ...$$

(13.26)

Shown in figure 13.32 is a receiver equalization architecture based on the FIR filter implementation. Alternatively, the receiver equalization circuit can also be constructed based on the IIR filter structures as shown in figure 13.33.

For such equalization circuit, we can derive the transfer function of the IIR equalizer as

$$H_{EQ}(z) = \frac{1}{H_{CH}(z)} = \frac{1}{1 + b_1 z^{-1} + b_2 z^{-2} + ...} \equiv \frac{1}{1 + \dfrac{a_1}{a_0} z^{-1} + \dfrac{a_2}{a_0} z^{-2} + ...}$$

(13.27)

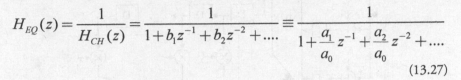

Fig. 13.32. The FIR-based equalization circuit for linear band-limited channel

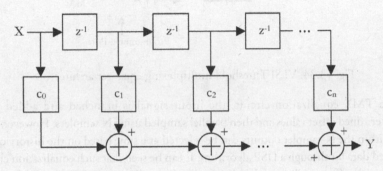

Fig. 13.33. The IIR-based equalization circuit for linear band-limited channel

The discrete-time equalization circuit can also be realized in full digital form for better design scalability. Shown in figure 13.34 is a VLSI implementation of the discrete-time equalizer structure based on the threshold multiplexing (TMX) algorithm.

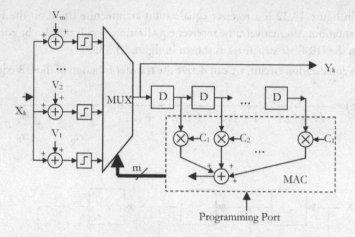

Fig. 13.34. VLSI Threshold multiplexing equalizer architecture

In the TMX equalization circuit, the input signal is branched and added with predetermined offset values and then parallel sampled using N samplers. However, only one within the N sampler outputs can be selected at a time based on the history of the received data bit through a DSP algorithm. It can be seen that such equalization circuit architecture offers similar to a decision feedback equalization (DFE) circuits, however, with highly improved operation speed. Shown in figure 13.35 is a VLSI-switched capacitor (SC) discrete-time equalization circuit implementation.

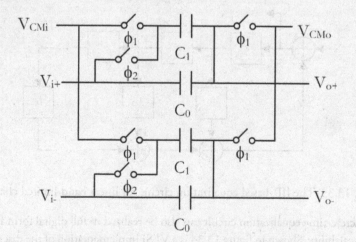

Fig. 13.35. Switched capacitor equalizer circuit

It can be derived that the above SC equalization circuit has a highpass frequency response similar to a continuous-time passive RC split path equalization circuit as shown in figure 13.36.

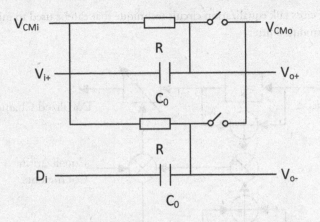

Fig. 13.36. Equivalent circuit of SC equalizer circuit

Since the inputs of high-speed I/O circuits are usually terminated with proper common-mode voltage, the input common-mode setting function of the above circuit can be eliminated and an alternative implementation of above circuit is shown in figure 13.37.

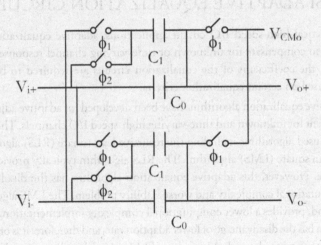

Fig. 13.37. Alternative SC equalizer circuit implementation

13.5 VLSI CROSS TALK EQUALIZATION CIRCUITS

In addition to the ISI effects, VLSI equalization circuit can also be used to compensate for other high-speed I/O circuit nonideality effects, such as the cross talk effects. It has been seen that the cross talk effects become very severe at very high data rate due mainly to the highpass characteristics of the cross talk effects. Shown in figure 13.38 is a VLSI two-channel cross talk equalization circuit technique that can be used to minimize the cross talk-introduced jitter.

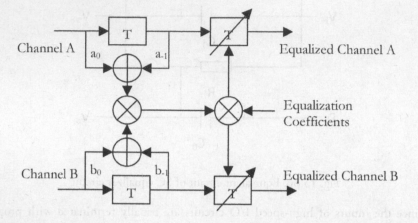

Fig. 13.38. VLSI cross talk equalization circuit

13.6 VLSI ADAPTIVE EQUALIZATION CIRCUITS

In some practical high-speed I/O circuit applications, adaptive equalization circuits are required to compensate for unknown or time-varying channel responses. In these applications, the coefficients of the equalization circuits are required to be adjusted adaptively based on certain equalization criteria.

Several adaptive equalization algorithms have been developed for adaptive adjustment of filter coefficient for unknown and time-varying high-speed I/O channels. There are two most widely used algorithms including the recursive least square (RLS) algorithm and the least mean square (LMS) algorithm. The RLS algorithm typically provides a faster converge rate. However, this adaptive equalization algorithm has the disadvantage of higher computational complexity and worse stability problem. The LMS algorithm, on the other hand, provides a lower computational complexity implementation. However, this algorithm has the disadvantage of lower adaption rate, and therefore, it is only suitable for a slow time-varying channel. A variation of LMS algorithm, known as the sign-sign LMS (SS-LMS), can provide further simplified circuit implementation, therefore, is very suitable for slow time-varying channel. This type of adaptive equalization circuit offers very low computational complexity at the cost of even reduced adaptation rate. In addition to the above equalization circuit techniques, the adaptive equalization technique can also be implemented with the ZF criterion.

In most high-speed I/O circuit applications, adaptive equalizations are usually implemented and associated with either FFE or DFE equalization architectures. A typical high-speed I/O equalization circuit usually has both the FFE and DFE circuits.

A conceptual circuit block diagram of the adaptive equalization circuit implementation is shown in figure 13.39, where the equalization core circuits can be implemented using the FFE or DFE equalization circuits as shown in figure 13.40 and figure 13.41. The equalization circuit coefficients are adjusted manually or adaptively to minimize the BER or to maximize the eye opening.

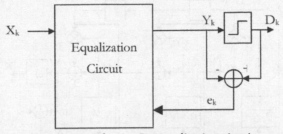

Fig. 13.39. Adaptive Rx equalization circuit

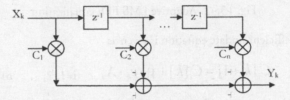

Fig. 13.40. FFE equalization circuit

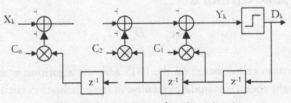

Fig. 13.41. DFE equalization circuit

However, in some applications, manual adjustment to optimize these parameters could be difficult when the number of weights is not small, or when there is significant variation in the circuit and the channel over time. As a result, adaptive equalization algorithms as the least-mean-square (LMS) and the zero-forcing (ZF) algorithm are required to address these problems.

13.6.1 LEAST-MEAN-SQUARE (LMS) EQUALIZATIONS

A FFE adaptive equalization circuit based on the LMS algorithm is shown in figure 13.42.

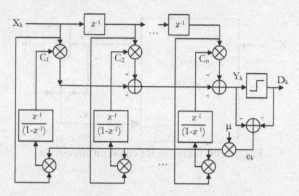

Fig. 13.42. Adaptive LMS FFE equalization

The weight coefficient update equation is given as

$$C_i[k+1] = C_i[k] + \mu \cdot e_k \cdot X_{k-i} \quad (i=1, 2, \dots, n) \qquad (13.28)$$

where $C_i[k]$ is the ith-tape coefficient during the kth-symbol period, μ is a constant that controls the adjustment rate of the weight, and e_k is the error at the equalizer output at the kth-symbol period expressed as

$$e_k = D_k - Y_k \qquad (13.29)$$

A DFE equalization circuit based on the LMS adaptive algorithm is shown in figure 13.43. The weight coefficient update equation of the equalization circuit is given as

$$C_i[k+1] = C_i[k] + \mu \cdot e_k \cdot D_{k-i} \quad (i=1, 2, \dots, n) \qquad (13.30)$$

It can be seen that the digital output D_k, instead of the unequalized input X_k, is used in the coefficient updating equation.

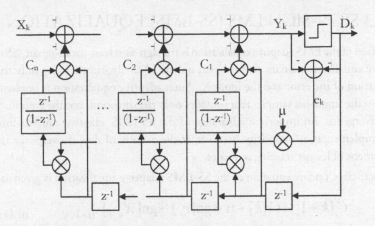

Fig. 13.43. Adaptive LMS DFE equalization

13.6.2 ZERO-FORCING (ZF) EQUALIZATION

A FFE adaptive equalization circuit based on the ZF algorithm is shown in figure 13.44.

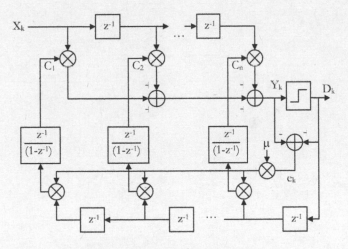

Fig. 13.44. Adaptive ZF FFE equalization

The weight coefficient update equations are similar to the LMS DFE equalization and are given as

$$C_i[k+1] = C_i[k] + \mu \cdot e_k \cdot D_{k-i} \quad (i=1, 2, \ldots, n) \qquad (13.31)$$

13.6.3 SIGN-SIGN LMS (SS-LMS) EQUALIZATION

A variation of the LMS adaptive equalization is the sign-sign least-mean-square (SS-LMS) adaptive equalization circuit that is based on the LMS algorithm, using only the sign information of the error and the input X_k. Such adaptive equalization is implemented based on the single-bit sampler that provides only the direction, not the amplitude, for the convergence. An important advantage of the SS-LMS adaptive equalization is its VLSI implementation simplicity, however, at the penalty of slow convergence and the requirement of longer training sequence.

The coefficient update equation of the SS-LMS adaptive equalization is given as

$$C_i[k+1] = C_i[k] + \mu \cdot \text{sgn}[e_k] \cdot \text{sgn}[X_{k-i}] \quad (i=1, 2, \ldots, n) \quad (13.32)$$

References:

[1] S. Armour, et al, "Performance Analysis of a Pre-FFT Equalizer Design for DVB-T," IEEE Trans. On Customer Electronics, Vol. 45, No. 3, August 1999.

[2] X. Lin, et al, "A High Speed Low-Noise Equalization Technique with Improved Bit Error Rate," IEEE 2002.

[3] A. J. Kim, et al, "Equalization and the Evolution of Gigabit Communications," 2003 IEEE GaAs Digest.

[4] J. L. Zerbe, et al, "Equalization and Clock Recovery for a 2.5-10 Gb/s 2-PAM/4-PAM Backplane Transceiver Cell," IEEE JSSC, Vol.38, No. 12, Dec. 2003.

[5] H. Witschnig, et al," Performance versus Effort for Decision Feedback Equalization—An Analysis based on SC/FDE Adapted to IEEE802.11a," 2004 IEEE.

[6] J. Buckwalter, et, al, "A 10Gb/s Data-Dependent Jitter Equalizer," IEEE 2004 Custom Integrated Circuits Conference.

[7] Y. Hur, et al, "Equalization and Near-end Crosstalk (NEXT) Noise Cancellation for 20-Gb/s 4-PAM Backplane Serial I/O Interconnections," IEEE Trans on Microwave Theory and Techniques, Vol.53, No. 1, Jan. 2005.

[8] Y. Tao, et, al, "A Signal Integrity-based Link Performance Simulation Platform," IEEE2005 Custom Integrated Circuits Conference.

[9] J. F. Buckwalter, et, al, "Analysis and Equalization of Data-Dependent Jitter," IEEE JSSC Vol. 41, No.3, March 2006.

[10] J. F. Buckwalter, et, al, "Phase and Amplitude Pre-Emphasis Techniques for Low Power Serial Links," IEEE JSSC Vol. 41, No.6, June 2006.

[11] H. J. Song, "Delay Locked Loop Based Circuit For Data Communication," US Patent #6466,615B1, 2002.

[12] H. J. Song, "Architecture and Implementation of Power and Area Efficient Receiver Equalization Circuit for High-Speed Serial Data Communication," SOCC2006.

[13] S. Hoyos, et al, "Mixed-Signal Equalization Architectures for Printed Circuit Board Channels," IEEE Transactions on Circuits and Systems—I: Regular Papers, Vol. 51, No.2, Feb 2004.

[14] L. Tong, et al, "Fast Blind Equalization Via Antenna Arrays," 1993 IEEE.

[15] S. Gondi, et al, "A 10-Gb/s CMOS Merged Adaptive Equalizer/CDR Circuit for Serial-Link Receivers," 2006 Symposium on VLSI Circuits Digest of Technical Papers.

[16] J. R. Schrader, et al, "Wireline Equalization using Pulse-Width Modulation," CICC2006.

[17] J. F. Bulzacchelli, et al, "A 10-Gb/s 5-Tap DFE/4-Tap FFE Transceiver in 90nm CMOS Technology," IEEE JSSC Vol. 41, No. 12, December 2006.

[18] W. J. Dally, et al, "Transmitter Equalization for 4-Gbps Signaling," IEEE Micro 1997.

[19] J. W. Lee, al, "A 2-Gbps CMOS Adaptive Line Equalizer," 2004 IEEE Asia-Pacific Conference on Advanced System Integrated Circuits, August, 2004.

[20] J. F. Buckwalter, et, al, "Cancellation of Crosstalk Induced Jitter," IEEE JSSC Vol. 41, No.3, March 2006.

[21] G. Esch, et al, "Near-Linear CMOS I/O Driver with Less Sensitivity to Process, Voltage, and Temperature Variations," IEEE Trans. VLSI Systems, Vol.12, No.11, November 2004.

[22] Y. Yan, et al, "Low Power High-Speed I/O Interfaces in 0.18um CMOS," IECS2003.

[23] M. Lin, et al, "Testable Design for Adaptive Linear Equalizer in High-Speed Serial Links," International Test Conference, 2006.

[24] E. Takahashi, et al, "8Gbps Paralleled Data Transmission with Adaptive I/O Circuit," IEEE, 2006.

[25] S. Kim, et al, "A 6-Gbps/Pin 4.2mW/Pin Half-Deuplex Pseudo-LVDS Transceiver," IEEE, 2006.

[26] D. Wei, et al, "Mostly Analog Disk Drive Read Channel with Practical Depth-of-Two Fixed Delay Tree Search," IEEE Tran Mangetics, vol. 38, no. 6, Nov. 2002.

[27] V. Balan, et al, "A 4.8-6.4 Gbps Serial Link for Back-Plane Applications using Decision Feedback Equalization," IEEE 2004.

[28] H. Wang, et al, "A Quad Multi-Speed Serializer/Deserializer with Analog Adaptive Equalization," IEEE VLSI Symposium, 2004.

[29] Y. Komatsu, et al, "Bi-Directional AC Coupled Interface with Adaptive Spread Spectrum Clock Generation," IEEE Asian Soiled-State Conference, Nov. 2007.

[30] J. F. Buckwalter, et al, "Analysis and Equalization of Data-Dependent Jitter," IEEE JSSC, vol 41, no. 3, March 2006.

[31] Y. Tomato, et al, "A 10-Gbps Receiver with Serials Equalizer and On-Chip ISI Monitor in 0.11-um CMOS," IEEE JSSC, vol. 40, no.4, April 2005.

[32] F. Wesis, et al, "Transmitter and Receiver Circuits for Serial Data Transmission over Lossy Copper Channels for 10Gb/s in 0.13um CMOS," IEEE.

[33] F. Yuan, et al, "A New Area-Efficient 4-PAM 10 Gb/s CMOS Serial Link Transmitter," IEEE 2006.

[34] K. Y. Chang, et al, "Level Selection Based 4_PAM Transmitter for Chip to Chip Communication," IEEE 2006.

[35] X. Lin, et al, "A High Speed Low-Noise Equalization Technique with Improved Bit Error Rate," IEEE 2002.

[36] J. L. Zerbe, et al, "Equalization and Clock Recovery for a 2.5-10-Gbps 2-PAM/4-PAM Backplane Transceiver Cell," IEEE JSSC, vol. 38, no. 12, Dec. 2003.

[37] H. Witschnig, et al, "Performance versus Effort for Decision Feedback Equalization—An Analysis based on SC/FDE Adapted to IEEE 802.11a," IEEE 2004.

[38] Y. Hur, et al, "Equalization and Near-End Crosstalk (NEXT) Noise Cancellation for 20-Gb/s 4-PAM Backplane Serial I/O Interconnections," IEEE Tran. Microwave Theory and Technique, vol. 53, no. 1, January 2005.

[39] S. Hoyos, et al, "Mixed-Signal Equalization Architectures for Printed Circuit Board Channel," IEEE Tran. Circuits and Systems-I: Regular Papers, vol. 51, no. 2, Feb. 2004.

[40] L. Tong, et al, "Fast Blind Equalization via Antenna Arrays," IEEE 1993.

[41] N. Ishii, et al, "Spatial and Temporal Equalization Based on an Adaptive Tapped-Delay-Line Array Antenna," IEEE PIMRC 1994.

[42] S. Gondi, et al, "A 10-Gb/s CMOS Merged Adaptive Equalizer/CDR Circuit for Serial-Link Receivers," IEEE VLSI Symposium, 2006.

[43] J. Hu, "A Clock Recovery Circuit for Blind Equalization of Multi-Gbps Serial Data Links," IEEE ISCAS 2006.

[44] M. Li, et al, "0.18um CMOS Backplane Receiver with Decision-Feedback Equalization Embedded," Electronics Letters, June 2006.

[45] E. Takahashi, et al, "8 Gbps Parallel Data Transmission with Adaptive I/O Circuit," IEEE 2006.

[46] A. J. Kim, et al, "Equalization and the Evolution of Gigabit Communications," 2003 IEEE GaAs Digest.

CHAPTER 14
BASIC VLSI HIGH-SPEED I/O CIRCUIT ARCHITECTURES

Various VLSI high-speed I/O circuits have been developed in the recent years targeted at high transmission data rates. These I/O circuit circuits can typically be classified into three basic architecture types, including the common clock I/O (or known as synchronous I/O) circuits, the forwarded clock I/O (or known as source synchronous I/O) circuits, and the embedded clock I/O (or known as data recovery I/O) circuits.

The common clock I/O circuits are based on the concept of electrically matched clock distribution to both the transmitter and receiver circuit from the same reference clock source. Common clock I/O circuits typically offer the timing plan simplicity. However, such I/O circuit suffers from the limitation for achieving low skew clock distributions to both the transmitter and receiver chips.

The forwarded clock I/O circuits are based on the concept of physical channel replica to send the data and reference clock together from transmitter to the receiver circuit. Such circuit technique can be effectively used to minimize the delay variation in the data transmission for high-speed data transmission at the penalty of added clock channel.

The embedded clock I/O circuits are based on the concept of clock and data recovery circuits for adaptive delay compensation of the transmitter, channel, and reference clocks. Such I/O circuit effectively eliminates the clock channel at some increased on-chip circuit design complexity.

14.1 TIMING CRITERIA OF VLSI HIGH-SPEED I/O CIRCUITS

VLSI high-speed I/O circuits are designed for reliable data transmission between IC chips at the high data rate. The general design criteria of VLSI high-speed I/O circuits can be described based on time-domain prototype circuit and SFG models. Under this model, the high-speed I/O design tasks are to generate a set of delay time parameters $\{\delta_{TX}, \delta_{DP}, \delta_{T/2}, \delta_{Tx}\}$ such that the maximum timing margin (MxTM) condition (or minimal delay phase tracking error E) specified in the following equations is achieved.

$$\overline{E} = \overline{\delta_{DP} + \delta_{TX} - \delta_{RX} - \delta_{T/2}} = 0 \tag{14.1}$$

$$|E| \equiv \left|\delta_{DP} + \delta_{TX} - \delta_{RX} - \delta_{T/2}\right| < E_{max} \tag{14.2}$$

where

$$E_{max} \equiv \frac{T - T_{Setup} - T_{Hold}}{2} \tag{14.3}$$

Shown in figure 14.1 is the SFG model for the general VLSI high-speed I/O circuit design approaches. Various VLSI high-speed I/O circuit architectures differ from each other on the implementation strategies to achieve the above delay timing equations.

In the common clock I/O architecture, the design approach is to balance the clock distribution delay skew between the transmitter and the receiver circuit such that the condition in delay constraint equation 14.2 can be achieved. In the forward clocking high-speed I/O circuit, the delay replica technique is introduced for the channel delay time in t_D such that the skew between the transited data and the receiver reference clock can be controlled. In the embedded clocking high-speed I/O circuit architecture, the receiver reference clock is regenerated based on the received data phase such that the time constraint equation 14.2 can be ensured.

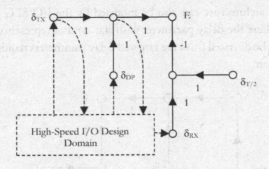

δ_{TX} E

1   1

1   -1

δ_{DP}

$\delta_{T/2}$

1

1

High-Speed I/O Design Domain

δ_{RX}

Fig. 14.1. General VLSI high-speed I/O circuits design model

14.2 COMMON CLOCK HIGH-SPEED I/O CIRCUITS

The common clock I/O circuits shown in figure 14.2 are based on the unified clock architecture, which directly distributes the clock to both the transmitter and the receiver circuits.

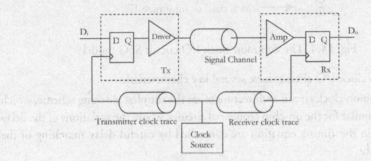

Fig. 14.2. The common clock I/O circuit structure

The common clock I/O circuits can be modeled using the prototype I/O circuit shown in figure 14.3.

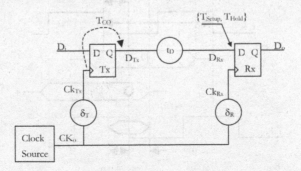

Fig. 14.3. The common clock I/O circuit prototype

Such I/O circuit architecture can also be modeled by the I/O SFG model as shown in figure 14.4, where the delay parameter with star symbol represents the actual delay values that might be deviated from the targeted delay parameters required in the timing constraint equation.

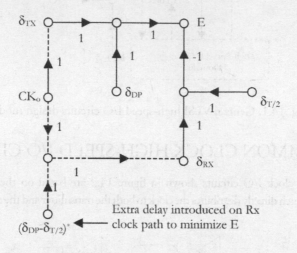

Fig. 14.4. The common clock I/O circuit SFG model

The common clock I/O circuits have several key characteristics:

- The common clock circuit architecture offers the simplest clocking scheme, which is very popular for the on-chip sequential circuits, where the variations of the delay element in the timing equation are controlled by careful delay matching of the clock paths.

- A single (i.e., common) clock source is used in common clock I/O circuit to synchronize both the transmitter and the receiver circuit operations.

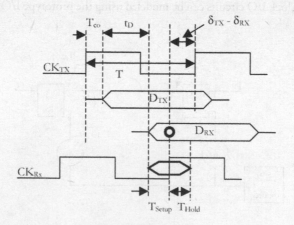

Fig. 14.5. Common clock I/O operation

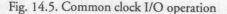

- Performance of common clock I/O circuits strongly depends on absolute delay values of the VLSI circuit components since delay parameters of each physically different circuit elements are usually poorly correlated.

$$\delta_{RX} = (T_{CO} + t_D + \delta_{TX} - \frac{T}{2})^*$$

(14.4)

- Major performance limitation of common clock I/O circuit is the large delay variations resulted from the poor correlation among delay parameters in the basic I/O timing constraint equation.

$$E = \Delta(\delta_{DP} + \delta_{TX} - \delta_{T/2})$$

(14.5)

- Maximum operation speed of a practical VLSI common clock I/O is typically limited by the uncorrelated delay variations (typically 5~10ns) in the basic I/O timing constraint equation, which is usually given as

$$f_{max} = \frac{1}{T_{min}} \sim \frac{1}{2\Delta E} \sim 10^2 \, Mhz$$

(14.6)

14.3 FORWARDED CLOCK I/O CIRCUITS

The forwarded clock I/O circuits shown in figure 14.6 are based on the physically matched channel for the compensation of the key delay parameters in the basic I/O timing equations, where the reference clocks are generated based on the transmitter clock with inserted replica channel delay the same as the data signal path.

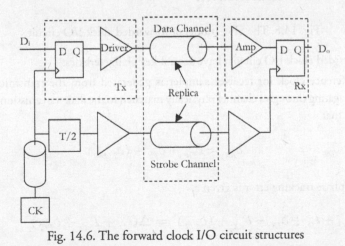

Fig. 14.6. The forward clock I/O circuit structures

A typical forward clock I/O circuit prototype circuit mode and SFG model are as shown in figure 14.7 and figure 14.8 respectively.

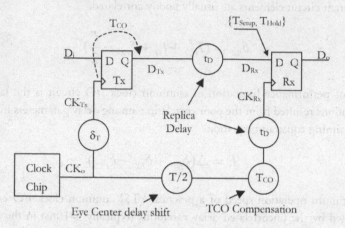

Fig. 14.7. The forwarded clock I/O circuit prototype

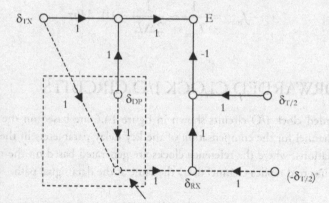

Fig. 14.8. The SFG model of forwarded clock I/O circuit

The forwarded clock I/O circuits have several key characteristics:

• A reference clock for receiver sampler is generated from the transmitter clock propagating through a channel physically matched to the data transmission channel, such that

$$\delta_{RX} \approx \delta_{DP} + \delta_{TX} - (\delta_{T/2})^* \tag{14.7}$$

The phase tracking error is given as

$$E = T_{CO} + t_D + \delta_{TX} - \delta_{RX} - (\delta_{T/2})^* = \Delta(t_D + T_{CO} + \delta_{T/2}) \tag{14.8}$$

- The forward clock circuit makes the receiver clock highly correlated to the transmitter clock. The effects of absolute channel delay variation can be effectively minimized.
- A half-clock period delay shift (eye centering) on the receiver reference clock path is inserted in either the transmitter side or the receiver side for the MxTM sampling operation.
- The half-frequency clocks are typically used to match the frequency response between the data and strobe channels.
- The maximum operation frequency of forward clock I/O is only limited by the uncorrelated delay variations in the basic I/O timing constraint equation, which is significantly smaller than the common clock I/O circuits.
- The requirement of an additional channel for sending clock from the transmitter to the receiver may increase the complexity and cost for the system implementation. Such penalty can usually be minimized by sharing clock reference among multiple data channels.
- Delay variation between the matched channel resulted from the pattern and frequency dependency channel response may increase the phase mismatch between the signal and the clock channel.
- Effective generation of the matched TCO and half-cycle eye-centering phase shift may result in significant design complexity penalties.

A practical high-speed I/O circuit using shared clock channel is shown in figure 14.9.

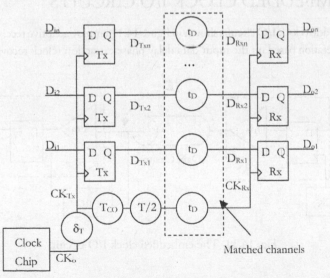

Fig. 14.9. The forwarded clock multibit I/O circuit prototype

Shown in figure 14.10 is a forwarded clock I/O circuit structure that is used to address the frequency-dependent jitter, the eye-centering phase shift, and the TCO compensation design requirements.

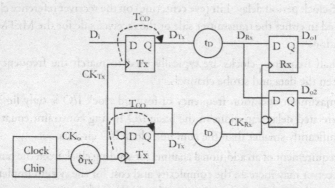

Fig. 14.10. The implementation of forwarded clock multibit I/O circuit

In this forwarded clock I/O circuit implementation, a half-speed clock is sent through the channel to match the highest data frequency through the channel. TCO matching is through the matched FF, which generates the half-speed clock. Eye-centering T/2 phase shift is realized by clocking the clock divider using the fall edge of the clock. However, due to the half-speed clocking, double-pump sampling scheme at the receiver must be used.

14.4 EMBEDDED CLOCK I/O CIRCUITS

The embedded clock I/O circuits shown in figure 14.11 employ adaptive receiver reference clock generation based on the input data delay phase extraction (clock recovery).

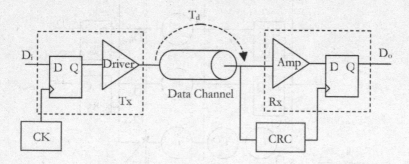

Fig. 14.11. The embedded clock I/O circuits

A typical embedded clock I/O circuit prototype model is shown in figure 14.12.

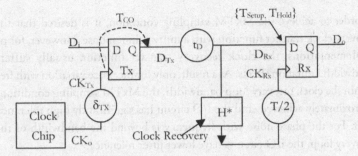

Fig. 14.12. The embedded clock I/O circuit prototype

Since clock is not explicitly sent to the receiver block, the receiver reference clock has to be extracted (or recovered) from the phase information of the received data stream usually based on a clock recovery circuit (CRC). Due to this reason, the embedded clock I/O circuits are also known as the data recover or clock and data recovery (CDR) I/O circuits.

The SFG model of the embedded clock I/O circuit architecture is shown in figure 14.13:

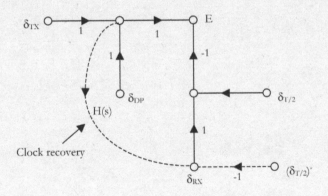

Fig. 14.13. The SFG model of forwarded clock I/O circuit

The embedded clock I/O circuits have several key characteristics:

- For the embedded clock I/O circuits, the receiver clock phase is regenerated (recovered) from the input data stream using a clock recovery loop with transfer function H as

$$\delta_{RX} = H(s)(\delta_{DP} + \delta_{TX}) - (\delta_{T/2})^*$$

(14.9)

- The phase tracking error of such I/O circuits can be expressed as

$$E = \delta_{DP} + \delta_{TX} - \delta_{RX} - \delta_{T/2} = (1 - H(s))(\delta_{DP} + \delta_{TX}) + \Delta\delta_{T/2} \quad (14.10)$$

- In order to achieve the MxTM sampling condition, it is desired that the clock recovery clock transfer function with a unity gain response. However, for practical implementations, the clock recovery transfer function usually suffers from bandwidth limitation effects. As a result, only for the phase variation with frequency within the clock recovery loop bandwidth, the MxTM sampling condition can be approximately achieved, and the I/O circuit has significantly high tolerance to the jitter. For the phase noise with frequency is beyond the bandwidth of the clock recovery loop, the I/O circuits have lower jitter tolerance.

- Compared with the forwarded clock I/O circuits, the penalty for dedicated clock channel is eliminated.

- The adaptive delay match makes the circuit highly reliable for high-speed I/O circuit operation beyond 1Gb/s.

- The requirement of certain clock recovery circuit for data recovery makes the design of the embedded clock I/O circuit design highly specialized.

References:

[1]   D. Chung, et al, "Chip-Package Hybrid Clock Distribution Network and DLL for Low Jitter Clock Delivery," IEEE JSSC, Vol. 41, No. 1, Jan 2006.

[2]   F. Zarkeshvari, et al, "High-speed Serial I/O Trends, Standards and Techniques," 2004 its International Conference on Electrical Engineering.

[3]   G. Balamurugan, et al, "Modeling and Mitigation of Jitter in Multi-Gbps Source Synchronous I/O Links," Proc. 21st International Conference on Computer Design (ICCD'03)

References

CHAPTER 15
VLSI HIGH-SPEED I/O TRANSMITTER CIRCUITS

VLSI high-speed I/O transmitter (Tx) circuits are used for signal conditioning for matching the characteristics of high-speed I/O channel, supporting the dedicated high-speed I/O link specifications, such as the signal swing, loading driving capabilities, bandwidths, slew rates, characteristic impedances, and data rate. Practical VLSI high-speed I/O transmitter circuit may typically include the circuit functions such as the parallel-in-serial-out (PISO) circuit, the transmitter impedance termination circuit, the equalizer circuit, and the predriver circuit. VLSI high-speed I/O transmitter circuit may also contain circuit features such as the design for testability (DFT) circuit and the power management (PM) circuit.

VLSI high-speed I/O transmitter circuits can be modeled in time-domain using a synchronization circuit element with the key timing parameter specified by the equivalent transmitter clock to output (TCO) delay.

Practical VLSI high-speed I/O transmitter circuits typically suffer from various delay time variation effects such as the transmitter reference clock delay variations, the datapath delay variations. Key design tasks of VLSI high-speed I/O transmitter circuits are to minimize the skew, jitter, and duty-cycle errors of transmitter reference clock buffer circuits; to minimize the transmitter TCO delay time variation under the PVT variations and loading conditions (capacitive loads, transmission-line and PCB trace discontinuities); to minimize the channel reflection, ringing, and cross talk effects through proper control of transmitter output impedance and slew rate; and to minimize the channel ISI effects through effective implementation of the transmitter equalization.

15.1 TIMING ASPECTS OF VLSI HIGH-SPEED I/O TRANSMITTERS

Under the SFG model shown in figure 15.1, VLSI high-speed I/O transmitter circuits contribute to the transmitter reference clock phase (δ_{T_x}) and the effective datapath delay (δ_{DP}) parameters.

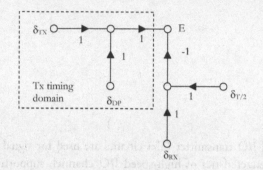

Fig. 15.1. Timing contribution of VLSI Tx circuit to I/O delay parameters

A typical VLSI high-speed I/O transmitter circuit usually includes the following key circuit components as shown in figure 15.2:

- The transmitter driver circuit to convert digital data bits to specific analog electrical signal, which is compatible to the channel characteristics. Major design parameters of the VLSI high-speed I/O transmitter deriver are the load driving capabilities and signal-to-noise ratio (SNR) (signal timing and voltage magnitude), channel impedance matching, channel equalization, noise tolerance, overvoltage, and ESD protections.

- The transmitter synchronization circuit to synchronize the high-speed I/O data transmission operation. The data transmission operation is synchronized using the transmitter reference clock to minimize the timing variation.

- The transmitter digital data encoder circuit to preprocess the data in the digital domain to ensure high reliability data transmission at targeted data rate. Transmitter data encoding usually includes the channel coding, such as the 8B/10B coding, the data scrambling, and the parallel to serial (PISO) conversion. Channel coding and scrambling operations are used for channel capacity maximization and error minimization. The PISO operation is used for rate matching between the low-speed digital core (parallel data) domain and the high-speed physical layer (serial) circuits.

- Practical VLSI high-speed I/O transmitter circuits will also include other supporting circuits, such as the clock distribution circuits, the voltage and current biasing generation circuits, the power management circuits, and the ESD protection circuits.

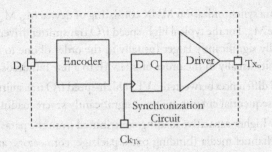

Fig. 15.2. Basic building components of VLSI high-speed I/O transmitter circuit

15.2 VLSI HIGH-SPEED I/O TRANSMITTER DRIVER CIRCUITS

VLSI high-speed I/O transmitter driver circuits can usually be viewed as the design extension of the on-chip sequential circuit as shown in figure 15.3 where data transfer within the circuit is controlled by the reference clock Ck. During the circuit operation, data signal stored at node A propagates to node B within the transmitter circuit at the rise edge of the reference clock, controlled by the digital MUX consisting of MOS device M_2 and M_3. Data at node B is amplified by the driver device M_D to generate the data transmitter output D_{TX}.

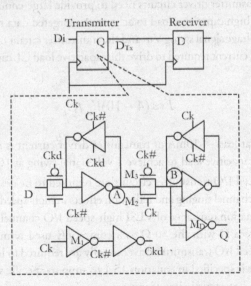

Fig. 15.3. Driver in typical VLSI sequential circuits

The key timing parameter of the transmitter circuit is the clock to output delay (TCO). The TCO of the transmitter circuit is mainly determined by the delay time of the clock buffer M_1, the data synchronization MUX consisting of device M_2, M_3, and the output data buffer device M_D. For the typical high-speed I/O transmitter driver circuit, the load capacitor is usually significantly larger (usually in the order of one to tens of pF) than the on-chip circuit (usually in the order of a few fF to a few hundreds of fF).

The fundamental differences between the VLSI high-speed I/O transmitter driver circuits and the on-chip sequential circuits are their significantly severe loading conditions:

- Significantly higher capacitive loading contributed from the parasitic capacitances of the I/O channel media (binding pads, package, connectors, and PCB traces), usually ranging from a few pF to a few hundreds of pF

- Significantly higher inductance and transmission line effects contributed from the parasitic inductances of binding pad, package, trace, connectors, and PCB traces, usually ranging from a few hundred pH to a few nH

- Significantly larger variations of the loading conditions depended on the applications

- Much severe circuit protection requirements resulting from the exposure of the I/O driver circuits to external circuits

As a result, dedicated transmitter driver devices, separating from the transmitter synchronization circuits, are usually required in VLSI high-speed I/O circuits. Practical VLSI high-speed I/O transmitter driver circuit designs usually follow the following considerations:

- VLSI I/O transmitter driver circuits need to provide large-enough current driving capability for high capacitive load to achieve the targeted data rate. For a specified transmitter voltage signal swing Vm and the transmitter circuit operation frequency f, the minimal current required to drive the capacitive load CL can be approximately expressed as

$$I \approx (4 \sim 10)V_m f C_L$$

(15.1)

For example, at least 4~10mA of transmitter driver current is usually required to drive 1pF of capacitive load to achieve 1 volt signal swing at 1Gb/s data rate.

- Most high-speed I/O transmitter circuits are required to be designed to minimize the inductive channel ringing and reflection effects in high-speed data transmission. Typical termination resistance of VLSI high-speed I/O channel ranges from a few tens to hundreds Ω with the 50 Ω as a commonly used termination resistance. VLSI high-speed I/O transmitter driver usually are required to have higher driving current than it is specified in equation 15.1 to support specific signal swing Vm at 50Ω channel characteristic impedance:

$$I \approx \frac{V_m}{50//50} = \frac{V_m}{25}$$

(15.2)

It can be seen that 20mA current is required for driving the load with 500mV nominal signal swing for a 50Ω termination impedance.

- Since the transmitter driver output currents are directly proportional to the signal voltage swings, low-voltage swing differential signaling (LVDS) techniques are widely adopted in practical VLSI high-speed I/O circuits, such as the USB2.0, the SATA, and the PCI Express I/O circuits, to minimize the driving current of the transmitter circuits. In addition, LVDS techniques are also used to improve the S/N ratio of the VLSI high-speed I/O circuits at high data rates. Shown in figure 15.4 are three basic LVDS configurations with different common-mode signal levels. In addition, various high-speed I/O standards may use slightly different signal voltage swing specifications.

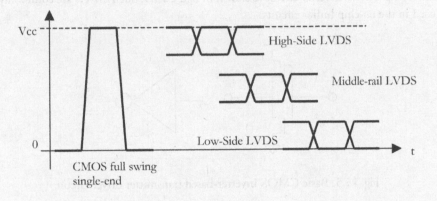

Fig. 15.4. Typical VLSI LVDS schemes

- In order to minimize the ringing and reflection effects of high-speed data transmission through the transmission line and parasitic LC-based channel, channel impedance termination and the transmitter output signal slew rate control techniques are typically required in the VLSI high-speed I/O transmitter driver circuits. The proper termination of the channel can be used to absorb the unused signal power during the signal transmission across the high-speed I/O channel. The slew rate control circuits, on the other hand, can be used to effectively reduce the out-of-band signal power transmission across the high-speed I/O channel. Both techniques can be used to improve the signal quality of the VLSI high-speed I/O circuits. Using the PCI Express transmitter as example, it is usually required that the output driver impedance is adaptively controlled to a target of 50Ω (single-ended) or 100Ω (difference) under the PVT condition with desired driver parasitic capacitance controlled to be less than 1pF.

- VLSI high-speed I/O transmitter driver circuits usually include the equalization function to immunize the bandwidth limitation and cross talk effects of high-speed I/O channel. The channel bandwidth limitation effects reduce the signal power in the I/O circuits, and the cross talk effects, on the other hand, can result in increased noise floor. Channel bandwidth limitation may also induce ISI effects. These effects may significantly degrade the signal-to-noise ratio in the VLSI high-speed I/O circuits. The preemphasis equalization techniques are commonly implemented in the VLSI high-speed I/O transmitter driver circuits for the channel bandwidth equalization.

15.2.1 CMOS VOLTAGE MODE TRANSMITTER DRIVER CIRCUITS

The simplest VLSI high-speed I/O transmitter driver structure directly employs full-swing CMOS inverter circuit as shown in figure 15.5. Such drivers are commonly used in the on-chip buffer circuits.

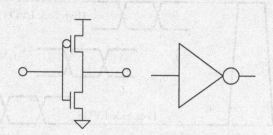

Fig. 15.5. Basic CMOS inverter-based transmitter driver circuit

Major advantage of such driver circuit structures is its design simplicity, the large load driving capability. For capacitive load C_L, the delay time of the driver can be approximately expressed as

$$t_d \propto \frac{C_L}{\beta(V_{cc} - (V_{TN} + |V_{TP}|))^\alpha} \quad \alpha = 1 \sim 2$$

(15.3)

Delay variation induced by the supply and substrate noise for such Tx circuit structure can be approximately expressed as

$$\frac{\Delta t_d}{t_d} = \alpha \frac{\Delta V_{cc} + \Delta(V_{TN} + |V_{TP}|)}{(V_{cc} - (V_{TN} + |V_{TP}|))}$$

(15.4)

Since delay variation of such type of driver circuit is approximately proportional to the total delay of the driver. Timing variation of such transmitter driver can usually be minimized by controlling the total delay time of the driver through minimizing the load capacitance, increasing the driver strength, and increasing the power supply voltage. However, these efforts are subject to the constraint of the transmitter output impedances that are required to terminate the channel. In addition, the effective supply voltage (Vcc and Vss) may also be regulated for minimizing the transmitter driver delay variation effects.

Random jitter (RJ) of the CMOS inverter-based transmitter driver is usually contributed by the thermal noise of the devices that can be approximately modeled as

$$\frac{\Delta t_{RMS}}{t_d} \approx \sqrt{\frac{2kT}{C_L}}\sqrt{1+\frac{2}{3}a_V}\,\frac{1}{V_{PP}} \tag{15.5}$$

It can be seen that RJ is inversely proportional to the signal peak voltage and the square root of the capacitive load value. Therefore, higher signal swing and reduced loading capacitance are usually desired for low RJ transmitter driver design.

Cross-coupling-introduced voltage noise in the CMOS voltage mode transmitter driver can be modeled with the circuit shown in figure 15.6. Such effect contributes to the cross-coupling noise that can be modeled using a noise transfer function. It can be seen that effective way for minimizing the cross talk noise is to minimize the cross-coupling capacitance, the reduced slew rate, and the signal swing of the signal.

$$\frac{\Delta V}{V_N} = \frac{RC_X s}{1+R(C+C_X)s} \tag{15.6}$$

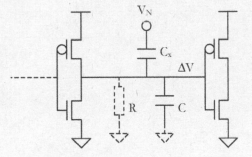

Fig. 15.6. Cross-coupling noise model of VLSI voltage mode driver

The reflection and ringing effects of VLSI voltage mode driver due to the inductive loads shown in figure 15.7 can be effectively minimized by proper termination of the high-speed I/O channel and by the signal slew rate control circuit techniques.

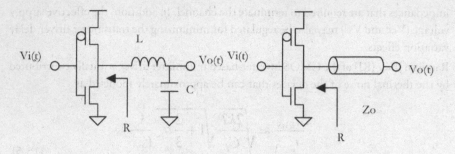

Fig. 15.7. Reflection and ringing effects in VLSI I/O driver circuits

The desired termination impedance of the transmitter driver is determined by the characteristic impedance of the channel as

$$R = Z_0 = \sqrt{L/C}$$

(15.7)

Shown in figure 15.8 are two VLSI circuit implementations of transmitter driver impedance control circuits. In these VLSI transmitter circuit structures, the equivalent output impedance of the transmitter driver circuits can be digitally controlled by controlling the driver device legs in the circuits.

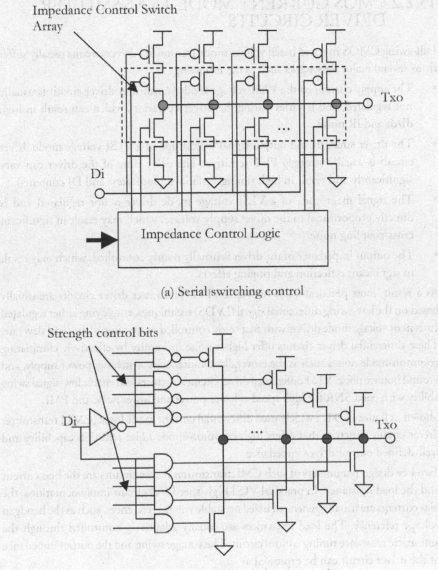

(a) Serial switching control

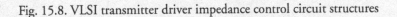

(b) Parallel switching control

Fig. 15.8. VLSI transmitter driver impedance control circuit structures

15.2.2 CMOS CURRENT MODE TRANSMITTER DRIVER CIRCUITS

Full-swing CMOS inverter-based voltage mode transmitter driver circuits usually suffer from several major drawbacks such as the following:

- The supply current to the VLSI voltage mode transmitter driver circuit is usually not regulated, and it varies during the driver operation, which can result in high dI/dt and IR noise.

- The driver strength and output signal magnitude of a VLSI voltage mode driver circuit is usually strongly PVT sensitive. The delay time of the driver can vary significantly and result in high timing variations (both skew and DJ contents).

- The signal magnitude of a VLSI voltage mode driver is not regulated and is directly proportional to the driver supply voltage, which may result in significant cross-coupling noise.

- The output impedance of the driver is usually poorly controlled, which may result in significant reflection and ringing effects.

As a result, most practical VLSI high-speed I/O transmitter driver circuits are usually based on the low-swing differential signal (LVDS) techniques, employing either regulated current or voltage mode drivers with adaptively controlled output impedance and slew rate. These differential driver circuits offer higher noise immunity by effectively eliminating common-mode noises such as the cross talk, simultaneous switching, power supply, and ground bounce noise. VLSI differential driver circuit structures also enable low-signal swing ability with good SNR for high speed at lower power and lower noise and EMI.

Shown in figure 15.9 is a widely used differential current mode logic (CML) transmitter driver circuit structure that offers high common-mode noise rejection capability and well-defined output driver impedance.

Two key design parameters of such CML transmitter driver circuits are the bias current and the load resistance. In practical VLSI high-speed I/O circuit implementations, the bias currents are usually generated based on stable voltage references, such as the bandgap voltage reference. The load resistances are usually adaptively controlled through the automatic resistance tuning control circuit. The voltage swing and the output impedance of the driver circuit can be expressed as

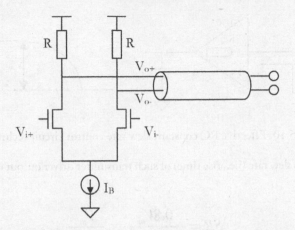

Fig. 15.9. VLSI CML transmitter driver circuit structure

$$V_m = I_B R \tag{15.8}$$

$$R_o = R \tag{15.9}$$

15.2.3 TRANSMITTER DRIVER SLEW RATE CONTROL CIRCUITS

The ringing effects of data signal in the high-speed I/O channel may significantly impact the I/O circuit performance. Most VLSI high-speed I/O circuits are designed to have specified output slew rate to limit the out-of-band signal contents in the transmitter output signals to ensure high-quality data transmission across the high-speed I/O channel. In these cases, the slew rate control technique effectively eliminates high-frequency components at the resonant frequency of the channel discontinuity, therefore minimizing the reflection and cross-coupling effects. Slew rate control circuit techniques are typically adopted in high-speed I/O transmitter driver circuit (such as the USB2.0 transmitter) for tight controlling of the transmitter slew rate to ensure the signal transmission quality.

High-speed I/O transmitter output signal slew rate control can be realized in various ways. Shown in figure 15.10 is a VLSI transmitter slew rate control circuit technique that is based on the control of the effective RC constant of the transmitter capacitive load.

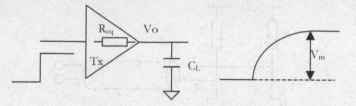

Fig. 15.10. Effective RC constant slew rate control circuit technique

The 10%-90% slew rate (i.e., rise time) of such transmitter driver output can be derived as

$$SR = \frac{0.8V_m}{2.2R_{Eq}C_L} \propto \frac{V_m}{R_{Eq}C_L}$$

(15.10)

It can be seen that the slew rate of the transmitter for a given load capacitance and signal swing can be obtained by properly selecting the transmitter effective output resistance of the driver circuit. However, the above slew rate control technique is highly sensitive to the load capacitance that usually is highly undefined in practical high-speed I/O applications. Shown in figure 15.11 is an improved transmitter slew rate control circuit technique that employs a Miller capacitor compensation technique to minimize the load capacitance sensitivity.

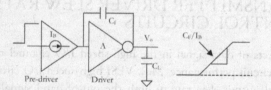

Fig. 15.11. Miller capacitor slew rate control circuit technique

In this slew rate control circuit, the slew rate is approximately independent of the load capacitance if the voltage gain of the driver and the feedback capacitance are high enough:

$$\begin{cases} SR = \dfrac{C_F}{I_B} \\ C_F A >> C_L \end{cases}$$

(15.11)

Shown in figure 15.12 is a VLSI slew rate control circuit implementation based on the Miller capacitor compensation technique.

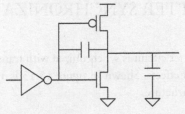

Fig. 15.12. Miller capacitor slew rate control circuit technique

The transmitter slew rate control circuit can also be implemented in digital circuit techniques as shown in figure 15.13, where accurate slew rate control is realized based on the sequential subdriver switching technique.

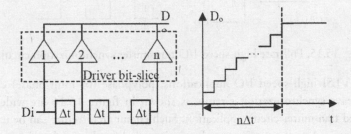

Fig. 15.13. Sequential switching slew rate control circuit architecture

15.3 I/O TRANSMITTER EQUALIZATION CIRCUITS

A typical VLSI high-speed I/O transmitter equalization circuit based on the transmitter de-emphasis technique employing a FIR filter structure to compensate for channel ISI effect is shown in figure 15.14.

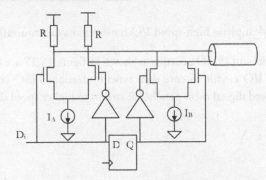

Fig. 15.14. VLSI high-speed I/O transmitter equalizer circuit

15.4 TRANSMITTER SYNCHRONIZATION CIRCUITS

High-speed I/O transmitter circuit is synchronized with transmitter reference clock to minimize delay variation effects. Shown in figure 15.15 is a full-rate high-speed I/O synchronization circuit structure.

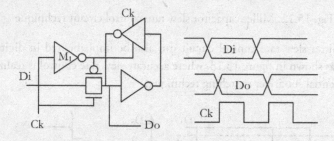

Fig. 15.15. Full-rate high-speed I/O transmitter synchronization circuit

In most VLSI high-speed I/O applications, polyphase (or multiphase) clocking or interleave synchronization circuits as shown in figure 15.16 are widely used high-speed transmitter circuit application. Such circuit architecture can be used very effectively to reduce the requirement for device speed for the synchronization circuit implementation.

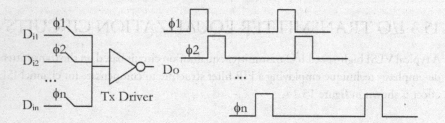

Fig. 15.16. Multiphase high-speed I/O transmitter synchronization circuit

The parallel-in-serial-out (PISO) circuits as shown in figure 15.17 are commonly used in the high-speed I/O circuit to core data synchronization. PISO circuit is used to convert the low-speed digital n-bit parallel bit stream to higher speed digital bit stream for transmission.

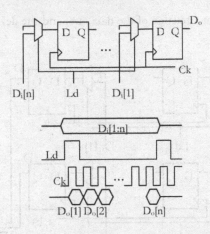

Fig. 15.17. Full-rate N-to-1PISO circuit

A commonly used PISO circuit structure is the half-rate PISO, which operates at half-of-the-clock frequency as shown in figure 15.18. Such a circuit implementation can be used to relax the speed rate requirement of the PISO digital circuit.

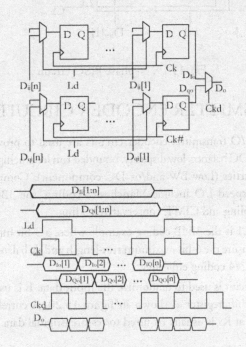

Fig. 15.18. Half-rate PISO circuit

The half-rate PISO circuit can be extended into general multiphase PISO circuit implementation shown in figure 15.19. In this circuit structure, each of the branch circuit only needs to operate at a fraction of the data rate, and the device speed requirement is highly reduced.

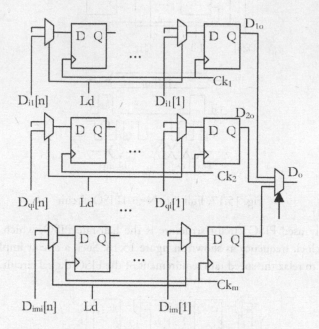

Fig. 15.19. N-phase PISO circuit

15.5 TRANSMITTER ENCODER CIRCUITS

VLSI high-speed I/O transmitter encoder circuits are used to provide desired coding properties, such as DC balance, low disparity, bounded run length, high coding efficiency, and spectral properties (Low BW and/or DC component). Commonly used coding schemes for high-speed I/O include Manchester coding, the 3B4B, 4B5B, 8B10B coding, the scrambling and CIMT, conservative coding.

Shown in table 15.1 is the 3B4B coding example where a 3-bit input data is mapped into 4-bit data to ensure the 6-bit maximum run-length and dc balancing. Such a coding scheme provides a 3/4 coding efficiency.

The scrambling circuit is used to randomize the input data. It is usually implemented using a feedback shift register as shown in figure 15.20. A correlated descrambling (reversing process) at Rx is usually required to restore original data.

Table 15.1. The 3B/4B coding

3B Input	4B output	
	Even Words	Odd Words
000	0011	
001	0101	
010	0110	
011	1001	
100	1010	
101	1100	
110	0110	1011
111	0010	1101
SyncA	0111	1110
SyncB	1000	0001

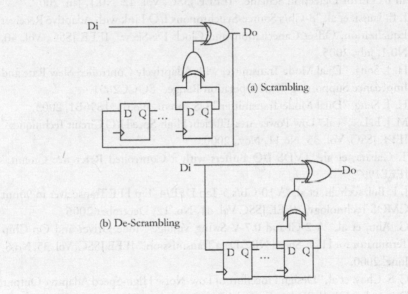

Fig. 15.20. VLSI scrambling/descrambling circuit structures

References:

[1]   H. S. Muthali, et al, "A CMOS 10 Gb/s SONET Transceiver," IEEE JSSC, Vol. 39, No.7, July, 2004.

[2]   S. Kim, et al, "A 6-Gbps/pin 4.2mW/Pin Half-Duplex Pseudo-LVDS Transceiver," IEEE2006.

[3]   J. H. Kim, et al, "A 4-Gb/s/Pin Low Power Memory I/O Interface Using 4-Level Simultaneous Bi-Direction Signaling," IEEEE JSSC, Vol. 40, No.1, July, 2005.

[4]   Y. S. Kim, et al, "A 4-Gb/s/Pin Current Mode 4-Level Simultaneous Bi-directional I/O with Current Mismatch Calibration," ISCAS 2006.

[5]   E. Takahashi, et al, "8Gbps Parallel Data Transmission with Adaptive I/O Circuit," IEEE2006.

[6]   K. H. Kim, "An 8 Gb/s/Pin 9.6ns Row-Cycle 288 Mb Data Rate SDRAM with an I/O Error Detection Scheme," IEEEE JSSC, Vol. 42, N0.1, Jan. 2007.

[7]   J. E. Jaussi, et al, "8-Gb/s Source-Synchronous I/O Link with Adaptive Receiver Equalization, Offset Cancellation, and Clock De-Skew," IEEE JSSC, Vol. 40, N0.1, July, 2005.

[8]   H. J. Song, "Dual Mode Transmitter with Adaptively Controlled Slew Rate and Impedance Supporting Wide Operation Range," SOCC2001.

[9]   H. J. Song, "Dual Mode Transmitter," US Patent, No.6531896B1, 2003.

[10] M. J. E. Lee, et al, "Low-Power Area-Efficient High-Speed I/O Circuit Techniques," IEEE JSSC, Vol. 35, No.11, Nov., 2000.

[11] T. Gabara, et al, "LVDS I/O Buffers with a Controlled Reference Circuit," IEEE1997.

[12] J. F. Bulzacchelli, et al, "A 10-Gb/s 5-Tap DFE/4-Tap FFE Transceiver in 90nm CMOS Technology," IEEE JSSC Vol. 41, No. 12, December 2006.

[13] G. Ahn, et al, "A 2-Gbaud 0.7-V Swing Voltage-Mode Driver and On-Chip Terminator for High-Speed NRZ Data Transmission," IEEE JSSC, Vol. 35, No.6, June. 2000.

[14] C. S. Choy, et al, "Design Procedure of Low-Noise High-Speed Adaptive Output Drivers," 1997 IEEE International Symposium on Circuits and Systems, July 9-12, 1997, Hong Kong.

[15] M. I. Montrose, et al, "Analysis on the Effectiveness of Clock Trace Termination Methods and Trace Lengths on a Printed Circuit Board," IEEE1996.

[16] S. Vishwanthalah, et al, "Dynamic Termination Output Driver for a 600MHZ Microprocessor," IEEE JSSC, Vol. 35, No. 11, Nov. 2000.

[17] W. G. Rivard, et al, "A 1.2um CMOS Differential I/O Systems Capable of 400Mbps Transmission Rates," IEEE.

[18] T. P. Thomas, et al, "Four-Way Processor 800MT/s Front Side Bus with Ground Reference Voltage Source I/O," 2002 Symposium on VLSI Circuits Digest of Technical Papers.

[19] T. F. Knight, et al, "A Self-Terminating Low-Voltage Swing CMOS Output Driver," IEEE JSSC Vol.23, No.2, April 1988.

[20] G. Esch, et al, "Near-Linear CMOS I/O Driver with Less Sensitivity to Process, Voltage, and Temperature Variations," IEEE Trans. VLSI Systems, Vol.12, No.11, November 2004.

[21] Y. Yan, et al, "Low Power High-Speed I/O Interfaces in 0.18um CMOS," IECS2003.

[22] I. S. Stievano, et al, "Parametric Macromodels of Differential Drivers and Receivers," IEEE Trans. Advanced Packaging, Vol. 28, No.2, May 2006.

[23] T. Matano, et, al, "A 1-Gb/s/pin 512-Mb DDRII SDRAM Using a Digital DLL and a Slew-Rate-Controlled Output Buffer," IEEE JSSC Vol.38, No. 5, May 2003.

[24] G. Mandal, et, al, "Low Power VLDS Transmitter with Low Common Mode Variation for 1Gb/s-Per Pin Operation," ISCAS 2005, I-1120-1123.

[25] M. D. Chen, et, al, "Low-Voltage Low-Power LVDS Drivers," IEEE JSSC, vol. 40, no. 2, Feb. 2005.

[26] L. Henrickson, et, al, "Low-Power Fully Integrated 10 Gb/s SONET/SDH Transceiver in 0.13um CMOS," IEEE JSSC, vol. 38, no. 10, Oct. 2003.

[27] C. C. Wang, et, al, "1.0 GBPS LVDS Transceiver Design For LCD Panels," AP-ASIC 2004.

[28] R. Nair, "Enhancing Signal Integrity in Cables: DVI to HDMI and Class-B Differential Signaling," ComLSI, Technology Preview-CBDS.

[29] A. B. Dowlatabadi, "A Robust, Load-Insensitive Pad Driver," IEEE JSSC, vol. 35, no. 4, April 2000.

[30] J. Parnklang, et, al, "Low Power Dissipation CMOS Output Driver Circuits," IEEE 2000.

[31] M. Hashimoto, et, al, "Low dI/dt Noise and Reflection Free CMOS Signal Driver," IEEE CICC 2989.

[32] J. H. Lou, et, al, "A 1.5-V Full-Swing Bootstrapped CMOS Large Capacitive-Load Driver Circuit Suitable for Low-Voltage CMOS VLSI," IEEE JSSC, vol. 32, no. 1, January 1997.

[33] Y. Yang, et, al, "CMOS Off-Chip Driver Design with Simultaneous Switching Noise and Switching Time Considerations," IEEE 1995.

[34] T. J. Gabara, et, al, "Digital Transistor Sizing Techniques Applied to 100K ECL CMOS Output Buffers," IEEE 1993.

[35] C. Svensson, et, al, "High Speed CMOS Chip to Chip Communication Circuit," IEEE.

[36] J. Povazanec, et, al, "On Design of Low-Noise Adaptive Output Drivers," IEEE Tran. Circuits and Systems—I: Fundamental Theory and Applications, vol. 46, no. 12, Dec. 1999.

[37] H. Hatamkhani, et, al, "Power Analysis for High-Speed I/O Transmitter," IEEE 2004 Symposium on VLSI Circuits Digest of Technical Papers.

[38]   G. Esch, et, al, "Near-Linear CMOS I/O Driver with Less Sensitivity to Process, Voltage, and Temperature Variations," IEEE Tran VLSI Systems, vol. 12, no. 11, Nov. 2004.

[39]   K. L. J. Wong, et, al, "A 27-mW 3.6-Gb/s I/O Transceiver," IEEE JSSC, vol. 39, no. 4, April 2004.

[40]   I. S. Stievano, et, al, "Parametric Macro models of Differential Drivers and Receivers," IEEE Tran. Advanced Packaging, vol 28, no. 2, May 2005.

CHAPTER 16
VLSI HIGH-SPEED I/O RECEIVER ANALOG FRONT END CIRCUITS

A VLSI high-speed I/O receiver (Rx) circuit can usually be partitioned into two major functional circuit blocks including a digital circuit clock (such as the data recover circuits (DRC) circuits) and an analog front-end (AFE) circuit. Receiver analog front-end (AFE) circuits are typically used to precondition the received data stream from the channel to ensure reliable receiving of the data. Key design challenges in receiver AFE are associated with the severe signal-to-noise ratio (SNR) degradation of the AFE input, due mainly to highly degraded input signal quality (both in swing and timing) as result of the channel loss effects. On the other hand, low signal magnitude of the AFE circuit also makes it highly sensitive to various circuit nonidealities, such as the reflection, ringing, ISI, voltage offset, and cross talk noise effects.

VLSI high-speed I/O receiver AFE circuits typically employ fully differential signaling technique to minimize the noise impacts. Receiver AFE circuits usually employ adaptive impedance termination and low input capacitance circuit techniques to minimize the channel reflection, ringing, and cross talk effects. Receiver AFE circuits also commonly employ wideband preamplifiers and sometimes the offset compensation circuits to improve the signal-to-noise ratio (SNR) and to minimize the voltage uncertainties of the high-speed sampler circuits. In some VLSI high-speed I/O circuit applications, receiver AFE circuits also employ receiver equalizers to compensate for the ISI effects.

16.1 TIMING ASPECTS OF HIGH-SPEED I/O RECEIVER CIRCUITS

Under the SFG model shown in figure 16.1, VLSI high-speed I/O receiver analog front-end (AFE) circuits typically contribute to part of the high-speed I/O circuit delay parameters including the eye-centering phase shift ($\delta_{T/2}$), the Rx clock path delay (δ_{Rx}), the effective datapath delay (δ_{DP}), and the maximum available timing window (E_{max}):

- Equivalent setup and hold times, voltage thresholds, and the delay variation of the Rx AFE circuits directly contribute to the timing variation of the eye-centering phase shifting. Setup and hold time variation and receiver threshold voltage variations in high-speed I/O circuits are significantly higher compared with the on-chip circuits, mainly resulting from the high channel attenuation, channel-induced ringing, reflection, and cross talk effects.

- Delay and delay variations of the clock distribution circuits also contribute to the Rx clock path delay parameters. For VLSI high-speed I/O circuits consisting of multiple transceiver ports, receiver reference clock repeating and distribution across these ports may result in significant timing variation.

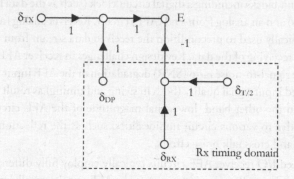

Fig. 16.1. Contribution of VLSI receiver circuit to I/O delay parameters

- The sampler circuits in the receiver AFE are usually part of the data recovery circuits that provide the input data stream and phase information capture. Such information is crucial for proper data recovery circuit loop operation.

The operation of the high-speed I/O receiver AFE circuits can be viewed as the extension of the VLSI on-chip sequential circuit as shown in figure 16.2. In such a circuit, the pass gate employing the device M_3 and M_4 serves as the equivalent sampling (i.e., synchronization) circuit of the receiver. Such sampling circuit is controlled by the clock signals Ck# and Ckd that are distributed using inverter buffer M_1 and M_2 from the input clock Ck. During the sampling operation, the input data D is first captured by the latch circuit formed by M_p and M_f within the receiver flip-flop circuit. The setup and hold times of such data receiver circuit are determined by the time duration for the sampling latch to reliably capture the input data.

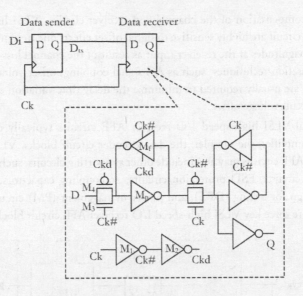

Fig. 16.2. Basic VLSI on-chip receiver circuit structure

The above basic sequential circuit suffers from two major performance limitations in VLSI high-speed I/O circuit applications, mainly due to the timing variation resulting from the delay variation of the input pass gate switch and the threshold variation of the inverter M_p associated with high noise floor, low useable input signal magnitude, and high MOS gate threshold variations. These effects are usually strong functions of the input signal magnitude and slew rate. In practical VLSI high-speed I/O circuits, signals at the receiver input are usually highly degraded in both the signal swing and the slew rate as a result of the channel loss and low-voltage signaling. In addition, there are usually significant ringing, reflection, and cross talk effects resulting from channel impedance discontinuity.

As a result, VLSI high-speed I/O receiver AFE circuits usually need to support the following circuit functions:

- Signal amplification. Amplification of received signal to compensate for the signal swing and slew rate degradations are highly desired for minimizing the receiver delay uncertainties resulting from coupling noise, sampler circuits voltage offsets, sampler circuit setup, and hold time and threshold uncertainties.

- Termination of channel. Channel termination can be used to effectively minimize the reflection, ringing, and cross talk effects for minimizing the delay variation of the received signals.

- Equalization of the channel. Receiver equalizations can be used to compensate for signal loss in high frequency that helps to enhance the swing and slew rate of the received signal. In addition, equalization can be used very effectively to minimize the high-speed I/O deterministic jitter (DJ) resulting from the channel ISI effects.

- Offset compensation of the channel and receiver circuits. VLSI high-speed I/O receiver circuit are highly sensitive to signal offset effects due to significantly lower signal magnitudes at the receiver input as result of the channel loss effects. Offset compensation techniques, such as AC signal coupling, offset voltage calibration circuits, are usually required to minimize the delay time variation effect resulting from circuit voltage offset.

The practical VLSI high-speed I/O receiver AFE circuits typically consist of the high-speed amplifier, the sampler, the data decoder circuit blocks. VLSI high-speed I/O receiver AFE circuits may also include other supporting circuits, such as impedance termination circuits, ESD protection circuits, ac coupling capacitors, Rx equalizer circuits, design for testing (DFT), and power management (PM) circuits. Shown in figure 16.3 are three key VLSI high-speed I/O receiver AFE circuit blocks.

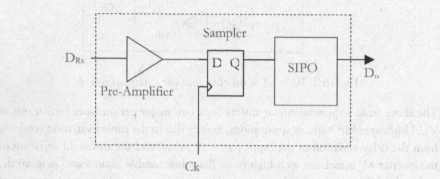

Fig. 16.3. Basic building blocks in high-speed I/O receiver AFE circuit

The received signals are usually first terminated with the impedance termination circuit. Then the signals are AC or DC coupled to the high-speed amplifier circuits. The amplified signals are sampled in the high-speed sampler circuits. The sampler outputs are synchronized to and rate converted for the core circuits employing the serial-in-parallel-out (SIPO) circuits. Offset calibration and equalization operations can be realized at the high-speed amplifier circuits or at the sampler circuits.

The key design goals of high-speed I/O Rx circuits are to achieve minimized setup and hold time window and delay timing variation effect under various PVT variations, cross-coupling noise, ringing, and reflection effects.

16.2 VLSI RECEIVER PREAMPLIFIER CIRCUITS

The high-speed amplifier circuits are typically used in VLSI high-speed I/O circuits to compensate for the input signal loss of received data stream.

16.2.1 SINGLE-END PREAMPLIFIER CIRCUITS

CMOS inverter shown in figure 16.4 provides a simple receiver preamplifier circuits. Such a circuit can be viewed as a "natural" design transition from the on-chip buffer circuits.

Fig. 16.4. Basic CMOS preamplifier

However, basic CMOS inverter suffers from the key problem of large voltage threshold variations under PVT conditions given as

$$\begin{cases} V_{iL} \approx V_{ThN} \\ V_{iH} \approx V_{cc} - |V_{THP}| \end{cases}$$

$$(16.1)$$

In addition, such amplifier circuit also suffers from high sensitivity to common-mode noise, such as the power supply noises and cross talk effects. The high dc current of the inverter for input signal close to the midrail of the supply also results in power dissipation and reliability issues.

16.2.2 DIFFERENTIAL PREAMPLIFIER CIRCUITS

Practical VLSI high-speed I/O receiver AFE circuits usually employ differential amplifier circuits to improve circuit performance. The bias regulated differential amplifier circuit structures as shown in figure 16.5 are commonly used in the high-speed I/O circuits. Such circuits offer high noise immunity since most noises are common-mode noise, which can be rejected by differential amplifier circuit structures. In addition, such amplifier structure provides highly defined input voltage threshold that provides an effective solution to the high dc current issue in basic CMSO inverter for midrail signal.

VLSI high-speed I/O differential receiver amplifier circuit can be implemented in different configurations. In the circuit structure shown in figure 16.5a, the single input signal is compared with a fixed threshold, which offers a pseudo-differential configuration for the data amplification. While the circuit shown in figure 16.5b offers the fully differential amplification of the input differential signal and provides improved performance over the single-end, pseudo-differential preamplifier structures.

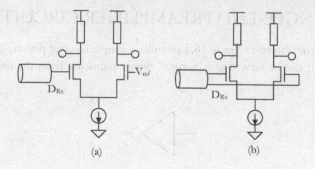

Fig. 16.5. Differential CMOS amplifier circuit structures

16.2.3 DC OFFSET CALIBRATION CIRCUITS

Since signal at the receiver input is highly attenuated due mainly to the high-speed I/O circuit channel loss, the dc offset in the receiver AFE circuit may significantly limit the high-speed I/O performance. As a result, offset calibration circuit is typically required in receiver AFE of high loss high-speed I/O channel to minimize the offset effects. Receiver dc offset calibration can be implemented in various ways. Shown in figure 16.6 is a VLSI dc offset calibration circuit that is based on the ac coupling of the signal. Such circuit can typically be used to eliminate the dc offset for signal input to the ac coupling capacitor.

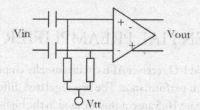

Fig. 16.6. Ac coupling-based dc offset cancellation circuit

However, the passive offset cancellation circuit will not be able to cancel the offset effect downstream circuits in the channel. In such case, the offset cancellation circuit needs to be implemented either in the preamplifier or in the sampler circuits.

Shown in figure 16.7 is a VLSI receiver offset compensation circuit that is based on the introduction of a bias current offset in the amplifier load to compensate for the offset of the circuit.

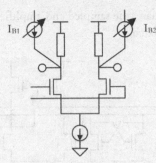

Fig. 16.7. VLSI offset calibration circuit structure

Offset calibration circuit can also be implemented within the sampler circuit. For offset calibration circuit shown in figure 16.8, the tunable capacitor loads are used in the sampler to compensate for the circuit offset effect.

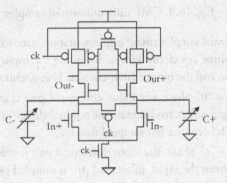

Fig. 16.8. VLSI offset-calibrated sampler circuit structure

16.3 VLSI HIGH-SPEED SAMPLER CIRCUITS

The amplified high-speed I/O data signals in receiver AFE are sampled using high-speed sampler circuit. The sampler circuits convert the received analog signals (or low-swing digital signals) into full-swing digital signals that are used for the DRC loop and core circuits. In the time-domain model, VLSI high-speed I/O sampler circuits serve as synchronization circuit elements that are controlled by the receiver reference clocks from the clock recovery circuits (CRC).

Shown in figure 16.9 is a differential CML flip-flop circuit structure that consists of two identical differential CML latch circuits. Such sampler circuit works as follows:

- During the Ck = "1" phase, the differential signal input is sampled by the input latch and the last sampled bit data is held in the second latch. During this state, the first latch serves as a differential amplifier circuit; such sampler circuit can be used to sampler low-swing input signals.

- During the Ck = "0" phase, the sampled and amplified data in the input latch is shifted into the output latch.

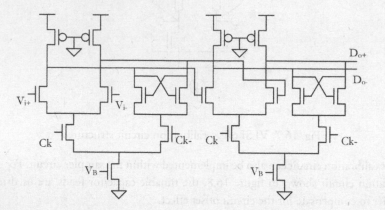

Fig. 16.9. CML fully differential sampler

An alternative differential sampler circuit implementation is shown in figure 16.10. Such sampler consists of three key circuit blocks, including the input differential pair (i.e., Gm), the input latch, and the output latch circuits. This circuit works as follows:

- During the Ck = "0" phase, the input differential pair is tristated and the input latch is precharged to Vcc to eliminate the residual charge of last sampled data bit. The output latch keeps the last sampled data bit.

- During the Ck = "1" phase, the input differential pair is enabled. The differential current output from the input differential pair is sampled by the input latch. The sampled data then directly transfer into the output latch. It can be seen that the differential pair in this phase serves as an amplifier; such sampler circuit can be directly used to sample the low-swing differential input signal.

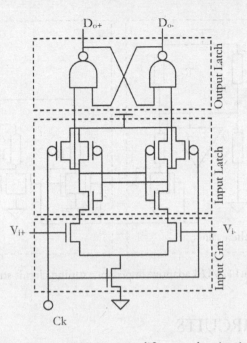

Fig. 16.10. VLSI sense amplifier sampler circuit

16.4 VLSI RECEIVER CHANNEL TERMINATION CIRCUITS

Receiver terminations are commonly required in high-speed I/O to ensure signal quality for minimizing the reflection from the channel. In order to compensate for the PVT variations, an adaptive control circuit is usually used to control the termination resistance for various operation conditions. Shown in figure 16.11 is a VLSI receiver termination circuit implementation that offers adaptively controlled receiver input impedance independent of the PVT conditions.

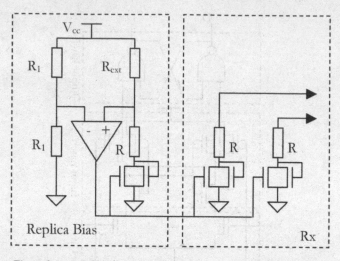

Fig. 16.11. VLSI adaptive impedance tuning circuit structure

16.5 SIPO CIRCUITS

VLSI high-speed I/O serial-in-parallel-output (SIPO) circuit serves as part of the synchronization circuits that synchronize data stream between the VLSI high-speed I/O receiver AFE and the core circuits. SIPO circuits also realize frequency domain conversion from high-speed serial AFE domain to low-speed parallel core domain circuits. Shown in figure 16.12 is a VLSI SIPO circuit implementation, where the high-frequency serial bit stream of frequency f is converted to N-bit parallel data stream of f/N frequency as shown figure 16.13.

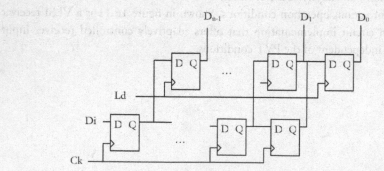

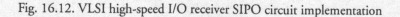

Fig. 16.12. VLSI high-speed I/O receiver SIPO circuit implementation

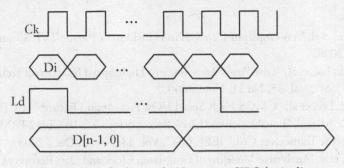

Fig. 16.13. VLSI high-speed I/O receiver SIPO logic diagram

References:

[1] S. Kim, et al, "A 6-Gbps/pin 4.2mW/Pin Half-Duplex Pseudo-LVDS Transceiver," IEEE2006.

[2] M. J. E. Lee, et al, "Low-Power Area-Efficient High-Speed I/O Circuit Techniques," IEEE JSSC, Vol. 35, No.11, Nov., 2000.

[3] M.J.E. Lee, et al, "CMOS High-Speed I/Os, Present and Future," ICCD2003.

[4] J. L. Zerbe, et al, "Equalization and Clock Recovery for a 2.5-10-Gb/s 2-PAM/4-PAM Backplane Transceiver Cell," IEEE JSSC, Vol. 38, No.12, Dec., 2003

[5] J. Lee, et al, "Analysis of Modeling of Bang-Bang Clock and Data Recovery Circuits," IEEE JSSC, Vol. 39, No.9, Sept., 2004.

[6] W. Ellersick, et al, "GAD: A 12 GS/s CMOS 4-Bit A/D Converter for Equalized Multi-Level Link," 1999 Symposium on VLSI Circuits Digest of Technical Papers.

[7] H. J. Song, "Architecture and Implementation of Power and Area Efficient Receiver Equalization Circuit for High-Speed Serial Data Communication," SOCC2006.

[8] I. S. Stievano, et al, "Parametric Macromodels of Differential Drivers and Receivers," IEEE Trans. Advanced Packaging, Vol. 28, No. 2, May 2006.

CHAPTER 17
VLSI HIGH-SPEED I/O DATA RECOVERY CIRCUITS

Precise timing recovery operation is among the most important operations of VLSI high-speed I/O circuits. Timing recovery in VLSI-embedded clock high-speed I/O circuits is usually realized using the data recovery circuits (DRC) that serve as a special phase-locked circuit family to adaptively control the receiver reference clock phase based on the received data stream such that the receiver sampling clock aligns to the center of the incoming data eye pattern. VLSI data recovery circuits estimate the most probable timing at which the incoming data stream should be sampled by comparing the estimated timing and the timing information embedded in the data transitions and determine the digital values based on the optimal sampling of signal.

The performance of the data recovery circuit is typically specified by its jitter tolerance and jitter transfer functions. Jitter tolerance is defined as the amplitude of sinusoidal jitter versus frequency a receiver DRC can handle for a given bit error rate (BER). Jitter tolerance represents the ability of the DRC to recover an incoming serial data correctly despite the applied jitter. Jitter tolerance of the DRC is usually specified using a jitter tolerance mask. Jitter transfer function is defined as the ratio of the DRC output jitter of a recovered clock to the input jitter versus frequency. Jitter transfer function represents the gain by which the DRC attenuates or amplifies jitter. Jitter tolerance and transfer function are usually directly related to the DRC loop frequency response. A reasonable loop bandwidth is usually desired for better jitter transfer function characteristics and loop stability.

17.1 BASIC DATA RECOVERY CIRCUIT OPERATIONS

The high-speed I/O circuit structure shown in figure 17.1 employs a data recovery circuit (DRC) to generate the receiver sampling clock that tracks the center of the input data eye pattern for optimal data sampling.

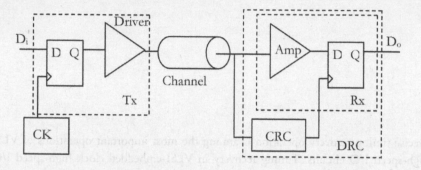

Fig. 17.1. Embedded clock high-speed I/O circuit architecture

VLSI data recovery circuits can be implemented in various ways employing VCDL, VCO, PI, or polyphase clock circuits. A typical high-speed I/O data recovery circuit structure shown in figure 17.2 consists of several key functional circuit blocks, including a phase detector (PD) circuit for incoming data delay phase extraction, a loop filter (LPF) circuit for loop response compensation, a phase adjustment unit (PAU) for optimal sampling clock phase generation, and a synchronization circuit for received data sampling and synchronization.

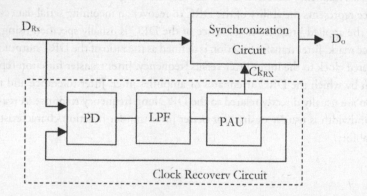

Fig. 17.2. Typical VLSI high-speed I/O DRC circuit structure

17.2 TIME-DOMAIN DRC SFG MODELS

A VLSI DRC can be mathematically modeled based on the time-domain SFG model as shown in figure 17.3, where a clock recovery circuit (CRC) serves as a phase domain signal processing circuit element for compensating the delay variations of the incoming data stream.

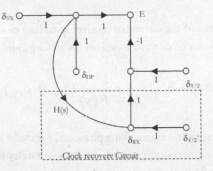

Fig. 17.3. SFG Model of high-speed I/O data recovery circuit

The CRC loop in the DRC can typically be modeled using a lowpass transfer function H(s) that has a unity dc gain and is characterized by a CRC loop bandwidth. A CRC circuit is bandwidth limited such that it will lose track of the incoming data signal phase for frequency beyond the CRC loop bandwidth.

Shown in figure 17.4 is a detailed implementation of the CRC loop that employs a loop filter F(s) with the general desired frequency characteristics as

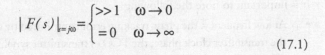

$$|F(s)|_{s=j\omega} = \begin{cases} \gg 1 & \omega \to 0 \\ = 0 & \omega \to \infty \end{cases} \tag{17.1}$$

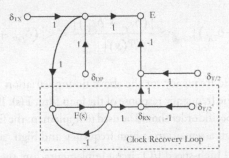

Fig. 17.4. SFG model of high-speed I/O data recovery circuits

The general operation of the VLSI high-speed I/O data recovery circuit is to approximate the MxTM condition by estimating the most probable delay time at which the incoming data stream should be sampled. Such operation can be expressed using the optimal regenerated receiver reference clock with a feedback loop based on the phase of the received data stream as

$$\delta_{RX} = \frac{F(s)}{F(s)+1}(\delta_{DP} + \delta_{TX}) - (\delta_{T/2})^* \rightarrow \delta_{DP} + \delta_{TX} - \delta_{T/2}$$

(17.2)

The tracking phase error of the estimated receiver sampling clock phase with respect to the MxTM condition can be derived based on the SFG model as

$$E \equiv \delta_{RX} - \delta_{DP} + \delta_{TX} - \delta_{T/2} = \frac{1}{1+F(s)}(\delta_{DP} + \delta_{TX}) + \Delta\delta_{T/2}$$

(17.3)

The loop transfer function and the tracking phase error transfer function with respect to the input phase uncertainties contributed by the datapath delay time and the transmitter reference clock can be derived as

$$H_{Loop}(s) \equiv \frac{\delta_{RX}}{(\delta_{DP} + \delta_{TX})}\Big|_{\Delta\delta_{T/2}=0} = \frac{F(s)}{F(s)+1}$$

(17.4)

$$H(s)_{eye\_centering} \equiv \frac{E}{\Delta\delta_{T/2}}\Big|_{(\delta_{DP}+\delta_{TX})=0} = 1$$

(17.5)

It is important to note the following:

- At low frequency, the phase tracking error contributed by the phase parameters from the transmitter clock phase, the TCO of transmitter synchronization circuit, and the effective datapath delay of the channel is highly attenuated by the DRC loop:

$$\Big|\frac{1}{F(s)+1}(\delta_{DP} + \delta_{TX})\Big|_{\omega\to 0} \approx \frac{|(\delta_{DP} + \delta_{TX})|}{|F(s)|_{\omega\to 0}} << |(\delta_{DP} + \delta_{TX})|$$

(17.6)

- The characteristic frequency for the above timing variation filtering operation is determined by the frequency response of the loop filter F(s). Based on such theory, integrations of specific order should be used to implement the filter F(s) for the DRC that provides high attenuation at high frequency and high gain at low frequency

- In practical VLSI high-speed I/O circuit implementation, the actual eye-centering phase shifting will usually deviate from its optimal value as specified by the MxTM condition resulting from circuit nonidealities, such as the nonsymmetry in the setup and hold times in the VLSI receiver sampler circuit, the static clock phase error and clock jitter, the ISI effect, etc. The impact of such delay variation can be modeled using the eye-centering phase error term as

$$\Delta\delta_{T/2} \equiv \delta^{*}_{T/2} - \delta_{T/2} \tag{17.7}$$

Alternatively, the high-speed I/O data recovery circuits can also be implemented based on the SFG shown in figure 17.5, where the input data signal phase is compared with the receiver clock with phase offset. Such a phase detection technique provides an alternative circuit implementation for the VLSI high-speed I/O circuit data phase detection.

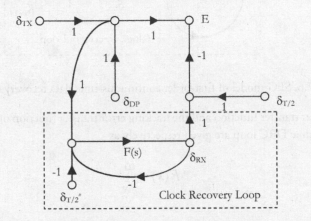

Fig. 17.5. Alternative SFG model of high-speed I/O data recovery circuits

The clock recovery circuit transfer function and the tracking phase error transfer function based on the above alternative DRC loop model can be expressed as

$$\delta_{RX} = \frac{F(s)}{F(s)+1}(\delta_{DP} + \delta_{TX}) - \frac{F(s)}{F(s)+1}(\delta_{T/2})^{*} \tag{17.8}$$

$$E = \frac{1}{1+F(s)}(\delta_{DP} + \delta_{TX}) + (\frac{F(s)}{F(s)+1}\delta^{*}_{T/2} - \delta_{T/2}) \tag{17.9}$$

Note that this DRC loop has similar phase error transfer functions as the previous one for low frequency up to the loop bandwidth. The major deviation between the two DRC implementations is the eye-centering phase error transfer function term.

17.2.1 FIRST-ORDER DRC LOOPS

Shown in figure 17.6 is a first-order continuous-time data recovery circuit loop employing a simple integrator loop filter.

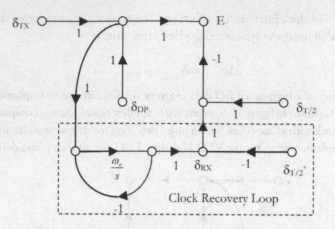

Fig. 17.6. SFG model of first-order continuous-time data recovery circuit

The loop filter transfer function and the tracking error transfer function of a first-order continuous-time DRC loop are given respectively as

$$F(s) = \frac{\omega_o}{s} \tag{17.10}$$

$$E = \frac{1}{1+F}(\delta_{DP} + \delta_{TX}) + \Delta\delta_{T/2} = \frac{s}{\omega_o + s}(\delta_{DP} + \delta_{TX}) + \Delta\delta_{T/2} \tag{17.11}$$

It can be seen that a first-order CRC loop provides a first-order lowpass response with the -3dB bandwidth ω_o determined by the loop filter time constant. The tracking phase error transfer function, on the other hand, has a first-order highpass response with the -3dB cutoff frequency specified by ω_o. As a result, the high-frequency jitter components in the input data stream beyond the loop bandwidth cannot be effectively attenuated.

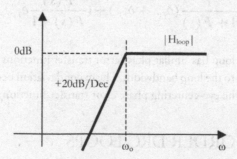

Fig. 17.7. Tracking phase error transfer function response of first-order DRC loops

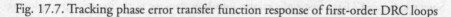

Shown in figure 17.8 is a typical first-order data recovery loop implementation employing a discrete-time integrator loop filter.

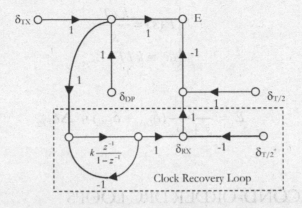

Fig. 17.8. SFG model of first-order DT data recovery circuit

The loop filter transfer function and the tracking error transfer function for the first-order discrete-time DRC loop are given respectively as

$$F(s) = k \frac{z^{-1}}{1 - z^{-1}}$$

(17.12)

$$E = \frac{1 - z^{-1}}{1 + (k-1)z^{-1}} \delta_i + \Delta \delta_{T/2}$$

(17.13)

The above z-domain system under the oversampling operation is equivalent to an s-domain system as shown in figure 17.9 as

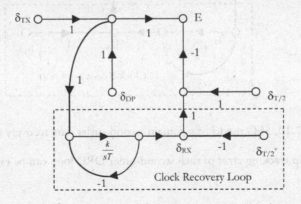

Fig. 17.9. s-domain equivalent SFG of first-order DT data recovery circuit

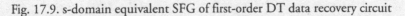

The equivalent s-domain loop filter transfer function and the loop tracking error transfer function for the first-order discrete-time DRC loop can be expressed as

$$\begin{cases} F(s) = \dfrac{k/T}{s} \\ \omega_o \equiv k/T \end{cases}$$

(17.14)

$$E = \frac{s}{s + \omega_o}(\delta_{DP} + \delta_{TX}) + \Delta\delta_{T/2}$$

(17.15)

17.2.2 SECOND-ORDER DRC LOOPS

A typical second-order continuous-time clock and data recovery loop can be realized employing a second-order loop filter in the form as

$$F(s) = \frac{\omega_o}{s}\left(\frac{\omega_o}{s} + \frac{1}{Q}\right)$$

(17.16)

The SFG model of such a data recovery system is shown in figure 17.10.

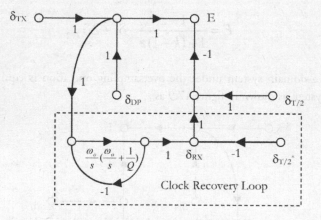

Fig. 17.10. SFG model of s-domain second-order data recovery circuit

The DRC loop tracking error of such second-order DRC loop can be expressed as

$$E = \frac{\left(\dfrac{s}{\omega_o}\right)^2}{\left(\dfrac{s}{\omega_o}\right)^2 + \dfrac{1}{Q}\left(\dfrac{s}{\omega_o}\right) + 1}(\delta_{DP} + \delta_{TX}) + \Delta\delta_{T/2}$$

$$(17.17)$$

It can be seen that a second-order DRC loop has a second-order highpass frequency response with respect to the tracking phase error as shown in figure 17.11.

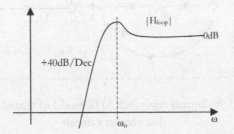

Fig. 17.11. Frequency response of the second-order CRC and DRC loops

It can also be seen that a second-order DRC loop may have the peaking effect at the characteristic frequency if the Q-factor is large than 1.

The second-order DRC loop can also be implemented in the discrete time form as shown in figure 17.12.

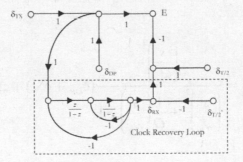

Fig. 17.12. SFG model of second-order DT data recovery circuit loop

The tracking phase error transfer function of such data recovery loop can be derived as

$$E = \left(1 - z^{-1}\right)^2 (\delta_{DP} + \delta_{TX}) + \Delta\delta_{T/2}$$

$$(17.18)$$

It can be seen that such a second-order discrete-time DRC loop also has a second-order highpass tracking error transfer function response.

Similarly under the oversampling operation condition, the above z-domain DRC loop can be approximated using an s-domain as shown in figure 17.13.

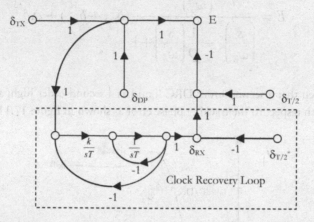

Fig. 17.13. S-domain equivalent SFG model of second-order DT
data recovery circuit

The equivalent clock recovery transfer and tracking phase error functions of such DRC loop can be expressed as

$$\delta_{RX} = (\delta_{DP} + \delta_{TX}) \frac{1}{\left(\dfrac{s}{\omega_o}\right)^2 + \dfrac{1}{Q}\left(\dfrac{s}{\omega_o}\right) + 1}$$

(17.19)

$$E = \frac{\left(\dfrac{s}{\omega_o}\right)^2}{\left(\dfrac{s}{\omega_o}\right)^2 + \dfrac{1}{Q}\left(\dfrac{s}{\omega_o}\right) + 1}(\delta_{DP} + \delta_{TX}) + \Delta\delta_{T/2}$$

(17.20)

where

$$\omega_o \equiv \sqrt{k}/T$$

$$Q \equiv \sqrt{k}$$

(17.21)

Shown in figure 17.14 is an alternative implementation of the second-order discrete-time DRC circuit loop.

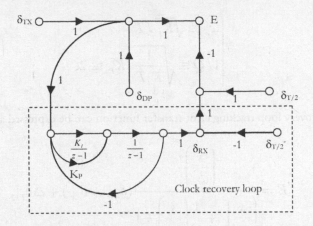

Fig. 17.14. SFG model of second-order DT data recovery circuit

The equivalent s-domain loop of the above clock recovery loop based on the oversampling condition is shown in figure 17.15.

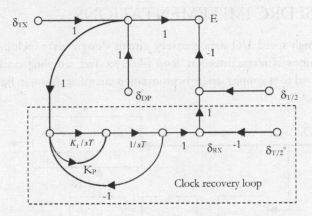

Fig. 17.15. The s-domain equivalent of second-order DT clock recovery loop

It can be seen that such a data recovery loop is the same as a second-order phase loop discussed in the previous chapter with the clock recovery transfer function given as

$$\frac{\delta_{Rx}^*}{(\delta_{DP} + \delta_{TX})} = \frac{\frac{1}{Q}\left(\frac{s}{\omega_o}\right) + 1}{\left(\frac{s}{\omega_o}\right)^2 + \frac{1}{Q}\left(\frac{s}{\omega_o}\right) + 1}$$

(17.22)

where

$$\begin{cases} \omega_o \equiv \sqrt{K_I} \, / \, T \\ 1/Q \equiv \sqrt{\dfrac{1}{K_I T}} \cdot K_P \equiv 2\zeta \end{cases}$$

(17.23)

The data recovery loop tracking error transfer function can be expressed as

$$E = \frac{\left(\dfrac{s}{\omega_o}\right)^2}{\left(\dfrac{s}{\omega_o}\right)^2 + \dfrac{1}{Q}\left(\dfrac{s}{\omega_o}\right) + 1}(\delta_{DP} + \delta_{TX}) + \Delta\delta_{T/2}$$

(17.24)

Consequently, the tracking error frequency response is similar to figure 17.11.

17.3 VLSI DRC IMPLEMENTATIONS

Key VLSI high-speed I/O data recovery circuit design tasks include the circuit implementations of phase detection, loop filter, receiver sampling clock generation circuit, received data sampler, and synchronization circuit as shown in figure 17.16.

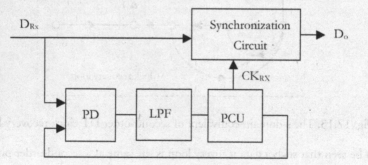

Fig. 17.16. Key VLSI high-speed I/O DRC circuit building blocks

Two typical SFG models of the high-speed I/O data recover circuits are shown in figure 17.17. These DRC circuit models can readily be implemented in the VLSI circuits by mapping the signal processing elements into their VLSI circuit forms.

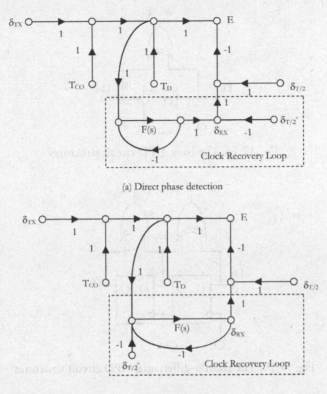

(a) Direct phase detection

(b) Phase shifted phase detection

Fig. 17.17. VLSI high-speed I/O data recovery circuit SFG models

17.3.1 PHASE DETECTOR CIRCUIT

Shown in figure 17.18 to figure 17.23 are several VLSI high-speed I/O data phase detector (DPD) circuit implementations. These phase detection circuits are used to compare the delay phase of the received data stream with respect to the receiver sampling clock. The outputs of the phase detector circuits are a sampled analog signal (i.e., PWM signal) for analog-based phase detector circuit architectures or digital signal (i.e., PCM) for digital-based phase detector circuit architectures.

For analog DPD, the output voltages are filtered using the s-domain loop filters for generating the analog control signal for the delay phase control circuit elements, such as the VCO, VCDL, and PI for the receiver sampling reference clock generations. For digital DPD, the output voltages are filtered using the z-domain loop filters for generating the digital control signal of receiver reference clock generations.

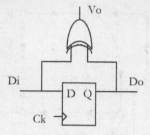

Fig. 17.18. The basic DPD circuit structures

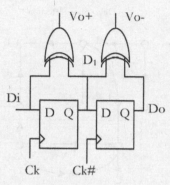

Fig. 17.19. The pseudo-differential DPD circuit structures

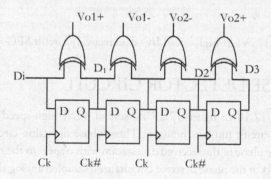

Fig. 17.20. The double pseudo-DPD circuit structures (I)

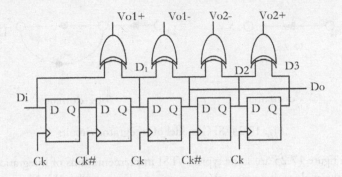

Fig. 17.21. The double pseudo-DPD circuit structures (II)

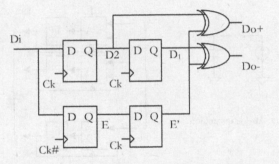

Fig. 17.22. The digital DPD circuit structure

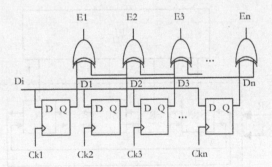

Fig. 17.23. The oversampling DPD circuit structures

17.3.2 LOOP FILTER

Analog or digital loop filters based on analog or digital integrators as shown in figure 17.24 are usually used in this VLSI high-speed I/O data recovery loop for DRC loop compensation.

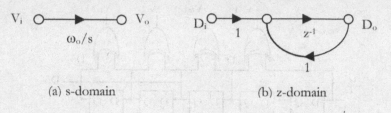

(a) s-domain   (b) z-domain

Fig. 17.24. SFG model of integrator circuits

Shown in figure 17.25 are some typical VLSI implementations of integration circuits based on the analog integration, the carry propagation parallel addition, and the shift register circuit structures.

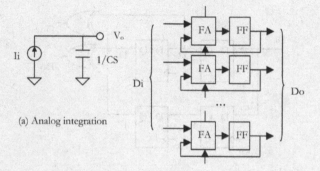

(a) Analog integration

(b) Carry propagation addition digital integration

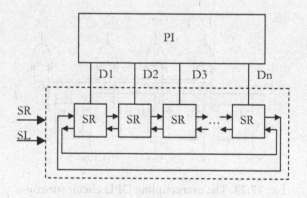

(c) Shift register digital integration

Fig. 17.25. VLSI integration circuit implementations

Digital integrator can be used as basic circuit elements to construct high order z-domain DRC loop filter as shown in figure 17.26 and figure 17.27.

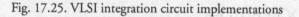

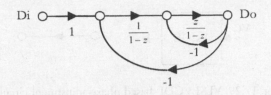

Fig. 17.26. SFG of second-order digital integration loop circuit

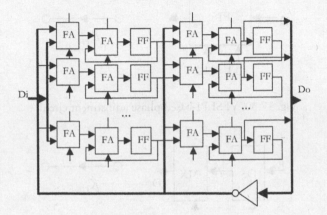

Fig. 17.27. VLSI second-order digital integration loop circuit

17.3.3 VOLTAGE-CONTROLLED DELAY CIRCUITS

Delay phase adjustment circuits in VLSI high-speed I/O circuits are used to control the receiver sampling clock phase such that it is aligned to the center of the incoming data eye pattern to achieve optimized receiver data capture in the presence of the PVT variations, loading conditions, and various noise source and jitter sources.

VLSI DRC delay phase adjustment circuit can be implemented using various VLSI circuit techniques, including the VCO, the VCDL, the PI, and the oversampling circuit technique as shown in figure 17.28, figure 17.29, figure 17.30, and figure 17.31.

Fig. 17.28. VLSI VCO-based phase adjustment circuit

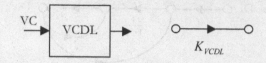

Fig. 17.29. VLSI VCDL-based phase adjustment circuit

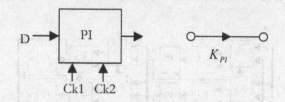

Fig. 17.30. VLSI PI-based phase adjustment circuit

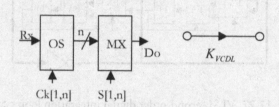

Fig. 17.31. VLSI oversampler-based phase adjustment circuit

17.3.4. DRC LOOP IMPLEMENTATIONS

Shown in figure 17.32 is a VCO-based first-order VLSI high-speed I/O DRC circuit implementation, where the phase detector is directly connected to the VCO with a usually added simple lowpass filter that has a much higher bandwidth than the DRC loop for attenuating the sampling noise.

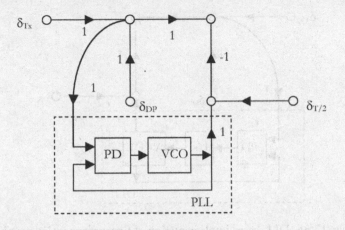

Fig. 17.32. PLL-based implementation of first-order continuous-time
data recovery circuit

The SFG model of this high-speed I/O DRC circuit loop is shown in figure 17.33.
The phase integration operation of the CRC loop in such circuit is realized by using
the integration operation of the VCO itself. As a result, the loop filter of the CRC can
be eliminated.

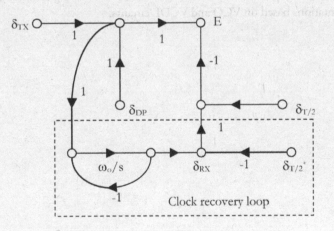

Fig. 17.33. SFG model of first-order continuous-time data recovery circuit

An equivalent DRC loop employing a VCDL-based first-order high-speed I/O circuit
implemented is shown in figure 17.34.

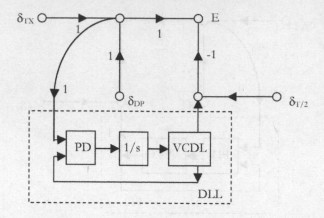

Fig. 17.34. DLL-based implementation of first-order continuous-time
data recovery circuit

In this circuit implementation, an analog integrator circuit is required since VCDL circuit does not offer the integration function. Such a circuit provides a first DRC loop that is functionally equivalent to the circuit shown in figure 17.32.

Alternatively, the above DRC loop can also be realized in digital forms employing VCO, VCDL, or PI circuits. Shown in figure 17.35 are two digital VLSI DRC circuit implementations based on VCO and VCDL circuits.

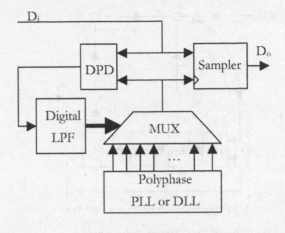

(a)

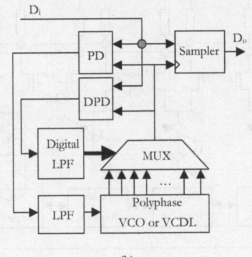

(b)

Fig. 17.35. Digital DRC circuit implementation employing VCO and VCDL

VLSI high-speed I/O digital DRC circuits usually employ VLSI PI circuits for phase adjustment. Shown in figure 17.36 is a VLSI digital DRC loop implemented using the PI circuit.

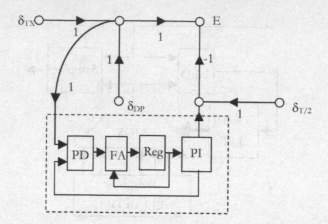

(a) PI based DRC circuit architecture

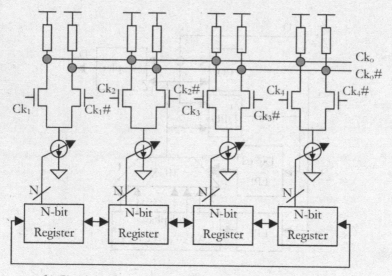

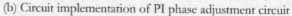

(b) Circuit implementation of PI phase adjustment circuit

Fig. 17.36. PI-based first-order continuous-time data recovery circuit

The SFG model of such first-order z-domain DRC loop is shown in figure 17.37. It can be seen that at the oversampling condition (i.e., the control clock frequency of the integrator and PI circuits are very much higher than the DRC loop bandwidth), such DRC circuit is equivalent to the VCO- and VCDL-based DRC loops.

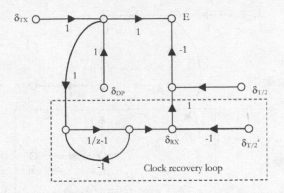

Fig. 17.37. SFG model of first-order discrete-time data recovery circuit

Note that for the high-speed I/O circuit data recovery circuit employing the phase interpolator, the digital-based implementation of the loop filter circuits are usually used for better process scalability and design robustness.

17.4 RECEIVER NOISE MARGIN MODELS

Receiver jitter tolerance (i.e., eye width) and eye height margin tests are two commonly used methods to specify the performance of the high-speed I/O receiver circuits.

Receiver jitter tolerance is used to specify the timing margin of the high-speed I/O receiver circuit. Receiver eye height margin is a parameter to specify the voltage margin of the high-speed I/O circuit. Both noise margins can be validated with the help of the receiver DRC circuits.

17.4.1 RECEIVER JITTER TOLERANCE

Receiver jitter tolerance (i.e., eye width) is usually defined as the maximum (single tone) periodic jitter (PJ) that a high-speed I/O receiver can handle with a certain BER. Consequently, receiver jitter tolerance of embedded clock I/O circuits also provides a measure of the frequency response of the receiver data recovery circuit loop.

Based on the embedded clock I/O circuit SFG model shown in figure 17.38, the tracking phase error equation can be expressed as

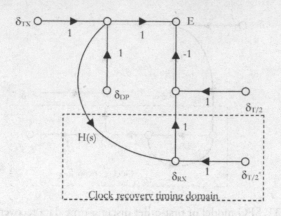

Fig. 17.38. SFG Model of high-speed I/O data recovery circuit

$$E = (1 - H(s))(\delta_{DP} + \delta_{TX}) + \Delta\delta_{T/2} \tag{17.25}$$

where H(s) is a lowpass transfer function representing the clock recovery circuit loop, $\Delta\delta_{T/2}$ is the tracking phase error that is contributed from the high-speed I/O circuit nonidealities, such as the static eye-centering phase shift error and the jitter components that cannot be effectively compensated by the DRC loop.

Receiver jitter tolerance for such system can be approximately expressed as

$$\left|\delta_i(j\omega)\right|_{max} = (E_{Max} - \Delta\delta_{T/2}) / \left|1 - H(j\omega)\right| \tag{17.26}$$

Note that such a jitter tolerance equation can usually have the following properties:

- At very low PJ frequency (i.e., within the bandwidth of the DRC loop transfer function), $|(1-H(j\omega)| \rightarrow 0$, due mainly to the lowpass CRC loop transfer characteristics, implying high jitter tolerance that can even be beyond the data UI.

$$\left|\delta_i(j\omega)\right|_{max} \underset{\omega \to 0}{=} (E_{Max} - \Delta\delta_{T/2}) / \left|1 - H(j\omega)\right| \rightarrow \infty \tag{17.27}$$

- At very high PJ frequency (i.e., beyond the bandwidth of the DRC loop transfer function), $|(1-H(j\omega)| \rightarrow 1$, due also to the lowpass CRC loop transfer characteristics, implying jitter tolerance that is limited by the jitter and phase error noise floor of the high-speed I/O receiver circuit.

$$\left|\delta_i(j\omega)\right|_{max} \underset{\omega \to \infty}{=} (E_{Max} - \Delta\delta_{T/2}) \tag{17.28}$$

- At PJ frequency near the bandwidth of the DRC loop, jitter tolerance curve decreases as result of the decrease of the DRC tracking capability limited by the DRC loop bandwidth.

As a result, the jitter tolerance response curve of the receiver is usually divided into two main regions with low-frequency performance limited by the DRC loop performance and the high-frequency performance limited by the inherent phase error and jitter performance of the receiver circuit itself. In addition, the boundary of the two regions will be shifted to higher frequency with higher DRC loop bandwidth and to lower frequency with higher DRC loop bandwidth.

Such behavior of the receiver jitter tolerance can be described using the jitter tolerance curve as shown in figure 17.39.

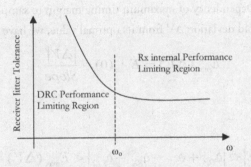

Fig. 17.39. High-speed I/O receiver jitter tolerance curve

In practical VLSI high-speed I/O circuit design, a jitter tolerance mask is usually used to specify the performance of the receiver performance. Such jitter tolerance mask is shown in figure 17.40.

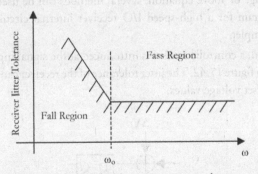

Fig. 17.40. High-speed I/O receiver jitter tolerance mask

17.4.2 RECEIVER EYE HEIGHT MARGIN

For VLSI high-speed I/O circuit, the available timing budget E_{max} in prototype circuit model depends on the threshold voltage of the comparator in the synchronization (sampler) circuit. As shown in figure 17.41, the effective high-speed I/O timing budget will be degraded when the threshold voltage is moved away from its optimized value.

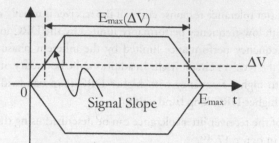

Fig. 17.41. Dependency of maximum timing margin to sampler threshold

For a given threshold deviation ΔV from its optimal value, we have that

$$E_{max}(\Delta V) = E_{max}(0) - \frac{|\Delta V|}{Slope} \qquad (17.29)$$

and that

$$\begin{cases} |E| \equiv |\delta_{DP} + \delta_{TX} - \delta_{RX} - \delta_{T/2}| < E_{max}(\Delta V) \\ \delta_{T/2} \equiv \dfrac{T - T_{setup} + T_{hold}}{2} \end{cases} \qquad (17.30)$$

By taking advantage of above equation, several methods can be used to construct the effective eye diagram for a high-speed I/O receiver internal circuit node that is dc coupled to the sampler.

In the first method, a controlled offset is introduced at the signal input to the sampler circuit as shown in figure 17.42. The jitter tolerance of the receiver can then be measured for each of the offset voltage values.

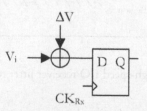

Fig. 17.42. Jitter tolerance-based internal eye diagram measurement

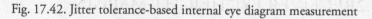

The jitter tolerance of the receiver by this method can then be used to approximately construct the internal eye diagram with respect to the receiver sampler clock by finding the slope of the signal as

$$\left|\delta_i\right|_{\max} \underset{\omega >> \omega_o}{=} E_{\max}(0) - \frac{|\Delta V|}{Slope} - \Delta\delta_{T/2}$$

(17.31)

In the second method, the BER is measured for the receiver circuit without the PJ injection in the input data. In such method, the static offset is purposely introduced into the eye-centering phase shift, and maximum offset without receiver failure for a given voltage signal dc offset is measured:

$$0 = E_{\max}(0) - \frac{|\Delta V|}{Slope} - (\Delta\delta_{T/2}(\Delta V))_{\max}$$

(17.32)

References:

[1] Hsiang-Hui Chang, et al, "Low Jitter and Multi-rate Clock and Data Recovery Circuit Using a MSADLL for Chip-to-Chip Interconnection," IEEE TRANSACTIONS ON CIRCUITS AND SYSTEMS—I: REGULAR PAPERS, VOL. 51, NO. 12, DECEMBER 2004

[2] Mehrdad Ramezaizi, et al," An Improved Bang-bang Phase Detector for Clock and Data Recovery Applications," IEEE2001.

[3] Michael H. Perrott, "Behavioral Simulation of Fractional-N Frequency Synthesizers and Other PLL Circuits," IEEE Design & Test of Computers

[4] Behzad Razavi," Challenges in the Design of High-Speed Clock and Data Recovery Circuits," IEEE Communications Magazine • August 2002

[5] Jafar Savoj, et al, "A 10-Gb/s CMOS Clock and Data Recovery Circuit with a Half-Rate Binary Phase/Frequency Detector," IEEE Journal Of Solid-State Circuits, Vol. 38, No. 1, January 2003

[6] Hideyuki Nosaka,et al, "A 10-Gb/s Data-Pattern Independent Clock and Data Recovery Circuit With a Two-Mode Phase Comparator," IEEE JSSC, VOL. 38, NO. 2, FEBRUARY 2003

[7] Seong-Jun Song, et al, "A 4-Gb/s CMOS Clock and Data Recovery Circuit Using 1/8-Rate Clock Technique," IEEE JOURNAL of SOLID-STATE CIRCUITS, VOL. 38, NO. 7, JULY 2003

[8] Bong-Joon Lee, et al, "A 2.5-10-Gb/s CMOS Transceiver With Alternating Edge-Sampling Phase Detection for Loop Characteristic Stabilization," IEEE JSSC, VOL. 38, NO. 11, NOVEMBER 2003

[9] Ruiyuan Zhang, et al, "Clock and Data Recovery Circuits with Fast Acquisition and Low Jitter," 2004 IEEE.

[10] S. I. Ahmed, et al, "An All-Digital Data Recovery Circuit Optimization Using MATLAB/SIMULINK," 2005 IEEE.

[11] H. J. Song," Algorithm, Architecture, and Implementation of Algorithmic delay-Locked Loop Based Data Recovery Circuit for High-Speed Serial Data Communication," 2001 IEEE SOCC.

[12] B. S. Song, et al, "NRZ Timing Recovery Technique for Band-Limited Channels," IEEE Journal of Solid-State Circuits, VOL.32, No. 4, April 1997.

[13] H. Tamura, et al., "Partial Response Detection Technique for Driver Power Reduction in High-Speed Memory to processor Communication," ISSCC-97, pp. 241-42

[14] H. Tamura, "CMOS High-Speed I/O—Background, Circuits, and Future Trends," IEEE 2004.

[15] R. Walker, "Clock and Data Recovery for Serial Digital Communication—(Plus Tutorial on Bang-Bang Phase-Locked Loops," Hewlett-Packard Company, Palo Alto, California

[16] H. J. Song, "Delay Locked-Loop Based Circuit for Data Communication," US Patent #6,466,615 B1, 2002.

[17] J. Savoj, et al, "A 10-Gb/s Clock and Data Recovery Circuits," IEEE 2000.

[18] R. J. Yang, et, al, "A 2000-Mbps ~2-Gbps Continuous-Rate Clock-and-Data-Recovery Circuit," IEEE Trans. Circuits and Systems, Vol. 53, No.4, April 2006.

[19] Y. Ohtomo, et, al, "A 12.5-Gbps Parallel Phase Detection Clock and Data Recovery Circuit in 0.13-um CMOS," IEEE JSSC, Vol. 41, No.9, September 2006.

[20] W. Dally et al," Transmitter Equalization for 4-GBPS Signal," IEEE Micro, January/February 1997.

[21] Y. Choi, et al, "Jitter Transfer Analysis of Tracked Over-sampling Techniques for Multigigbit Clock and Data Recovery," IEEE TRANSACTIONS ON CIRCUITS AND SYSTEMS—II: ANALOG and Digital signal processing, VOL. 50, NO. 11, November 2003.

[22] J. Lee, et al, "Analysis and Modeling of Bang-Bang Clock and Data Recovery Circuits," IEEE JOURNAL of SOLID-STATE CIRCUITS, VOL. 39, NO. 9, September 2004.

[23] B. Kim, et, al, "A 30-Mhz Hybrid Analog/Digital Clock Recovery Circuit in 2-um CMOS," IEEE JSSC Vol. 25, No. 6, March 1990.

[24] J. Lee, et al, "A 40-Gb/s Clock and Data Recovery Circuit in 0.18um CMOS Technology," IEEE JOURNAL of SOLID-STATE CIRCUITS, VOL. 38, NO. 12, September 2004.

[25] Y. D. Choi, et, al, "Jitter Transfer Analysis of Tracked Oversampling Techniques for Multigigabit Clock and Data Recovery," IEEE Tran. Circuits and Systems—II: Analog and Digital Signal Processing, vol. 50, no. 11, Nov. 2003.

[26] E. de Vasconcelos, et, al, "A 0.7um CMOS Clock Recovery Circuit for 622Mb/s SDH System," IEEE 1998.

[27] R. F. Rad, et, al, "A 33-mW 8-Gb/s CMOS Clock Multiplier and CDR for Highly Integrated I/Os," IEEE JSSC Vol. 39, No. 9, Sept. 2004.

[28] J. Lee, et, al, "Modeling of Jitter in Bang-Bang Clock and Data Recovery Circuits," IEEE CICC 2003.

[29] B. J. Lee, et, al, "A 2.5-10-Gb/s CMOS Transceiver with Alternating Edge-Sampling Phase Detection for Loop Characteristic Stabilization," IEEE JSSC Vol. 38, No. 11, Nov. 2003.

[30] J. K. Kwon, et, al, "A 5-Gb/s 1/8-Rate CMOS Clock and Data Recovery Circuit," IEEE ISCAS 2004.

[31] H. T. Ng, et, al, "A Second-Order Semi digital Clock Recovery Circuit Based on Injection Locking," IEEE JSSC Vol. 38, No. 12, Dec. 2003.

[32] H. H. Chang, et, al, "Low Jitter and MultiMate Clock and Data Recovery Circuit using a MSADLL for Chip-to-Chip Interconnection," IEEE Tran. Circuits and Systems—I: Regular Papers, vol. 51, no. 12, Dec. 2004.

[33] C. Pease, et, al, "Practical Measurement of Timing Jitter Contributed from a Clock-and-Data Recovery Circuit," IEEE Tran. Circuits and Systems—I: Regular Papers, vol. 52, no. 1, January 2005.

[34] D. W. Hong, et, al, "BER Estimation for Serial Links Based on Jitter Spectrum and Clock Recovery Characteristics," IEEE ITC 2004.

[35] D. J. Lee, et, al, "A Quad 3.125Gbps Transceiver Cell with All Digital Data Recovery Circuits," IEEE 2005 Symposium of VLSI Circuits.

[36] H. Djahanshahi, et, al, "Differential CMOS Circuits for 622-Mhz/933-Mhz Clock and Data Recovery Applications," IEEE JSSC Vol. 35, No. 6, Jun 2000.

[37] S. Wang, et, al, "Design Considerations for 2nd-Order and 3rd-Order Bang-Bang CDR Loops," IEEE CICC 2005.

[38] S. I. Ahmed, et, al, "Behavioral Test Benches for Digital Clock and Data Recovery Circuits using VerilogA," IEEE CICC 2005.

[39] M. Y. He, et, al, "A CMOS Mixed-Signal Clock and Data Recovery Circuit for OIF CEI-6G+Backplane Transceiver," IEEE JSSC Vol. 41, No. 3, March 2006.

[40] Y. Miki, et, al, "A 50-mW/Ch 2.5-Gb/s Data Recovery Circuit for the SFI-5 Interface with Digital Eye-Tracking," IEEE JSSC Vol. 39, No. 4, April 2004.

[41] R. Y. Zhang, et, al, "Clock and Data Recovery Circuits with Fast Acquisition and Low Jitter," IEEE 2004.

[42] A. L. Coban, et, al, "A 2.5-3.125 Gb/s Quad Transceiver with Second Order Analog DLL based CDRs," IEEE CICC 2004.

[43] M. H. Perrott, et, al, "A 2.5-Gb/s Multi-Rate 0.25-um CMOS Clock and Data Recovery Circuit Utilizing a Hybrid Analog/Digital Loop Filter and All-Digital Referenceless Frequency Acquisition," IEEE JSSC Vol. 41, No. 12, Dec. 2006.

[44] F. Lambrecht, et, al, "Accurate System Voltage and Timing Margin Simulation in CDR based High Speed Designs," 2006 IEEE Electrical Performance of Electronic Packaging.

[45] M. Lin, et, al, "Production-Oriented Interface Testing for PCI-Express by Enhanced Loop-Back Technique," IEEE ITC 2005.

CHAPTER 18
VLSI JITTER FILTER CIRCUITS

Timing jitter and phase errors may interact with VLSI circuits. Based on the characteristics of the circuits, jitter may experience frequency selection, amplification, attenuation effects. Practical VLSI signal processing circuits can be divided into two major circuit families with respect to their interaction with the timing jitters, including the delay (or phase) domain signal processing circuit family and the filterlike signal processing circuit family.

VLSI phase-locked loop (PLL, DLL, DRC, etc.) and other time-domain signal processing circuits belong to the first circuit family that are usually directly specified, and design in delay or phase signal forms and their signal transfer functions are the same as their jitter transfer function. VLSI analog filter circuits, clock buffer circuits, interconnects, transmitter circuit, receiver AFE circuits, and the I/O channel belong to the second circuit family that are usually specified in voltage (or current) domain and their signal transfer functions are significantly different from their jitter transfer functions.

Timing jitter of high-speed I/O circuits can be explicitly or implicitly impacted by the phase domain or filterlike VLSI signal processing circuits respectively. The effects can be expressed using their jitter transfer functions.

The interactions of VLSI circuits to the jitter and phase errors can be used to construct a family of jitter and phase error processing circuits (known as jitter filters) for compensating the delay variations of the VLSI high-speed I/O circuits.

18.1 VLSI JITTER TRANSFER FUNCTION MODEL

Timing jitter of the clock or data signals may experience certain amplification, attenuation, or frequency selection effects as the signals propagate through the VLSI high-speed I/O circuits. The impacts of VLSI circuits to jitter can typically be modeled using the jitter transfer function as shown in figure 18.1

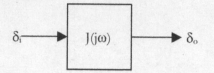

Fig. 18.1. Jitter transfer function

where δi and δo are the input and output jitter of the circuit respectively.

Under the I/O prototype circuit model, jitter transfer characteristics can be expressed by a frequency (or s-) domain jitter transfer function as shown in figure 18.2.

$$J(s) \equiv \frac{\delta_o(s)}{\delta_i(s)}$$

(18.1)

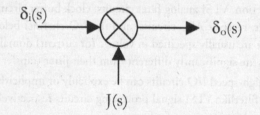

Fig. 18.2. Graphical expression of s-domain jitter transfer function

The impact of jitter transfer function on the clock paths within the prototype I/O circuit can be modeled as a circuit diagram shown in figure 18.3.

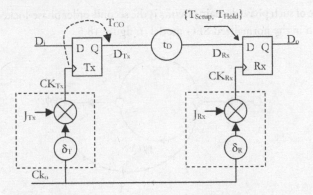

Fig. 18.3. The I/O prototype circuit model including clock jitter transfer functions

Based on the prototype I/O circuit model, the output jitters at the transmitter and receiver of the high-speed I/O circuit as influenced by the VLSI circuit are modeled with the jitter transfer functions of the two clock paths respectively. In practical VLSI high-speed I/O circuits, these jitter transfer functions are contributed from the clock distribution network, the clock PLL, and the data recovery circuits.

Shown in figure 18.4 is the modified I/O circuit SFG model to include the jitter transfer function effects of the clock and the datapaths,

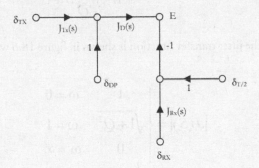

Fig. 18.4. I/O circuit jitter transfer function SFG model

where $JTx(s)$ represents the jitter transfer function of the transmitter clock path, $JRx(s)$ the jitter transfer function of the receiver clock path, and $JD(s)$ the transfer function of the data channel respectively.

18.1.1 PHASE DOMAIN CIRCUITS

Jitter transfer functions of VLSI phase domain circuits, such as the PLL, DLL, and DRC circuits, are the same as the signal transfer functions. This type of circuit, in practical applications, can be used for phase noise shaping or jitter filtering operations.

One example of such phase domain circuits is the second-order phase-locked loop (PLL) circuit shown in the normalized SFG model in figure 18.5.

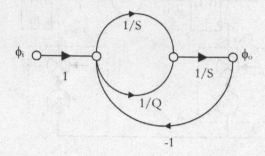

Fig. 18.5. SFG of normalized second-order PLL circuit

The jitter transfer function of the PLL circuit, which is the same as its signal transfer function, can be directly expressed as

$$J(S) \equiv \frac{\varphi_o}{\varphi_i} = \frac{\frac{1}{Q}S + 1}{S^2 + \frac{1}{Q}S + 1}$$

(18.2)

The Bode plot of the jitter transfer function is shown in figure 18.6 with several special points given as

$$|J(S)| = \begin{cases} 1 & \omega = 0 \\ \sqrt{1+Q^2} & \omega = 1 \\ 0 & \omega = \infty \end{cases}$$

(18.3)

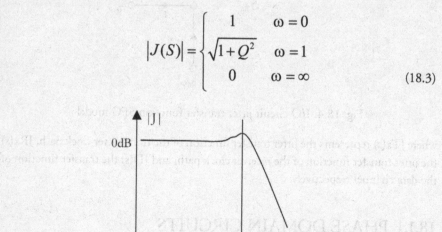

Fig. 18.6. Bode plots of second-order normalized PLL jitter transfer function

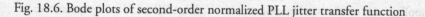

18.1.2 FILTERLIKE CIRCUITS

The filterlike VLSI signal processing circuits, such as amplifier, buffers, receiver, and the channel, can in general be modeled as VLSI filters. This type of circuits can typically be specified by signal transfer functions H(s) as shown in figure 18.7 with respect to the input and output voltage (current) signals. The signal transfer function provides a model for characterizing the circuit's interaction with the magnitude and the phase of the signal.

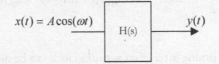

Fig. 18.7. Signal transfer function of filterlike VLSI circuits

For a given VLSI filterlike circuit with the signal transfer function as

$$H(s) = \frac{Y(s)}{X(s)}$$

(18.4)

The steady state output of an ideal single-tone signal from the circuit can be expressed in terms of the signal transfer function as

$$y(t) = A|H(j\omega)|\cos(\omega t + \angle H(j\omega))$$

(18.5)

For a nonideal clock signal, the timing jitter of the clock can be mathematically modeled using phase modulation (PM) sinusoidal clock signal as

$$x(t) = A\cos(\omega t + \beta_x(\omega_m)\cos(\omega_m t))$$

(18.6)

where ω_m is the modulation (or jitter) frequency and $\beta(\omega_m)$ is the phase noise (or jitter) magnitude at the frequency ω_m. Under this model, the output clock signal can be expressed based on similar PM sinusoidal signal as

$$y(t) = B\cos(\omega t + \beta_y(\omega_m)\cos(\omega_m t) + \varphi)$$

(18.7)

The frequency domain jitter transfer function of the filter circuit can be defined as the frequency-dependent ratio of the output to the input phase noise magnitude at specific frequency ω_m as

$$J(j\omega_m) \equiv \frac{\beta_y(\omega_m)}{\beta_x(\omega_m)}$$

(18.8)

For small phase noise (or jitter) signal, the PM clock signal models in equation 18.6 can be expressed into an alternative signal form as

$$
\begin{aligned}
x(t) &\approx A\cos(\omega t + \beta_x(\omega_m)\cos(\omega_m t)) \\
&= A[\cos(\omega t)\cdot\cos(\beta_x(\omega_m)\cos(\omega_m t)) - \sin(\omega t)\cdot\sin(\beta_x(\omega_m)\cos(\omega_m t))] \\
&\approx A\cos(\omega t) - A[\beta_x(\omega_m)\cdot\cos(\omega_m t)]\cdot\sin(\omega t) \\
&= A\cos(\omega t) - \frac{A\beta_x(\omega_m)}{2}[\sin((\omega+\omega_m)t) + \sin((\omega-\omega_m)t)]
\end{aligned}
$$

$$(18.9)$$

It can be seen that the timing jitter effect within a clock can be modeled using a pair of added adjacent tones next to the ideal the signal tone as shown in figure 18.8.

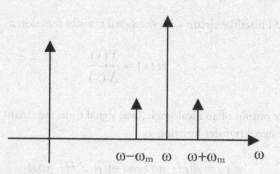

$$\omega-\omega_m \quad \omega \quad \omega+\omega_m \qquad \omega$$

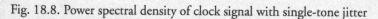

Fig. 18.8. Power spectral density of clock signal with single-tone jitter

Based on such modeling approach, the single-tone phase noise (jitter) can be equivalently expressed as the ratio of the magnitude of the added adjacent tone versus the ideal signal tone.

In most cases, it is more convenient to express the signal in the complex form as

$$x^*(t) = Ae^{(j\omega t + j\beta_x(\omega_m)\cos(\omega_m t))}$$

$$(18.10)$$

For small timing jitter values, the above equation can be approximately expressed as

$$x^*(t) \approx Ae^{j\omega t} + \frac{jA\beta_x(\omega_m)}{2}\left(e^{j(\omega+\omega_m)t} + e^{j(\omega-\omega_m)t}\right)$$

$$(18.11)$$

18.2 JITTER TRANFER FUNCTION OF FILTERLIKE CIRCUITS

The impact of filterlike circuit to jitter can be analyzed based on the frequency selectivity of the ideal clock tone versus the noise tones as shown in figure 18.9.

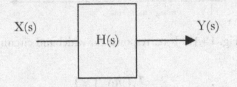

$$X(s) \quad\longrightarrow\quad \boxed{H(s)} \quad\longrightarrow\quad Y(s)$$

Fig. 18.9. Signal transfer function of filterlike circuits

It can be seen in the appendix that the jitter transfer function of the general filter for a small jitter can be approximately expressed as

$$
\begin{cases}
J(j\omega_m) \approx \sqrt{\left|\rho^+\right|^2 + \left|\rho^+\right|^2 + 2\left|\rho^+\right|\left|\rho^+\right|\cos(\theta)} \\[2mm]
\rho^+ \equiv \dfrac{H(j(\omega+\omega_m))}{2H(j\omega)} \\[2mm]
\rho^- \equiv \dfrac{H(j(\omega-\omega_m))}{2H(j\omega)} \\[2mm]
\theta \equiv \angle H(j(\omega+\omega_m)) + \angle H(j(\omega-\omega_m)) - 2\angle H(j\omega)
\end{cases}
\tag{18.12}
$$

where ρ^+ and ρ^- represent the ratios of the jitter signal tones with respect to the signal tone. $\cos(\theta)$ represents the interference between the two jitter signal tones.

18.2.1 VLSI WIDEBAND CIRCUITS

For VLSI wideband circuits with the bandwidth much higher than the signal and jitter frequency as shown in figure 18.10,

we have that

$$
\begin{cases}
\rho^+ = 1 \\
\rho^- = 1 \\
\theta = 0
\end{cases}
\tag{18.13}
$$

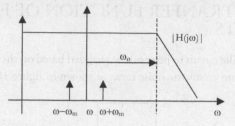

Fig. 18.10. Jitter response for wideband circuits

$$J_J(j\omega_m) \approx 1 \tag{18.14}$$

It can be seen that wideband circuits also represent wideband jitter response, and the timing jitter will not be impacted by the wideband circuits.

In general, for lowpass filter with in-band clock frequency, the Bode plot of the jitter transfer function can be approximately given as figure 18.11.

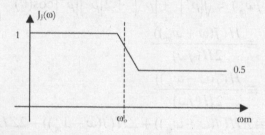

Fig. 18.11. Jitter transfer characteristics of wideband circuits

The variation of jitter response at high frequency is based on the fact that the high-side jitter tone is attenuated at high frequency.

18.2.2 BAND-LIMITED CIRCUITS

For the VLSI band-limited filterlike circuit with the clock signal frequency located beyond the passband of the circuit as shown in figure 18.12, the parameters in the jitter transfer function are given as

$$\begin{cases} \rho^+ \equiv \left| \dfrac{H(j(\omega + \omega_m))}{H(j\omega)} \right| \ll 1 \\[4mm] \rho^- \equiv \left| \dfrac{H(j(\omega - \omega_m))}{H(j\omega)} \right| \gg 1 \end{cases} \tag{18.15}$$

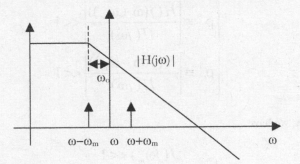

Fig. 18.12. Jitter response for out-of-band clock

As a result, we have that

$$J(j\omega_m) \approx \left| \frac{H(j(\omega - \omega_m))}{H(j\omega)} \right| \gg 1$$

(18.16)

It can be seen that jitter will be amplified passing through the band-limited circuit. In general, the Bode plot of out-of-band clock passing through the lowpass filterlike can be approximately plotted as figure 18.13. Note that the 20dB/dec increase in the jitter transfer function is based on the fact that ρ^- increases at 20dB/dec rate for a first-order filterlike circuit.

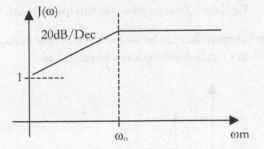

Fig. 18.13. Jitter transfer characteristics of lowpass filter

18.2.3 BANDPASS CIRCUITS

For in-band clock passing through the bandpass filter as shown figure 18.14, with signal tone located at the center of the passband, we have that

$$\begin{cases} \rho^{+} \equiv \left| \dfrac{H(j(\omega + \omega_m))}{H(j\omega)} \right| << 1 \\[4mm] \rho^{-} \equiv \left| \dfrac{H(j(\omega - \omega_m))}{H(j\omega)} \right| << 1 \end{cases}$$

(18.17)

Consequently, we have

$$J(j\omega_m) << 1$$

(18.18)

Note that equation 18.18 leads to an important concept of jitter filtering that can be used in practical VLSI circuit design for minimizing the timing jitter in the high-speed I/O circuits.

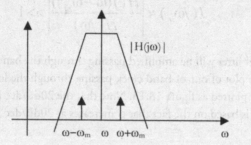

Fig. 18.14. Jitter response for bandpass circuits

It can be seen that bandpass filter can be used to attenuate the timing jitter effects. The Bode plot of the jitter of an in-band clock can be plotted as

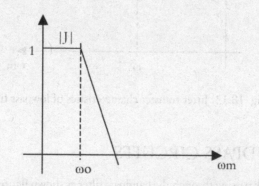

Fig. 18.15. Jitter transfer characteristics of bandpass filter

18.3 DUTY-CYCLE DISTORTION MODELS

The above jitter and jitter transfer function model can also be used to analyze the duty-cycle distortion (DCD) effects in VLSI clocking circuits, where the DCD amplification effects are commonly observed in the band-limited VLSI circuit or PCB channel.

18.3.1 SINGLE-TONE DCD MODEL

A clock signal of frequency ω with duty-cycle distortion (DCD) can be modeled using a special PM signal as

$$x(t) \approx A\sin(\omega t + \alpha\cos(\omega t))$$

(18.19)

It can be seen that the phase modulation tone due to the DCD effect has the highly correlated frequency of the ideal clock. As a result, the rising and falling edges of the clock signal are deviated from the ideal edges in the opposite direction. Consequently, the nonideal clock signal is deviated from its ideal waveform given as

$$\begin{cases} x(t)\big|_{t=0} \approx A\sin(\alpha) \\ x(t)\big|_{t=T/2} \approx A\sin(\alpha) \end{cases}$$

(18.20)

We may introduce a duty-cycle distortion parameter of the clock as

$$DCD \equiv \frac{t_{High\_Side} - t_{Low\_Side}}{2T} = \frac{\alpha}{2\pi}$$

(18.21)

On the other hand, the small-magnitude DCD effect can also be expressed in the frequency domain as

$$x(t) \approx A\sin(\omega t + \alpha\cos(\omega t))$$
$$= A[\sin(\omega t) \cdot \cos(\alpha\cos(\omega t)) + \cos(\omega t) \cdot \sin(\alpha\cos(\omega t))]$$
$$\approx A\sin(\omega t) + A\alpha(\cos(\omega t))^2 = A\sin(\omega t) + \frac{A\alpha}{2}[1 - \sin(2\omega t)]$$
$$\approx \frac{A\alpha}{2} + A\sin(\omega t)$$

$$(18.22)$$

It can be seen that the DCD effect can be mathematically modeled as an added dc offset in the clock signal. Note that in the above model method, the high frequency signal tone due to the DCD is purposely ignored since most VLSI circuits of interest are typically band limited.

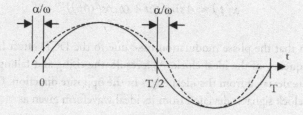

Fig. 18.16. PM DCD model

Equation 18.22 can be rewritten as

$$x(t) \approx \frac{A\alpha}{2} + A\sin(\omega t) = V_{dc} + V_{ac}\sin(\omega t)$$

$$(18.23)$$

It can be seen that DCD can be expressed as the ratio of the dc offset versus the ideal signal magnitude as

$$DCD \equiv \frac{t_{High\_Side} - t_{Low\_Side}}{2T} \approx \frac{1}{2\pi}\left|\frac{V_{dc}}{V_{ac}}\right|$$

$$(18.24)$$

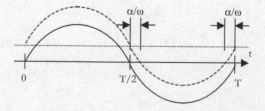

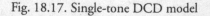

Fig. 18.17. Single-tone DCD model

18.3.2 DCD TRANSFER FUNCTION

For the lowpass VLSI circuit, the dc offset signal will have higher gain compared with the idea clock signal tone. As a result, the DCD effect will be amplified.

$$(DCD)_o \approx \left| \frac{H(j\omega)|_{\omega=0}}{H(j\omega)|_{\omega=\omega}} \right| (DCD)_i > (DCD)_i$$

(18.25)

Such DCD amplification effect has been widely observed in VLSI high-speed I/O circuits.

18.3.3 TIME-DOMAIN DCD MODEL

DCD amplification effect of the band-limited channel can also be modeled in time-domain. For the lowpass RC filter circuit shown in figure 18.18, with the 3dB bandwidth given as

$$\omega_{-3dB} = 1/RC$$

(18.26)

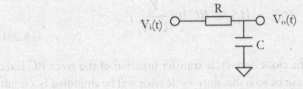

Fig. 18.18. Lowpass filter model for bandwidth-limiting circuit

For a nonideal clock signal of clock period T and high-side duty-cycle t_1, the output waveform $V_o(t)$ can be expressed as a signal $f_1(t)$ for rising side and $f_2(t)$ for the falling side as shown in figure 18.19. In such equation, Y_1 and Y_2 are the initial values of the rise and fall waveforms respectively.

$$\begin{cases} f_1(t) = 1 - (1 - Y_1)e^{\frac{-t}{RC}} \\ Y_1 = \dfrac{e^{t_1/RC} - 1}{e^{T/RC} - 1} \\ Y_2 = 1 - (1 - Y_1)e^{\frac{-t_1}{RC}} \\ f_2(t) = Y_2 e^{\frac{-(t-t_1)}{RC}} \end{cases}$$

(18.27)

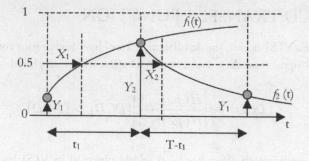

Fig. 18.19. Output clock waveform of bandwidth-limited circuit

The 50% magnitude delay time of the rising and falling signals can be expressed respectively as

$$\begin{cases} X_1 = 0.7RC + RC\ln(1 - Y_1) \\ X_2 = 0.7RC + RC\ln(Y_2) \end{cases}$$

$$(18.28)$$

The output duty-cycle as function of the input duty-cycle and the time constant can be expressed as

$$(t_1)_o = (t_1)_i + RC\ln(\frac{Y_2}{1 - Y_1})$$

$$(18.29)$$

Shown in figure 18.20 is the clock duty-cycle transfer function of the given RC filter for various T/RC values. It can be seen that duty-cycle error will be amplified as a result of the circuit lowpass filtering effect.

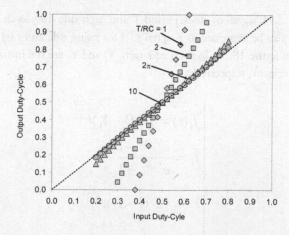

Fig. 18.20 DCD transfer function of RC lowpass filter

Shown in figure 18.21 is the plot of amplification gain of the DCD versus the T/RC parameters. It can be seen that significant DCD amplification can be expected for lowpass circuit.

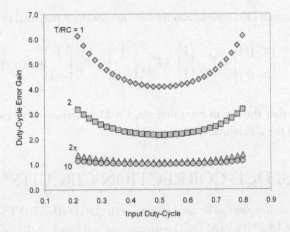

Fig. 18.21. DCD transfer gain of RC lowpass filter

18.3.4 DIFFERENTIAL DCD MODEL

The single-tone DCD model can also be extended into the differential clock circuits.

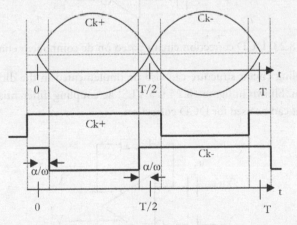

Fig. 18.22. Single-tone differential DCD model

For differential clock signals with common-mode dc component as shown in figure18.22, the DCD parameter in the differential clock signal can be modeled using the single-tone expression as

$$\begin{pmatrix} x_+(t) \\ x_-(t) \end{pmatrix} \approx V_{dcc} \begin{pmatrix} 1 \\ 1 \end{pmatrix} + V_{ac} \begin{pmatrix} 1 \\ -1 \end{pmatrix} \sin(\omega t)$$

(18.30)

In general case, such DCD model should also include the differential dc component as

$$\begin{pmatrix} x_+(t) \\ x_-(t) \end{pmatrix} \approx V_{dcc} \begin{pmatrix} 1 \\ 1 \end{pmatrix} + V_{dcd} \begin{pmatrix} 1 \\ -1 \end{pmatrix} + V_{ac} \begin{pmatrix} 1 \\ -1 \end{pmatrix} \sin(\omega t)$$

(18.31)

It can be seen that the key to minimize the DCD, therefore, is to eliminate the dc components in the differential clock signals.

18.3.5 VLSI DCD CORRECTION CIRCUITS

Various circuit techniques can be used to correct the DCD effect in VLSI clock signals. Shown in figure 18.23 is a VLSI DCD correction circuit based on the single-tone DCD model. In this circuit, the dc component of the input clock signal can be eliminated using an ac coupling circuit technique. As a result, the DCD effect in the input clock can be corrected.

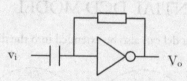

Fig. 18.23. DCD correction circuit based on dc component eliminating

The ac coupling circuit structure can also be implemented in the differential circuit configuration. Shown in figure 18.24 is a VLSI ac coupling differential buffer circuit structure that can be used for DCD correction.

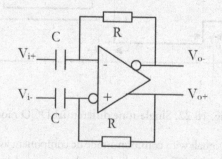

Fig. 18.24. VLSI differential DCD correction circuit

Based on the common-mode rejection of the differential amplifier circuit as shown in figure 18.25, these circuits can also be used to correct the DCD effect for the differential clock signals.

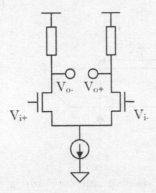

(a) Differential buffer

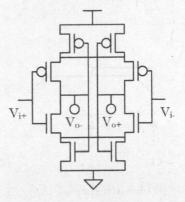

(b) Symmetrical differential buffer

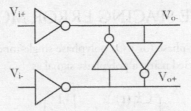

(c) Pseudo Differential buffer

Fig. 18.25. VLSI differential DCD correction circuits

18.4 POLYPHASE CLOCK PHASE SPACING ERROR

Polyphase clock signals with multiple clocks of same frequency and magnitude and phase uniformly distributed across the entire clock period are commonly used in VLSI high-speed I/O for data recovery, SIPO, PISO, and other operations.

Phase spacing error can be defined as the deviation of the polyphase clock phases from their ideal values. Phase spacing error may directly impact the performance of the high-speed I/O circuit. Shown in figure 18.26 is a high-speed I/O data recovery circuit example that is based on the edge sampling using Ck_1 for input data phase detection and the data sampling using Ck_2 for optimized data capture. If there is a phase spacing error, the sampling point will be deviated from its optimal position and result in timing error in the $\delta_{T/2}$ term of the timing equation shown below.

$$\begin{cases} -E_{max} < E \equiv \delta_{DP} + \delta_{TX} - \delta_{T/2} - \delta_{RX} < E_{max} \\ E_{max} \equiv \dfrac{T - (T_{setup} + T_{hold})}{2} \end{cases}$$

(18.32)

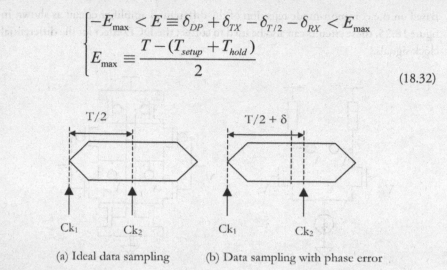

(a) Ideal data sampling   (b) Data sampling with phase error

Fig. 18.26. Impact of phase spacing error to high-speed I/O circuits

18.4.1 I/Q PHASE SPACING ERROR MODEL

An ideal complex form 4-phase (or I/Q) polyphase single-tone clock signals shown in figure 10.27 can be modeled using a polyphase signal as

$$\begin{pmatrix} Ck_1(t) \\ Ck_2(t) \\ Ck_3(t) \\ Ck_4(t) \end{pmatrix} = V_+ \begin{pmatrix} 1 \\ j \\ -1 \\ -j \end{pmatrix} e^{j\omega t}$$

(18.33)

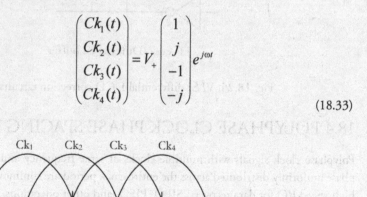

Fig. 18.27. Single-tone 4-phase polyphase clock signals

Such 4-phase clock signals in the frequency domain can be expressed using a polyphase signal with frequency located at the positive frequency ω as shown in figure 18.28.

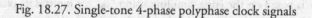

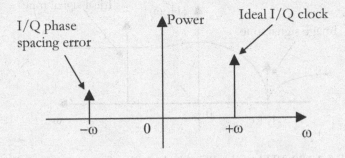

Fig. 18.28. Power spectral density of ideal single-tone I/O clock

It can be proved that the phase spacing error for I/Q clocks is equivalent to an added signal tone at the image frequency of the ideal I/Q clock. As a result, the phase spacing error correction for I/Q clocks can be effectively realized by eliminating the image signal tone in the frequency domain.

18.4.2 I/Q PHASE ERROR CORRECTION CIRCUITS

Shown in figure 18.29 is an RC polyphase filter circuit that can be used to correct the I/Q phase error of the I/O clock signals.

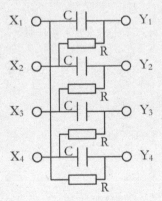

Fig. 18.29. VLSI RC-passive polyphase filter

Such polyphase offers the property of image signal elimination as shown in figure 18.30, desired for I/Q phase error correction.

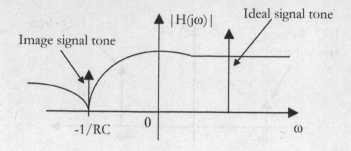

Fig. 18.30. VLSI passive-RC polyphase filter frequency response

APPENDIX 18

A18.1:JITTER TRANSFER FUNCTION OF FILTER CIRCUITS

When a clock signal x(t) passes through a filter with signal transfer function H(s), the output signal can be expressed in time-domain as

$$x(t) = \cos(\omega t + \beta_x (\omega_m) \cos(\omega_m t)) \qquad (18.34)$$

$$
\begin{aligned}
y(t) &\approx |H(j\omega)|\cos(\omega t + \angle H(j\omega)) \\
&\quad - \frac{\beta_x(\omega_m)}{2}[|H(j(\omega+\omega_m))|\sin((\omega+\omega_m)t + \angle H(j(\omega+\omega_m)))] \\
&\quad - \frac{\beta_x(\omega_m)}{2}[|H(j(\omega-\omega_m))|\sin((\omega-\omega_m)t + \angle H(j(\omega-\omega_m)))] \\
&= |H(j\omega)|\{\cos(\omega t + \angle H(j\omega)) \\
&\quad - \frac{\beta_x(\omega_m)}{2}[\frac{|H(j(\omega+\omega_m))|}{|H(j\omega)|}\sin((\omega+\omega_m)t + \angle H(j(\omega+\omega_m)))] \\
&\quad - \frac{\beta_x(\omega_m)}{2}[\frac{|H(j(\omega-\omega_m))|}{|H(j\omega)|}\sin((\omega-\omega_m)t + \angle H(j(\omega-\omega_m)))]\}
\end{aligned}
$$

$$(18.35)$$

The terms represent the jitter are given as

$$
\begin{aligned}
y(t) &\approx |H(j\omega)|\{\cos(\omega t + \phi_o) \\
&\quad - J_J^+ \sin((\omega+\omega_m)t + \phi_m^+ + \phi_o) - J_J^- \sin((\omega-\omega_m)t + \phi_m^- + \phi_o)\} \\
&= |H(j\omega)|\{\cos(\omega t + \phi_o) - [J_J^+ \cos(\omega_m t + \phi_m^+) + J_J^- \cos(\omega_m t - \phi_m^-)]\sin(\omega t + \phi_o) \\
&\quad - [J_J^+ \sin(\omega_m t + \phi_m^+) - J_J^- \sin(\omega_m t - \phi_m^-)]\cos(\omega t + \phi_o)\} \\
&\approx |H(j\omega)|\{\cos(\omega t + \phi_o) - [J_J^+ \cos(\omega_m t + \phi_m^+) + J_J^- \cos(\omega_m t - \phi_m^-)]\sin(\omega t + \phi_o)\}
\end{aligned}
$$

$$(18.36)$$

where

$$
\begin{cases}
\phi_o \equiv \angle H(j\omega) \\
\phi_m^+ \equiv \angle H(j(\omega + \omega_m)) - \angle H(j\omega) \\
\phi_m^- \equiv \angle H(j(\omega - \omega_m)) - \angle H(j\omega)
\end{cases}
$$

$$
\begin{cases}
J_J^+ \equiv \dfrac{\beta_x(\omega_m)}{2} \left| \dfrac{H(j(\omega + \omega_m))}{H(j\omega)} \right| \\[4mm]
J_J^- \equiv \dfrac{\beta_x(\omega_m)}{2} \left| \dfrac{H(j(\omega - \omega_m))}{H(j\omega)} \right|
\end{cases}
$$

$$(18.37)$$

For small output jitter, and ignore the overall initial phase ϕ_o for simplicity since it does not impact the jitter analysis, the above equation can be further simplified as

$$
y(t) \approx |H(j\omega)|\{\cos(\omega t)
$$
$$
-[J_J^+ \cos(\omega_m t + \phi_m^+) + J_J^- \cos(\omega_m t - \phi_m^-)]\sin(\omega t)
$$
$$
= |H(j\omega)|\{\cos(\omega t)
$$
$$
-[J_J^+ (\cos(\omega_m t)\cos(\phi_m^+) - (\sin(\omega_m t)\sin(\phi_m^+))
$$
$$
-J^- (\cos(\omega_m t)\cos(\phi_m^-) + (\sin(\omega_m t)\sin(\phi_m^-))]\sin(\omega t)
$$
$$
= |H(j\omega)|\{\cos(\omega t)
$$
$$
-[(J_J^+ \cos(\phi_m^+) + J_J^- \cos(\phi_m^-))\cos(\omega_m t)
$$
$$
+(J_J^+ \sin(\phi_m^+) - J_J^- \sin(\phi_m^-))\sin(\omega_m t)]\sin(\omega t)
$$

$$(18.38)$$

let

$$
\begin{cases}
\alpha \equiv \sqrt{(J_J^+ \cos(\phi_m^+) + J_J^- \cos(\phi_m^-))^2 + (J_J^+ \sin(\phi_m^+) - J_J^- \sin(\phi_m^-))^2} \\[3mm]
\gamma \equiv \tan^{-1}\left(\dfrac{J_J^+ \sin(\phi_m^+) - J_J^- \sin(\phi_m^-)}{J_J^+ \cos(\phi_m^+) + J_J^- \cos(\phi_m^-)} \right)
\end{cases}
$$

$$(10.39)$$

We have that

$$
y(t) \approx |H(j\omega)|\{\cos(\omega t) - \alpha \cos(\omega_m t + \gamma) \cdot \sin(\omega t)\}
$$
$$
\approx |H(j\omega)|\{\cos(\omega t) - \alpha \cos(\omega_m t + \gamma) \cdot \sin(\omega t)\}
$$

$$(18.40)$$

From above equations, we have that

$$\beta_y(\omega_m) \approx 2\alpha = 2\sqrt{\left(J_J^+\right)^2 + 2J_J^+ J_J^- \cos(\phi_m^+ + \phi_m^-) + \left(J_J^-\right)^2}$$

$$= \frac{\beta_x(\omega_m)}{|H(j\omega)|} \sqrt{\begin{array}{l} \left(H(j(\omega+\omega_m))\right)^2 + \left(H(j(\omega-\omega_m))\right)^2 \\ +2\left|H(j(\omega+\omega_m))H(j(\omega-\omega_m))\right|\cos(\angle H(j(\omega+\omega_m)) \\ +\angle H(j(\omega-\omega_m)) - 2\angle H(j\omega)) \end{array}}$$

$$\text{(18.41)}$$

The jitter transfer function is given as

$$J(j\omega_m) \equiv \frac{\beta_y(\omega_m)}{\beta_x(\omega_m)} \tag{18.42}$$

$$\begin{cases} J(j\omega_m) \approx \sqrt{\left|\rho^+\right|^2 + \left|\rho^+\right|^2 + 2\left|\rho^+\right|\left|\rho^+\right|\cos(\theta)} \\ \rho^+ \equiv \frac{H(j(\omega+\omega_m))}{H(j\omega)} \\ \rho^- \equiv \frac{H(j(\omega-\omega_m))}{H(j\omega)} \\ \theta \equiv \angle H(j(\omega+\omega_m)) + \angle H(j(\omega-\omega_m)) - 2\angle H(j\omega) \end{cases} \tag{18.43}$$

References:

[1]   M. Li, "Transfer Functions for the Reference Clock Jitter in A Serial Link: Theory and Applications," ITC2004.

[2]   J. Buckwalter, et al, "Predicting Data-Dependent Jitter," IEEE Trans. Circuits and Systems—II. Express Briefs, Vol.51, No. 9, Sept. 2004.

[3]   B. Analui, "Data-Dependent Jitter in Serial Communications," IEEE Trans. Microwave Theory and Techniques, Vol.53, No. 11, Nov. 2005.

[4]   K. K. Kim, et al, "On the Modeling and Analysis of Jitter in ATE Using Matlab," DFT2005.

[5]   C. K. Ong, et al, "Random Jitter Extraction Technique in a Multi-Gigahertz Signal," DATE2004.

[6]   G. Balamurugan, et al, "Modeling and Mitigation of Jitter in High-Speed Source Synchronous Inter-Chip Communication Systems,"ICCD2003.

[7]   M. Shimanouchi, et al, "An Approach to Consistent Jitter Modeling for Various Jitter Aspects and Measurement Methods," ITC2001.

[8]   J. Walker, et al, "Modeling of the Synchronization Process Jitter Spectrum with Input Jitter," IEEE Trans. Communications, Vol.47, No.2, Feb. 1999.

[9]   M.-J. E. Lee, et al, "Jitter Transfer Characteristics of Delay-Locked Loops—Theories and Design Techniques," IEEE JSSC, Vol.38, No.4, April 2003.

[10]   H. J. Song, "A general Method to VLSI Polyphase Filter Analysis and Design for Integrated RF Applications," IEEE SOCC 2006.

[11]   H. J. Song, et, al, "I/Q-Channel Mismatch Transfer and Amplification Effects and Applications to the Measurement and Calibration of Integrated VLIF RF Receivers," IEEE SOCC2006.

CHAPTER 19
SIGNAL INTEGRITY AND POWER DELIVERY DESIGN

VLSI high-speed I/O circuits are facing increasingly severe design challenges related to the signal integrity (SI) and power delivery (PD) in high-speed data transmission. Major signal integrity and power delivery issues in high-speed I/O circuits are related to ringing, reflection, delay variation (or jitter), cross talk, dI/dt, and IR noise effects.

These signal integrity and power delivery issues previously appeared only in the PCB circuit interconnects that are related to the nonideal effects of the metal routing, such as the parasitic resistance effects at low frequency, the parasitic capacitance effects at midrange frequency, and the parasitic inductance effects at high frequency. These issues are becoming increasingly problematic in the VLSI high-speed I/O circuits due mainly to the increasing data rate, clock frequencies, signal slew rate, and interconnect complexity.

Signal integrity and power delivery issues are now among the major concerns in both high-speed I/O PCB and VLSI on-chip circuit design that are main root causes of pattern dependent or soft failures, which are difficult to debug, resulting in either wrong data captured due to poor data signal quality or data captured at a wrong time due to poor clock signal timing.

19.1 RINGING EFFECTS IN HIGH-SPEED I/O CIRCUITS

Signal ringing occurs when signal travels through the underdamped, lumped, or distributed LRC circuits. Ringing effects in the high-speed I/O channel circuits are caused by the access energies stored in the circuit during the data transmission, which do not dissipate properly. Ringing effects can cause significant timing and voltage variation that degrades the I/O circuit performance or even result in circuit failure.

19.1.1 LRC CIRCUIT RINGING EFFECTS

The ringing effects can be demonstrated using the LRC network shown in figure 19.1.

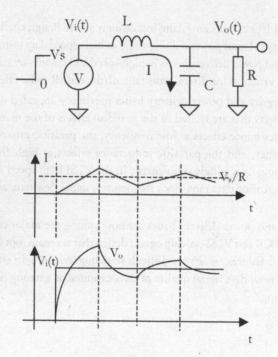

Fig. 19.1. Ringing effects in LRC circuits

It can be seen that such ringing effects can be described in time-domain as follows:

- When a step voltage input Vi(t) is applied to the circuit, a current will be induced in the inductor to supply the capacitor and the resistor loads.
- For a low dissipation (i.e., high Q) circuit, the current in the inductor will increase to a value higher than the steady state current of the circuit when the capacitor is charged to the steady state voltage value. As a result, the current in the inductor

will continue to charge the capacitor to a voltage higher than the steady state value before the inductor current reaches the steady state value.

- Since the voltage on the capacitor is higher than the steady state voltage value when the inductor current reaches steady state, the capacitor will back-charge the voltage source through the inductor until the capacitor voltage reaches the steady state voltage.

- When the voltage on the capacitor reaches the steady state again, the current in the inductor is lower than the steady state value. As a result, the current will continue to discharge the capacitor.

- The above process continues until the access energy is finally dissipated into the resistive load, resulting in the ringing of the output voltage.

The ringing effects can also be modeled in frequency domain using the frequency transfer function of the circuit as

$$H(s) \equiv \frac{V_o(s)}{V_i(s)} = \frac{1}{\left(\dfrac{s}{w_o}\right)^2 + \dfrac{1}{Q}\left(\dfrac{s}{w_o}\right) + 1}$$

$$(19.1)$$

where

$$\begin{cases} w_o \equiv 1/\sqrt{LC} \\ Q \equiv R\sqrt{\dfrac{C}{L}} \end{cases}$$

$$(19.2)$$

The gain response of the circuit based on equation 19.1 can be calculated and is shown in figure 19.2.

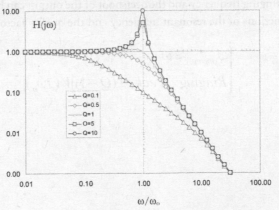

Fig. 19.2. Gain frequency response of the LCR network

For a given general voltage input signal $V_i(t)$, the output waveform can be calculated using the Fourier transformation method as

$$V_o(j\omega) = H(j\omega)V_i(j\omega) = \dfrac{1}{\left(\dfrac{s}{\omega_o}\right)^2 + \dfrac{1}{Q}\left(\dfrac{s}{\omega_o}\right) + 1} \cdot V_i(j\omega)$$

(19.3)

The output voltage waveforms can be analyzed based on the following cases:

- If $Q < 0.7$, all input signal contents with frequency higher than the bandwidth of the LRC circuit will be attenuated due to the band-limiting effect of the circuit. The output voltage signal slew rate will be limited to the time constant set by the bandwidth of the circuit and no ringing can occur.

- If $Q > 1$ and the input signal is band limited to frequency within ω_o, no ringing will happen since there is no excitation energy at the clock resonant frequency.

- If $Q > 1$ and the input signal contains significant power at ω_0 frequency, the output signal is given as

$$V_o(j\omega) = H(j\omega)V_i(j\omega) = [H(j\omega)\big|_{|\omega-\omega_o|\leq\Delta\omega} + H(j\omega)\big|_{|\omega-\omega_o|>\Delta\omega}] \cdot V_i(j\omega)$$

$$= [H(j\omega)\big|_{Q\sim1} \cdot V_i(j\omega)] + [H(j\omega) - H(j\omega)\big|_{Q\sim1}]\big|_{|\omega-\omega_o|\leq\Delta\omega} \cdot V_i(j\omega)$$

$$\approx [H(j\omega)\big|_{Q\sim1} \cdot V_i(j\omega)] + [H(j\omega) - 1]\big|_{|\omega-\omega_o|\leq\Delta\omega} \cdot V_i(j\omega_o)$$

(19.4)

In the above equation, the first term provides a slew rate limited output waveform. The second term provides a significant gain for signal at the resonant frequency ω_o of the circuit. This term appears as ringing in the output signal as shown in figure 19.3. For high Q circuits, the ringing frequency and the overshoot of the output can be approximately expressed as functions of the resonant frequency and the quality factor as

$$\begin{cases} \omega_{ringing} \approx \omega_o \\ Ringing\_Peak \approx (Q-1)|V(j\omega_o)| \end{cases}$$

(19.5)

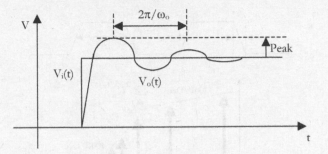

Fig. 19.3. Ringing effects in high Q LRC circuit

It is important to note that

- signal ringing effect is caused by the low loss (high Q) circuit, coupled with signal content at resonant frequency of the circuit;
- one way to eliminate circuit ringing is to properly damp the LRC circuit (i.e., low Q design); and
- alternatively, ringing of LRC circuit can also be eliminated by using slew rate control circuit technique that eliminates the excitation signal content at the resonant frequency ω_0.

It can be seen that for the lumped LRC circuit, the critical resistor value for avoiding the ringing effect is approximately given as

$$R \approx \sqrt{\frac{L}{C}}$$

(19.6)

The principle of slew rate control-based ringing elimination can be explained by using the power spectrum of a typical digital signal as shown in figure 19.4 as follows:

- The harmonics of signal within a characteristics frequency (known as the knee frequency) is typically attenuated at -20dB/Dec rate.
- The high-frequency harmonic contents beyond knee frequency are strongly attenuated at -40dB/Dec rate.
- Due to the antisymmetrical property of the signal, there are no even order harmonics frequency contents.

where the knee frequency of the digital signal can be expressed in terms of the signal rise time as

$$f_{Knee} = 0.35\frac{1}{t_r}$$

(19.7)

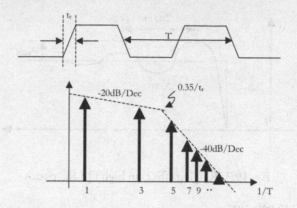

Fig. 19.4. Typical digital signal waveform and knee frequency

It can be seen that when the input signal is slew rate limited, the high-frequency signal content beyond the knee frequency is highly attenuated. If the resonant frequency ω_o of the RLC circuit is significantly higher than the knee frequency of the signal, ringing effects are eliminated since there is no excitation energy at the circuit resonant frequency.

In practical VLSI circuit, damping of the LRC circuit can also be realized by terminating the source with proper value resistor. Shown in figure 19.5 is LRC circuit with the resistor in series with the voltage source.

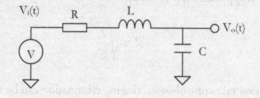

Fig. 19.5. Parallel termination of RCL circuit

The s-domain transfer function of the circuit can be derived as

$$H(s) \equiv \frac{V_o(s)}{V_i(s)} = \frac{1}{\left(\dfrac{s}{\omega_o}\right)^2 + \dfrac{1}{Q}\left(\dfrac{s}{\omega_o}\right) + 1}$$

(19.8)

The resonant frequency and quality factor can be derived as

$$\begin{cases} \omega_o \equiv 1/\sqrt{LC} \\ Q \equiv \dfrac{1}{R}\sqrt{\dfrac{L}{C}} \end{cases}$$

$$\text{(19.9)}$$

It can be seen that critical damping (or impedance matching) can be reached when the resistor is selected as

$$R \approx \sqrt{\frac{L}{C}}$$

$$\text{(19.10)}$$

19.1.2 TRANMISSION LINE RINGING EFFECT

Signal ringing effects also occur in the distributed LRC (or transmission line) circuits, where transmission lines can be viewed as infinite segments of many small LRC segments connected in series. Ringing occurs if the transmission line is not properly terminated (or damped).

For a general transmission line circuit structure shown in figure 19.6, the ringing effects in general depend on the transmission line delay time (t_d), the rise time of the signal, the load and the driver impedance, as well as the transmission line impedance. Such ringing effects can usually be analyzed using the reflection diagram as follows:

- First, ringing is always associated with unmatched (underdamped) circuit since when the transmission line is matched at either the load or the source, at least one of the reflection coefficients (ρ_i or ρ_L) is zero and signal ringing cannot occur.

Fig. 19.6 Ringing effects in transmission line circuits

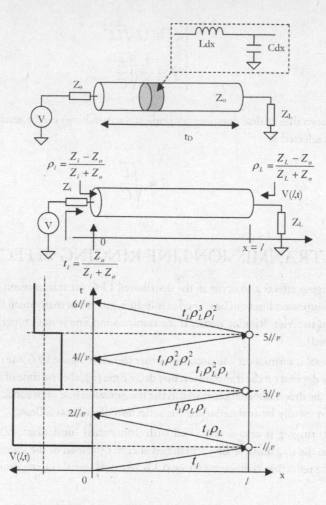

Fig. 19.6. Ringing effects in transmission line circuits

- Second, when the rise time of the signal is much longer than the transmission line delay, ringing can be minimized. This effect was well explained in the lumped LRC circuit case where long rise time means the knee frequency of the signal is significantly reduced to below the resonant frequency of the circuit. For the transmission line circuit, this effect can be explained by the effects that when the transmission delay is much shorter than the rise time, the equivalent step input associated with the delay of the transmission line is significantly reduced and therefore can be ignored. In the practical design, the rule of thumb is to limit the transmission line delay to within 1/6 of the rise time of the signal for ringing effect minimization:

$$t_{rise} > 6 \cdot t_D \qquad (19.11)$$

19.2 DELAY EFFECTS IN HIGH-SPEED I/O CIRCUITS

Signal passing through the channel in the high-speed I/O circuit suffers from significant delay effects. In practical high-speed I/O circuits, the actual delay depends on the transmission line length, the signal rise time, the load/source impedance, the observation points (end or middle), and the logic thresholds. It is important to note that:

- Transmission line length is the most important parameter for the channel delay. Its contribution to the channel delay can be expressed as

$$t_d = l\sqrt{LC} \tag{19.12}$$

  A classic rule of thumb for delay of an inside (strip-line) trace on PCB is 180ps/inch.

- Rise/fall time (i.e., slew rate) of the signal may impact the delay of the channel as the time for signal to rise and pass the given logic threshold is approximately proportional to the rise time of the signal. In practical case, degraded rise time may significantly increase the delay and delay variation of the signal and result in poor jitter performance.

- Load and source impedance usually impact the delay of the channel through either the RC constant or the ringing of the signal.

- Depending on the termination conditions, delay time at the middle and end points may be significantly different. For example, for the open load transmission line, the end point may pass threshold voltage earlier than the middle point. As a result, the end point may even have a shorter delay time than the middle point.

- Logic threshold sets where the signal is interpreted to "0" and "1." Variation of the logic threshold may result in the variation of the delay of the circuit.

19.3 CROSS TALK EFFECTS IN HIGH-SPEED I/O CIRCUITS

Cross talk may cause high-speed I/O circuit performance degradations. Typical cross talk effects of VLSI circuits include increased noise levels, unplanned spikes, jitter on data edges, reflections of undesired signals. Level of cross talk in high-speed I/O circuit depends on design parameters, such as line length, rise time, line spacing, load/source impedance, trace orientation (same way or opposite), and reference plane geometry.

Cross talk effects occur during the switching of the attacker lines. The changes of the attacker line signals will couple or "cross talk" a voltage into the adjacent victim line through both the mutual capacitance and the inductance effects as shown in figure 19.7. The cross talk voltage usually appears on the victim line as a narrow pulse with a width equal to the pulse rise time on the attacker lines.

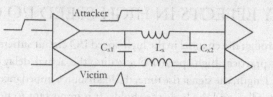

Fig. 19.7. Cross talk effects in signal routing

The cross talk effects can usually be minimized by reducing the switching signal slew rate, the signal swing, and by proper signal routing. The common rule of thumb to control cross talk by signal routing is to maintain spacing between traces of 4:1 (horizontal spacing to vertical height above ground) as shown in figure 19.8 since the cross talk typically inversely depends on the square of spacing:

$$S > 4H \qquad (19.13)$$

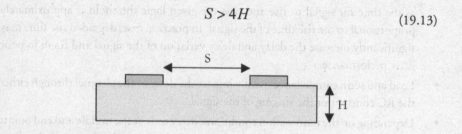

Fig. 19.8. The 4:1 rule for minimizing cross talk effects in transmission line circuits

Cross talk in VLSI circuits may also be caused by the discontinuity such as the ground plane slot in the current return path as shown in figure 19.9. Such a reference-plane slot can usually be seen on dense backplanes, which passes through fields of connector pins. In order to minimize cross talk due to reference plane slotting, it is usually required that the solid lane clear out around each pin to maintain continuity of the plane between all pins.

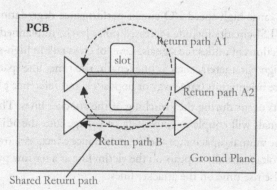

Fig. 19.9. Cross talk effects due to signal return path sharing

19.4 POWER DELIVERY-INDUCED NOISES

The shared power delivery network among different circuit blocks may introduce dI/dt and IR noises through the power (Vcc and ground) line coupling effects.

Shown in figure 19.10 is a lumped circuit model of a typical VLSI circuit power delivery network. In such a network, the on-chip switching current (usually known as $I_{cc}(t)$) induced noise is coupled with the equivalent impedance Z of the power delivery network to introduce the on-chip supply voltage noise, and that can impact the performance of the VLSI circuits.

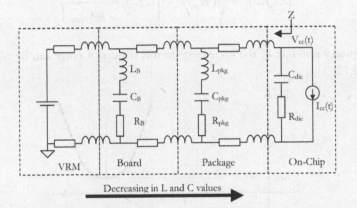

Fig. 19.10. Lumped power delivery circuit model

The frequency response of such power delivery circuit can be modeled based on several characteristic frequency ranges of the power delivery network as follows:

- At very high frequency, the impedance contributions form the package and PCB can be ignored since their impedance is usually at orders of magnitude higher than the on-chip impedance of the LRC circuit components. The power delivery circuit can be approximately simplified to a circuit shown in figure 19.11:

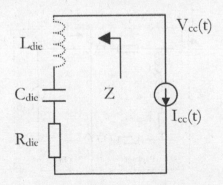

Fig. 19.11. Equivalent power delivery circuit at very high frequency

451

The equivalent power delivery network impedance in such frequency range can be expressed as

$$Z(s) = R_{die} + sL_{Die} + \frac{1}{sC_{Die}}$$

(19.14)

or

$$|Z(j\omega)| = \sqrt{\left(R_{die}\right)^2 + \left(\omega L_{Die} - \frac{1}{\omega C_{Die}}\right)^2}$$

(19.15)

The frequency response of the impedance in this frequency range can be plotted as figure 19.12.

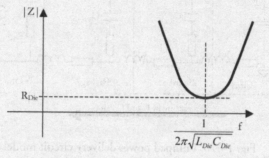

Fig. 19.12. High-frequency power delivery circuit impedance plot

- At slightly lower frequency within the characteristic frequency range of the package, the package trace and the package decoupling capacitors dominate the power delivery circuit impedance. The power delivery circuit in this frequency range can be simplified to the circuit shown in figure 19.13.

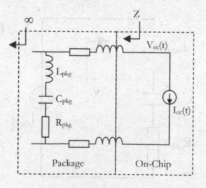

Fig. 19.13. Equivalent power delivery circuit near package resonant frequency

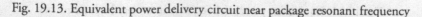

The equivalent power delivery network impedance in such frequency range is given as

$$Z(s) = R'_{pkg} + sL'_{pkg} + \frac{1}{sC_{pkg}}$$

(19.16)

or

$$|Z(j\omega)| = \sqrt{\left(R'_{pkg}\right)^2 + \left(\omega L'_{pkg} - \frac{1}{\omega C_{pkg}}\right)^2}$$

(19.17)

where R'_{pkg} and L'_{pkg} are the equivalent package resistance and inductance.

The frequency response of the impedance in this frequency range can be plotted as figure 19.14.

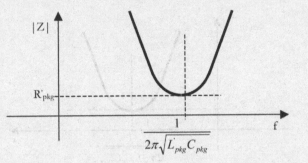

Fig. 19.14. High-frequency power delivery circuit impedance plot

• At frequency within the characteristic frequency range of the PCB, the PCB trace and the board decoupling capacitors dominate the power delivery circuit impedance, and the circuit can be simplified as circuit shown in figure 19.15.

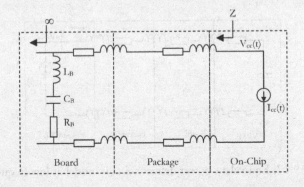

Fig. 19.15. Equivalent power delivery circuit near PCB resonant frequency

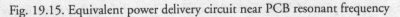

The equivalent power delivery network impedance in such frequency range can be expressed as

$$Z(s) = R'_{VRM} + sL'_{VRM}$$

(19.18)

or

$$|Z(j\omega)| = \sqrt{\left(R'_{PCB}\right)^2 + \left(\omega L'_{PCB} - \frac{1}{\omega C_{PCB}}\right)^2}$$

(19.19)

where R'_{PCB} and L'_{PCB} are the equivalent PCB resistance and inductance.

The frequency response of the impedance in this frequency range can be plotted as figure 19.16.

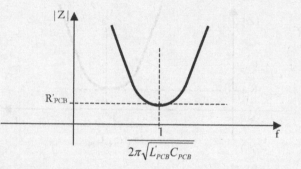

Fig. 19.16. High-frequency power delivery circuit impedance plot

- At low frequency, the voltage regulator module (VRM) dominates the power delivery circuit impedance. The circuit can be simplified as figure 19.17.

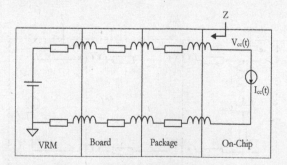

Fig. 19.17. Equivalent power delivery circuit model at low frequency

The equivalent power delivery network impedance in such frequency range is given as

$$Z(s) = R'_{VRM} + sL'_{VRM} \tag{19.20}$$

or

$$|Z(j\omega)| = \sqrt{\left(R'_{VRM}\right)^2 + \left(\omega L'_{VRM}\right)^2} \tag{19.21}$$

where R'_{VRM} and L'_{VRM} are the equivalent VRM resistance and inductance, which are the total serial resistance and inductance of the power delivery circuit of the VRM to the integrated circuits.

The frequency response of the impedance in this frequency range can be plotted as figure 19.18.

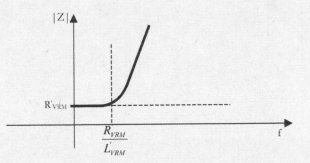

Fig. 19.18. Low-frequency power delivery circuit impedance plot

As the summary of the all operation frequency range from dc, the power delivery network impedance can be plotted as the form shown in figure 19.19.

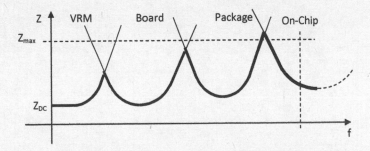

Fig. 19.19. Typical impedance profile of power delivery network

At each frequency, the voltage noise components is determined by the Icc(t) current components and the impedance at that frequency.

Impedance peaking effect can usually occur between two decoupling circuit resonant frequencies. Such impedance peaking effect can result in increased power delivery noise at these frequencies.

References:

[1] R. Goya, "Managing Signal Integrity," IEEE Spectrum March 1994.

[2] L. Gal, "On-Chip Crosstalk—the New Signal Integrity Challenge," IEEE 1995 Custom Integrated Circuit Conference.

[2] R. H. G. Cuny, et al, "SPICE and IBIS Modeling Kits, The Basis for Signal Integrity Analysis," 1996 IEEE.

[2] N. Jain, et al, "SI Issues Associated with High-speed Packages," 1997 IEEE/ CPMT Electronic Packaging Technology Conference.

[3] J. Kim, et al, "Via and Reference Discontinuity Impact on High-Speed Signal Integrity," IEEE

[4] A. K. Palit, et al, "Analysis of Crosstalk Coupling Effects between Aggressor and Victim Interconnect Using Two-Port Network Model," 2004 IEEE.

[7] A. E. Engin, et, al, "Modeling and Analysis of the Return Path Discontinuity Caused by Vias using the 3-Conductor Model," 2003 IEEE.

[8] C. Halford, "Serial I/O Layout Shifts Signal Integrity Design," IEE Electronics System and Software, Oct/Nov. 2005.

[9] K. Srinivasan, et al," Enhancement of Signal Integrity and Power Integrity with Embedded Capacitors in High-Speed Packages," Proc. 7th International Symposium on Quality Electronic Design (ISQED'06).

[10] J. Zhang, et al, "Effect of Shield Insertion on Reducing Crosstalk Noise between Coupled Interconnects," 2004 IEEE.

[11] J. Zhang, et, al, "Decoupling Techniques and Crosstalk Analysis for Coupled RLC Interconnects," ISCAS 2004.

[12] Y. Eo, et al, "A New On-Chip Interconnect Crosstalk Model and Experimental Verification for CMOS VLSI Circuit Design," 2000, IEEE.

[13] Y-L. Li, et al, "Design and Performance Evaluation of Microprocessor Packaging Capacitors Using Integrated Capacitor-Via-Plane Model," IEEE Trans. Advanced Package, Vol. 23, No, 3, Aug. 2000.

[14] Y-M Lee, et al, "Power Grid Transient Simulation in Linear Time Based on Transmission-Line-Modeling Alternating-Direction-Implicit Method," IEEE Trans on Computer-Aided Design of Integrated Circuits and Systems, Vol. 12, No. 11, Nov. 2002.

[15] Y. Eo, et al, "A Compact Multilayer IC Package Model for Efficient Simulation, Analysis and Design of High-Performance VLSI Circuits," IEEE Trans on Advanced Packaging Vol.26, No.4 Nov. 2003.

[16] W. Cui, et al, "Power Delivery Design and Analysis on a Network Processor Board," IEEE

[17] I. Kantorovich, et al, "In-Situ Measurement of Impedance of Die Power Delivery System," 2004 IEEE.

[18] M. Popovich, et al, "Decoupling Capacitors for Multi-Voltage Power Distribution Systems," IEEE Trans on LVSI systems, Vol.14, No.3, March 2006.

[19]   C-T Tsai, "Signal Integrity Analysis of High-Speed, High-Pin-Count Digital Packages," IEEE.

CHAPTER 20
TDR AND VNA MEASUREMENT TECHNIQUES

The time-domain reflectometer (TDR) has long been a standard measurement tool for characterizing and troubleshooting high-speed I/O circuits. The vector network analyzer (VNA), on the other hand, is becoming a more commonly used signal integrity measurement tool as a result of increasing differential signaling speed and the need for more accurate characterization and modeling of differential interconnects (such as cables, connectors, and printed circuit boards).

The TDR measurements are based on the time-domain stimulation of the devices under test (DUT) with voltage steps. The time delay and the signal magnitude of the reflection of the voltage step traveling through the DUT can be captured to extract circuit parameters such as the length and impedance information of the DUT.

On the other hand, the measurements with VNA are implemented in the frequency domain, where the DUT are characterized at each frequency of interest, one point at a time. The magnitude and phase shift from the DUTs are measured relative to the incident signals. The phase shifts are then related to the length of the DUTs: the longer the DUTs, the larger the phase shifts. Also, the higher the frequencies, the larger the phase shifts.

20.1 TDR MEASUREMENT BASIS

Shown in figure 20.1 is the conceptual setup for the TDR measurement. It includes a step voltage signal generator, a source impedance, a known transmission line, and the device under test (DUT).

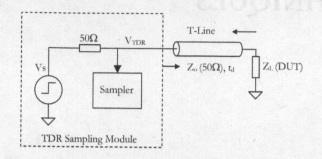

Fig. 20.1. TDR measurement setup

For a step voltage input, the voltage at sampler input is given by the incident wave from the source and the reflected wave back from the load as shown in figure 20.2.

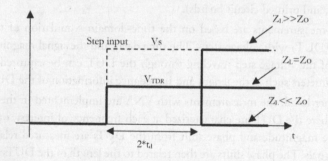

Fig. 20.2. TDR measurement waveforms

Mathematically, it can be seen that TDR measurements are based on a series of impedance ratios. Most TDR instruments will perform the necessary ratio calculations internally and display a numerical result. For the general load impedance Z_L, the reflection and incident wave ratio can be mathematically expressed as

$$\rho \equiv \frac{V_{reflection}}{V_{incident}} = \frac{Z_L - Z_o}{Z_L + Z_o}$$

(20.1)

On the other hand, based on the measured reflection coefficient, the load impedance can be calculated as

$$Z_L = Z_o \frac{1+\rho}{1-\rho}$$

(20.2)

20.2 TDR ELECTRICAL SIGNATURES

Based on equation 20.1, each load under test can be related to a reflection waveform. These waveforms provide the signatures of the type of load under test.

20.2.1 SHORT CIRCUIT LOAD

For short circuit load, we have that

$$\rho \equiv \frac{V_{reflection}}{V_{incident}} = \frac{0 - Z_o}{0 + Z_o} = -1$$

(20.3)

It can be seen that the incident and the reflection waves cancel each other after the return delay of the known T-line. The signature based on TDR for short circuit load is shown in figure 20.3.

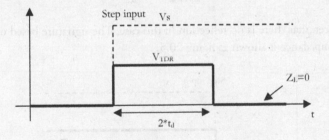

Fig. 20.3. TDR signature for impedance short

20.2.2 OPEN CIRCUIT LOAD

For open circuit load, we have that

$$\rho \equiv \frac{V_{reflection}}{V_{incident}} = \frac{\infty - Z_o}{\infty + Z_o} = 1$$

(20.4)

It can be seen that the incident and the reflection waves add up after the return delay of the known T-line. The signature based on TDR for open impedance is shown as figure 20.4.

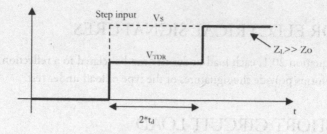

Fig. 20.4. TDR signature for impedance open

20.2.3 MATCHED IMPEDANCE

For matched circuit load, we have that

$$\rho \equiv \frac{V_{reflection}}{V_{incident}} = \frac{Z_o - Z_o}{Z_o + Z_o} = 0$$

(20.5)

It can be seen that there is no reflection in this case. The signature based on TDR for matched impedance is shown as figure 20.5.

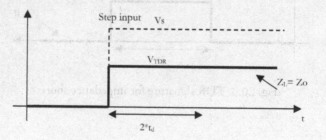

Fig. 20.5. TDR signature for match impedance

20.2.4 GENERAL RESISTIVE LOADS

For a general resistive circuit load, we have that

$$\rho \equiv \frac{V_{reflection}}{V_{incident}} = \frac{R - Z_o}{R + Z_o} \tag{20.6}$$

The reflection coefficient is a real number with the sign determined by the value of the load resistor. For high-value resistor, the reflection coefficient is positive, and for-low value resistor, the reflection coefficient is negative. The TDR for general resistive impedance is shown as figure 20.6.

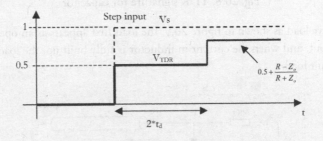

Fig. 20.6. TDR signature for resistor

20.2.5 CAPACITIVE AND INDUCTIVE LOADS

For capacitive load as shown in figure 20.7, the load first appears as a short circuit to the step input, and when the capacitor is fully charged, the load appears as an open circuit load.

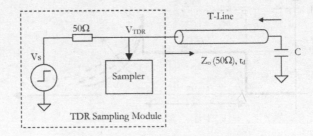

Fig. 20.7. TDR for capacitive loads

As a result, the TDR signature for a capacitor is shown in figure 20.8.

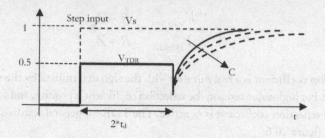

Fig. 20.8. TDR signature for capacitor

For inductive load as shown in figure 20.9, the load first appears as an open circuit to the step input, and when the current in inductor is fully built up, the load appears as a short circuit load.

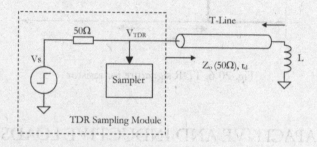

Fig. 20.9. TDR for inductive loads

As a result, the TDR signature for an inductor is then given in figure 20.10.

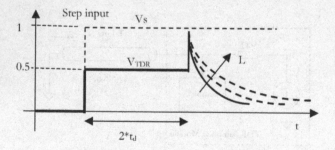

Fig. 20.10. TDR signature for capacitor

20.2.6 T-LINE DISCONTINUITIES

Signal in transmission line shown in figure 20.11 will be reflected at the capacitive discontinuities. Such a load appears as a short circuit for the step input. The load then appears as an open circuit load after the capacitor is charged up.

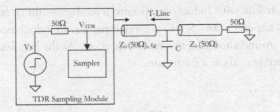

Fig. 20.11. TDR for capacitive discontinuity

The TDR signature for transmission line with such capacitor discontinuity is shown in figure 20.12.

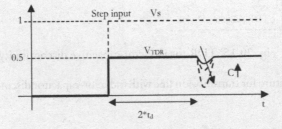

Fig. 20.12. TDR signature for capacitor

For transmission line with inductive discontinuity as shown in figure 20.13, the load first appears as an open circuit to the step input, and then it appears as a short circuit load when the current in the inductor is fully built up.

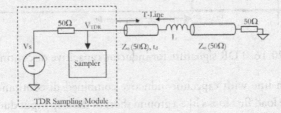

Fig. 20.13. TDR for inductive discontinuity

The TDR signature for transmission line with inductor discontinuity is shown in figure 20.14.

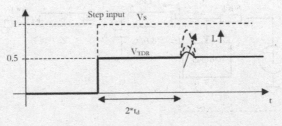

Fig. 20.14. TDR signature for inductive discontinuity

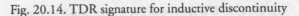

For transmission line with inductive-capacitor load discontinuity as shown in figure 20.15, the load appears as an open to the step input due to inductance effect, and then it appears as a ground short due to capacitance effect. As the line reaches steady state, the discontinuity appears as a short load.

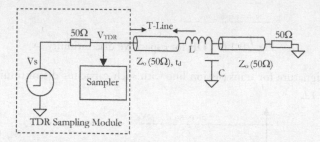

Fig. 20.15. TDR for inductor-capacitive discontinuity

The TDR signature for transmission line with inductor-capacitor discontinuity is shown in figure 20.16.

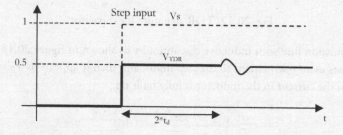

Fig. 20.16. TDR signature for inductor-capacitive discontinuity

For transmission line with capacitor-inductive combined discontinuity as shown in figure 20.17, the load first looks like a ground short to the step input due to capacitance effect, and then it appears as an open due to inductance effect. As the line reaches steady state, the discontinuity appears as a short load.

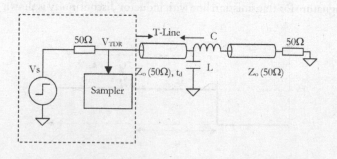

Fig. 20.17. TDR for capacitive-inductor discontinuity

The TDR signature for transmission line with capacitor-inductor discontinuity is shown in figure 20.18.

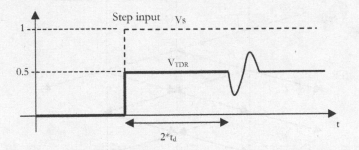

Fig. 20.18. TDR signature for capacitor-inductor discontinuity

20.2.7 TDR MULTIPLE REFLECTION EFFECTS

Signal in transmission line will be reflected at discontinuities. For the multisegment trace with different impedance, reflections occur at all trace boundaries with the TDR signature as shown in figure 20.19.

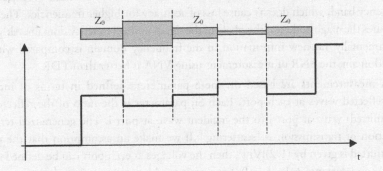

Fig. 20.19. TDR signature for segmented transmission lines

Such reflection effects can also be described graphically based on the reflection diagram as shown in figure 20.20.

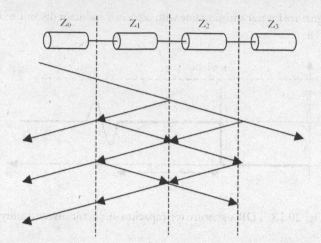

Fig. 20.20. Graphical method for TDR signature of segmented transmission lines

20.3 VNA MEASUREMENT BASIS

The VNA technique uses a single-tone frequency to sweep across a desired frequency range. The source power is typically leveled in a VNA and is constant over the entire frequency band, which doesn't cause loss of accuracy for higher frequencies. The VNA measures the magnitude and the phase of the signal in the frequency domain. Although, fundamentally, no new information in the frequency domain is compared with the time-domain, the SNR of measurement using VNA is better than TDR.

VNA measurements are based on the S-parameters defined in terms of incident and reflected waves at each port. Each S_{ij} parameter is the ratio of the reflected (or transmitted) wave at port j to the incident wave at port i. The generalized term for reflection or transmission is "scattering." If we make an assumption that the power transmitted is given by $(1/2)|V_i^+|^2$, then the voltages at each port can be defined as $V = V^+ + V^-$ and currents as $I = I^+ + I^-$. For a reciprocal junction (such as an interconnect), the scattering matrix is symmetrical, i.e., $S_{21} = S_{12}$. For a 4-port, the picture is more complicated, even though a 4-port is a direct extension of a 2-port definition.

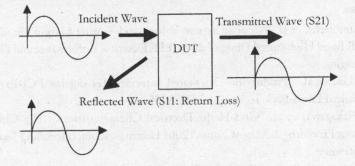

Fig. 20.21. VNA measurement principle

The main conceptual difference between the TDR and the VNA techniques is in the use of a wideband steplike source for TDR versus a narrowband sine wave generator for VNA. Additionally, TDR is a transient measurement (all transitions are observable), whereas VNA is a steady-state measurement—all transitions are lumped together, and the measurement is performed at a single frequency, with narrowband filtering employed to minimize the effects of noise.

References:

[1] F. Lambrechi, et al, "Accurate System Voltage and Timing Margin Simulation in CDR Based High-Speed Design," 2006 IEEE Electrical Performance and Electronic Packaging.

[2] M. Lin, et al, "Production-Orientated Interface Testing for PCI-Express by Enhanced Loop-Back Technique," ITC2005.

[3] K. Nakagawa, et al, "Giga-Herhz Electrical Characteristics of Flip-Chip BGA Package Exceeding 2,000pinCounts," 2004 Electronic Components and Technology Conference.

[4] "VNA-Based System Tests the Physical Layer," Application Note 1382-7, Agilent

CHAPTER 21
VLSI HIGH-SPEED I/O REFERENCE AND BIASING CIRCUITS

Voltage and current bias and reference circuits are critical supporting components in VLSI high-speed I/O circuits. The bias circuits are used to set the proper current or voltage operation points for various VLSI analog circuits within the high-speed I/O circuits. The voltage and current reference circuits are used to set the critical circuit operation specifications, such as the swings of the transmitter signals, the preemphasis levels, and the squelch circuit threshold levels of the high-speed I/O circuits. The resistive reference circuits, on the other hand, set the impedance values, such as the channel termination resistance of the high-speed I/O circuits.

Other important reference circuits are the timing reference circuit for clocking generation employing the PLL, DLL, or PI circuits for data transmission synchronization and data recovery and the temperature reference circuits for temperature sensing and thermal protection.

21.1 VLSI CURRENT BIAS CIRCUIT STRUCTURES

A basic VLSI current bias circuit structure is shown in figure 21.1. Such current bias circuit is commonly used in applications that do not need high accuracy current bias.

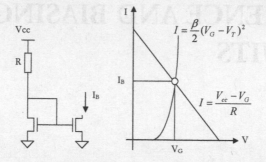

Fig. 21.1. The simple VLSI current bias circuit

The bias current of such circuit can be approximately expressed as

$$I_B \approx \frac{V_{CC} - V_T}{R}$$

$$(21.1)$$

Such a bias circuit is simple and highly reliable. However, it is highly sensitive to PVT variations, resulting mainly from the dependency of supply voltage (V_{cc}), device parameters (R and V_T), and temperature (PVT) variations.

The PVT sensitivity of the basic current bias circuit can be mathematically expressed using the sensitivity parameters as

$$S_{V_{cc}}^{I_B} \equiv \frac{\partial I_B}{I_B} / \frac{\partial V_{cc}}{V_{cc}} = \frac{V_{cc}}{V_{cc} - V_T}$$

$$(21.2)$$

$$S_{V_T}^{I_B} \equiv \frac{\partial I_B}{I_B} / \frac{\partial V_T}{V_T} = -\frac{V_T}{V_{cc} - V_T}$$

$$(21.3)$$

$$S_R^{I_B} \equiv \frac{\partial I_B}{I_B} / \frac{\partial R}{R} = -1$$

$$(21.4)$$

In most practical VLSI circuit applications, especially low power applications, the basic current bias circuit can usually be modified to include the enable control function as shown in figure 21.2.

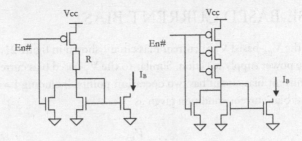

Fig. 21.2. Basic VLSI current bias circuits with enable control

21.1.1 VT-BASED CURRENT BIAS

The power supply voltage sensitivity of the VLSI current bias circuit can be improved by employing the so-called V_T-based current bias circuit structure shown in figure 21.3.

Such a VLSI current bias circuit consists of two current paths that are connected in the negative feedback configuration. This circuit has two operation points, including $I = 0$ (undesired operation point) and the desired operation point as

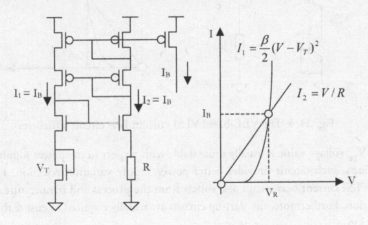

Fig. 21.3. The VT-based VLSI current bias circuit structures

$$I \approx \frac{V_T}{R}$$

(21.5)

It can be seen that it provides near-zero supply voltage sensitivity. In practical VLSI circuit applications, certain start-up circuits are usually required to ensure the circuit operates on the desired operation point. Based on equation 21.5, such circuit suffers from the bias current variations mainly contributed from the process and the temperature sensitivity of the V_T and R values.

21.1.2 VBE-BASED CURRENT BIAS

Alternatively, the V_{BE}-based VLSI current bias circuit shown in figure 21.4 can be used to improve the power supply rejection. Similar to the V_T-based bias current generation circuit, such current bias circuit has two operation points, including $I = 0$ (undesired) and the desired bias current condition given as

$$I \approx \frac{V_{BE}}{R}$$

(21.6)

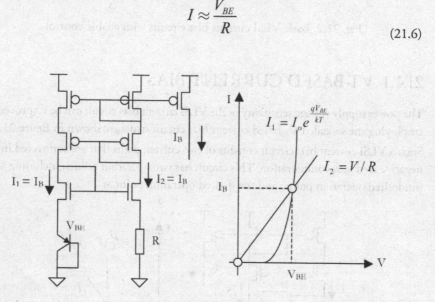

Fig. 21.4. The VBE-based VLSI current bias circuit structures

Since V_{BE} voltage value is usually more stable with respect to the power supply voltage variations, such circuit provides better power supply variation rejection. However, such VLSI current bias circuit also suffers from the process and temperature sensitive limitation. Furthermore, the start-up circuits are usually required to ensure the proper operation.

One useful feature of the V_{BE}-based current bias circuit is the negative temperature coefficient resulting from the negative temperature coefficient of the PN junction in the circuit. Such property can be used for temperature compensation operations.

21.1.3 CONSTANT GM CURRENT BIAS

The VLSI constant Gm current bias circuit structures are among the most widely used VLSI bias structures that can offer constant Gm values for MOS devices using such bias circuits. A constant-Gm current bias circuit is shown in figure 21.5.

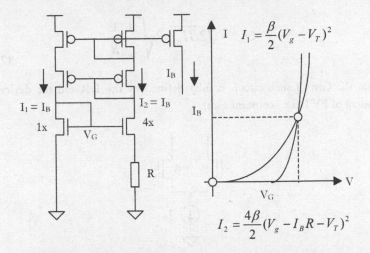

$$I_1 = \frac{\beta}{2}(V_g - V_T)^2$$

$$I_2 = \frac{4\beta}{2}(V_g - I_B R - V_T)^2$$

Fig. 21.5. VLSI constant Gm current bias circuit structure

Such current bias circuit also has two current paths connected in negative feedback that forms two operation points. One of them is undesired and is given as $I_B = 0$. The desired operation point can be derived using the following procedures.

Assuming all devices are in saturations, the bias current IB for the two paths is given as

$$\begin{cases} I_B = \frac{\beta_n}{2}(V_G - V_T)^2 \\ I_B = 4\frac{\beta_n}{2}(V_G - V_T - I_B R)^2 \end{cases} \Rightarrow \begin{cases} \sqrt{I_B} = \sqrt{\frac{\beta_n}{2}}(V_G - V_T) \\ \sqrt{I_B}/2 = \sqrt{\frac{\beta_n}{2}}(V_G - V_T - I_B R) \end{cases} \quad (21.7)$$

Finding the difference of above two equations yields

$$\Rightarrow \sqrt{I_B}/2 = \sqrt{\frac{\beta_n}{2}}(I_B R) \quad (21.8)$$

Solving for I_B, we have that

$$\Rightarrow I_B = \frac{1}{2\beta_n R^2} \quad (21.9)$$

When this bias circuit is applied to the circuit shown in figure 21.6, the Gm of the circuit is given as

$$G_m = \sqrt{2\beta I_B} = \sqrt{\frac{\beta}{\beta_n}} \frac{1}{R}$$

(21.10)

Note that the Gm of such circuit is fully defined by the 1/R and the device ratio independent of PVT (i.e., constant Gm).

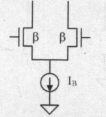

Fig. 21.6. VLSI differential pair

21.2 VLSI VOLTAGE REFERENCE CIRCUIT STRUCTURES

The basic voltage divider circuit shown in figure 21.7 can be used as voltage reference for VLSI circuit applications that do not require high accuracy.

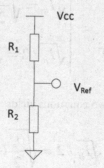

Fig. 21.7. The voltage divider-based VLSI voltage reference circuit structure

The voltage reference value based on above circuit can be expressed as

$$V_{Ref} = \frac{R_2}{R_1 + R_2} V_{CC}$$

(21.11)

Note that such a voltage reference is process and temperature independent.

21.2.1 VLSI BANDGAP VOLTAGE REFERENCE CIRCUITS

Bandgap voltage references are by far the most accurate on-chip VLSI voltage reference circuits. Bandgap voltage reference circuits are inherently PVT insensitive that are based on three basic VLSI circuit techniques including the following:

- The temperature coefficients compensation circuit technique that yields near-zero temperature sensitivity
- The self-bias circuit technique that yields very low supply voltage insensitivity
- The ratio-based circuit implementation technique that yields very low circuit design insensitive to VLSI manufacture process variations

Shown in figure 21.8 is a basic VLSI bandgap reference circuit implementation, where the ratio-based design approaches are used for the resistors R, R_1, and BJTs.

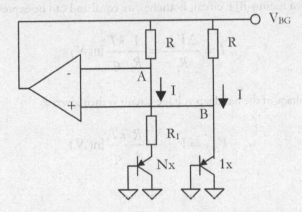

Fig. 21.8. VLSI bandgap voltage reference circuit structure

Bandgap reference circuits rely on the negative temperature coefficient of forward biased PN junction. The voltage current relationship of the BJT transistor shown in figure 21.9a can be expressed as

$$I = I_o e^{\frac{q V_{BE}}{kT}}$$

(21.12)

Consequently, the voltage difference of two matched forward biased BJT transistors as shown in figure 21.9b has a positive temperature coefficient that is proportional to the absolute temperature (PTAT) as

$$\Delta V_{BE} \equiv V_{BE2} - V_{BE1} = \frac{kT}{q}\ln(N_1 / N_2)$$

(21.13)

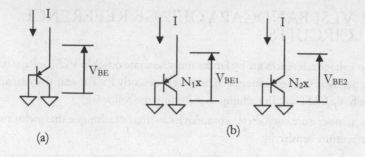

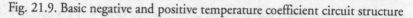

Fig. 21.9. Basic negative and positive temperature coefficient circuit structure

When such circuit structure is used in a negative feedback look as shown in figure 21.10, the two voltages at nodes A and B will be forced to be equal and the voltage across the resistor R_1 is equal to the V_{BE} difference of the two BJTs. As a result, the currents passing through the two resistor-BJT circuit branches are equal and can be expressed as

$$I = \frac{\Delta V_{BE}}{R_1} = \frac{1}{R_1}\frac{kT}{q}\ln(N)$$

(21.14)

The output voltage of the bandgap voltage circuit is then given as

$$V_{BG} = V_{BE} + \frac{R}{R_1}\frac{kT}{q}\ln(N)$$

(21.15)

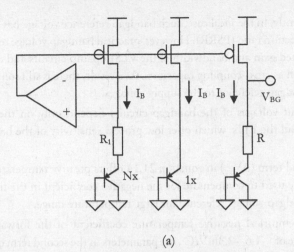

(a)

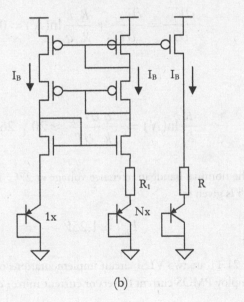

(b)

Fig. 21.10. Alternative VLSI bandgap voltage reference circuit structures

The above bandgap circuit structures suffer from the limitation of requiring supply voltage greater than the bandgap voltage of 1.25V. Shown in figure 21.11 is a VLSI bandgap voltage reference circuit structure that can be operated with sub-1.0V supply voltage.

This voltage reference has a few important properties:

- Power supply voltage independent. Based on the high gain opamp in negative feedback, the power supply voltage is eliminated from the output voltage equation.

Consequently, in the ideal case, such bandgap reference voltage has infinite power supply rejection ratio (PSRR). However, practical bandgap voltage references suffer from limited gain and bandwidth of the VLSI opamp circuits and other nonideal effects such as cross-coupling capacitors. As a result, there is still some dependency of the bandgap reference to the supply voltages.

- The output voltages of the bandgap circuits depend only on the ratios of the resistors and the BJTs, which offer low process sensitivity of the bandgap voltage reference.

- The second term (ΔV_{BE}) in equation 21.14 offers positive temperature coefficient that can be used to compensate for the negative coefficient in the first term (V_{BE}) in the bandgap voltage reference for large temperature range.

Based on the empirical negative temperature coefficient of the forward biased PN junction voltage of ~-1.6~-2.3mV/C°, the parameters in the second term of the bandgap voltage can be derived targeting at zero temperature coefficients as

$$\frac{\partial V_{BG}}{\partial T} = \frac{\partial V_{BE}}{\partial T} + \frac{R}{R_1}\frac{k}{q}\ln(N) = 0$$

(21.16)

that is,

$$\frac{R}{R_1}\ln(N) = -\frac{q}{k}\frac{\partial V_{BE}}{\partial T} \approx 20 \sim 26$$

(21.17)

Consequently, the nominal bandgap reference voltage at 27C° from equations 21.13, 21.14, and 21.15 is given as

$$V_{BG} \approx 1.25V$$

(21.18)

Shown in figure 21.11 are two VLSI circuit implementations of the bandgap voltage reference that employ PMOS current drivers or current mirror circuits.

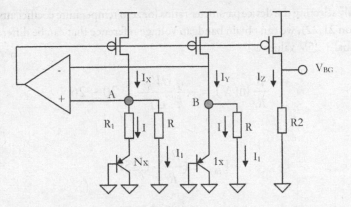

Fig. 21.11. Sub-1V VLSI bandgap voltage reference circuit

In such low-voltage bandgap circuit, the negative connected opamp circuit forces the two input node voltages A and B to be the same. The voltage across the resistor R_1 is then given as

$$\Delta V_{BE} = \frac{kT}{q} \ln(N) \tag{21.19}$$

The currents I_1 and I can be derived as

$$\begin{cases} I = \dfrac{1}{R_1} \dfrac{kT}{q} \ln(N) \\ I_1 = \dfrac{V_{BE}}{R} \end{cases} \tag{21.20}$$

When the PMOS transistors are selected to be identical, the current passing through the resistor R_2 is given as

$$I_Z = I_Y = I_X = I_1 + I_2 = \frac{V_{BE}}{R} + \frac{1}{R_1} \frac{kT}{q} \ln(N) \tag{21.21}$$

The output voltage of such reference is then a bandgap voltage that is given as

$$V_{BG} = \frac{R_2}{R} [V_{BE} + \frac{R}{R_1} \frac{kT}{q} \ln(N)] \tag{21.22}$$

By properly selecting the device parameter ratios for zero temperature coefficients (similar to equation 21.22), we can obtain bandgap voltage reference that can be different from the nominal 1.25V values as

$$\frac{R}{R_1} \ln(N) = -\frac{q}{k} \frac{\partial V_{BE}}{\partial T} \approx 20 \sim 26$$

$$(21.23)$$

$$V_{BG} \approx (\frac{R_2}{R}) \cdot 1.25V$$

$$(21.24)$$

In this bandgap voltage reference circuit, the minimal power supply voltage is only limited by the proper operation of the BJT, which usually requires ~700mV, and the saturation voltage of the PMOS device, which is typically around 100~200mV. As a result, this circuit may operate at supply voltage below 1V.

21.3 VLSI PTAT REFERNCE CIRCUITS

One of the practical applications of the VLSI bandgap circuits is for the generation of the proportional to absolute temperature (PTAT) current references that are widely used in applications such as the temperature sensing, thermal protection in VLSI high-speed I/O circuits. For the circuit shown figure 21.11, the current in the braches can be expressed as

$$I_B = \frac{kT}{Rq} \ln(N)$$

$$(21.25)$$

Note that this current is proportional to the absolute temperature. If the current is passed through a matched resistor of value R, the output voltage is directly proportional to the absolute temperature for accurate temperature sensing:

$$V = \frac{kT}{q} \ln(N)$$

$$(21.26)$$

References:

[1] N. Talebbeydokhti, et al, "Constant Transconductance Bias Circuit with An On-Chip Resistor," ISCAS 2006.

[2] A. Tekin, et al, "A Bias Circuit Based On Resistorless Bandgap Reference in 0.35um SOI CMOS," 2004 IEEE.

[3] N. Maghari, et al, "A Dynamic Start-up Circuit for Low Voltage CMOS Current Mirrors with Power-Down Support," 2005 IEEE.

[4] S. Pavan, "A Fixed Transconductance Bias Technique for CMOS Analog Integrated Circuits," ISCAS 2004.

[5] R. Dehghani, et al, "A New Low Voltage Precision CMOS Current Reference with No External Components," IEEE Trans. Circuit and Systems—II: Analog and Digital Signal Processing, Vol. 50, No. 12, Dec. 2003.

[6] G. Isano, et al, "A Replica Biasing For Constant-Gain CMOS Open-Loop Amplifiers," 1998, IEEE.

[7] M. A. P. Pertiji, et al, "A CMOS Smart Temperature Sensor with a 3σ Inaccuracy of +/-0.1oC from -55oC to 125oC," IEEE JSSC, Vol 40, No. 12, Dec. 2005.

INDEX

CPSIA information can be obtained
at www.ICGtesting.com
Printed in the USA
LVHW091909010323
740691LV00019B/699/J